Principles and Practices of
POND AQUACULTURE

Principles and Practices of POND AQUACULTURE

Edited by
JAMES E. LANNAN
R. ONEAL SMITHERMAN
GEORGE TCHOBANOGLOUS

OREGON STATE UNIVERSITY PRESS
CORVALLIS, OREGON

The paper in this book meets the guidelines for permanence and durability of the Committee on Production Guidelines for Book Longevity of the Council on Library Resources.

Library of Congress Cataloging in Publication Data
Main entry under title:

Principles and practices of pond aquaculture.

"Prepared as part of the Aquaculture Collaborative Research Support Program".
Includes bibliographies and index.
1. Fish-culture. 2. Fish ponds. I. Lannan, James E., 1935- . II. Smitherman, R. Oneal, 1937- .
III. Tchobanoglous, George, 1935- . IV. Aquaculture Collaborative Research Support Program.
SH159.P67 1986 639.311 84-2272
ISBN 0-87071-341-8

Printed in the United States of America

This book has been adapted for general readership from a report prepared for the U.S. Agency for International Development as part of the Title XII Collaborative Research Support Program (CRSP) in Pond Dynamics/Aquaculture (Specific Support Grant No. AID/DSAN-G-0264).

CONTENTS

FIGURES

PART 1 - PRINCIPLES OF POND AQUACULTURE

PART 2 - POND CULTURE PRACTICES

PART 3 - MODELLING OF POND CULTURE SYSTEMS

A Survey of the Mathematical Models Pertinent to Fish Production and Tropical Pond Aquaculture

TABLES

PART 1 - PRINCIPLES OF POND AQUACULTURE

PART 2 - POND CULTURE PRACTICES

Pond Production Systems: Fertilization Practices in Warmwater Fish Ponds

Pond Production Systems: Feeds and Feeding Practices in Warmwater Fish Ponds

Pond Production Systems: Water Quality Management Practices

FOREWORD

The title chosen for this book is intended to recognize the curious blend of art and science underlying every successful aquaculture system. The term "principles" denotes the physical, chemical, and biological processes occurring in pond systems and the interactions between them. The term "practices", on the other hand, denotes the fish cultural activities related to the design, management, and operation of pond culture systems.

Aquaculture has been practiced as a highly developed art form for a very long time. As the information about aquaculture expands there is increasing interest worldwide in moving from the practice of aquaculture as an art form towards a true agricultural technology.

This book had its origin during the planning of an international collaborative research program in aquaculture. For the purpose of program planning, it was necessary to summarize the diverse literature in a single volume. A secondary goal was that this summary would be useful to fish culturists around the world, and especially to those working in remote areas and less developed countries where access to the technical literature is limited.

Although the reports included in this volume address a broad range of topics relating to aquaculture principles and practices, they are intended to collectively address two questions crucial to the development of an aquaculture technology. The first question addresses the nature of the variation observed in pond production. The manifestation of this variation is all too familiar to experienced fish culturists. The typical scenario starts with the study of a pond that consistently produces satisfactory yields. The study of the pond includes careful measurement of a variety of physical, chemical, and biological variables. After evaluation of these data, attempts are made to establish additional ponds to be managed in exactly the same way. Although these are intended to be replicates of the successful pond, their performance is often observed to deviate greatly from the original.

This variation is also evident in aerial photographs of farm pond complexes to which similar pond management practices have been uniformly applied. In spite of this standardization there is frequently a dramatic color gradation ranging from olive brown through the magic green color to nearly yellow or, in some cases, red. It is clear that there are subtle differences regulating the development of biological communities from pond to pond, but the nature of this regulation remains obscure.

The second question relates to the economic efficiency of pond culture systems. Are contemporary pond management practices the most efficient approach to fish production? In order to answer this question, it is necessary to develop quantitative production functions to facilitate economic analyses of the various strategies or combinations thereof. It is not presently possible to develop these functions without making numerous and often tenuous assumptions, because the dynamic mechanisms regulating the productivity of the ponds are poorly understood.

The answers to both questions are clearly related to understanding pond dynamics. In this context, it is convenient to distinguish between two sources of fish nutrients: supplemental feeds and food organisms that develop in the pond. In terms of pond management, these sources are not mutually exclusive and, in fact, the most efficient system may involve some combination of the two. Regarding them as separate management approaches is a useful simplifying assumption because it will not be possible to optimize a system combining both approaches in the absence of reliable production functions for each approach considered separately.

In the one case, the major proportion of nutrients consumed by the fish originate externally in the form of prepared rations. These rations may range from raw grains to complex formulated feeds. An alternate strategy is to stimulate the growth of food organisms within the pond. In this case the availability of food resources limits the fish biomass that the pond can support. Fish yield can be increased by improving the environment for the biological community, including the food organisms consumed by the fish. This may be accomplished, for example, by the addition of lime to increase alkalinity and improve buffering, and by the judicious addition of organic and inorganic fertilizers.

The effect of fertilizer addition may be rationalized by considering the functional relationship between the short interval growth rate (rate of change of fish biomass with time) and stocking density (Hepher 1978). At lower stocking densities growth is at a maximum because food resources in the pond are not limiting. This relationship continues until a stocking density is reached where food resources start to become limiting. At this point, termed the critical standing crop, the growth rate starts to decrease. It continues to decrease as stocking density is increased until a stocking density is reached where the growth rate is equal to zero, i.e., the fish biomass neither increases nor decreases with time. This point is termed the carrying capacity of the pond.

The critical standing crop is a useful concept in pond management because at stocking levels below the critical standing crop there is no advantage to be gained from either supplemental feeding or fertilizer addition. However, both the critical standing crop and the carrying capacity can be extended to higher stocking densities by either practice. This increase is finite; both practices serve to increase the availability of food, but at some level the pond will become limited by other environmental variables, for example the availability of dissolved oxygen.

Understanding the mechanisms determining this functional relationship requires a rather complete understanding of the dynamic interactions occurring within a pond. It follows that under a given set of conditions the critical standing crop and carrying capacities of a given pond are unique properties of that pond and do not necessarily apply to other ponds.

In order to optimize the production of farm ponds, it will be necessary to develop predictive models that express both qualitatively and quantitatively how the pond will respond to different management practices.

The management of supplemental feeding in farm ponds is relatively less complicated than that of fertilizer addition. It focuses mainly upon satisfying the environmental and dietary requirements of the fish. Assuming the environmental requirements of the species are reasonably well understood, a few basic water quality determinations usually suffice to determine if a particular farm pond will provide a suitable environment. Environmental variables such as temperature, alkalinity, and the like must remain within the allowable limits for the applicable species, and the water must be free of toxic substances. The dissolved oxygen concentrations must remain at satisfactory levels to support aerobic respiration of the fish at all times, and metabolite concentrations, principally ammonia nitrogen, must not rise to toxic levels.

Satisfying the nutritional requirements of the fish involves both qualitative and quantitative considerations. The ration must provide all of the nutrients required by the species concerned. Rations are usually evaluated in a series of feeding trials conducted over a sufficient period of time to permit evaluation of growth, food conversion, and general fish health, including evidence of dietary deficiencies.

The amount of feed required under different conditions of temperature and fish size can be estimated from an empirically determined energy budget (for example see Brett et al. 1969). From this energy budget, functional relationships between growth rate and temperature for fish of given size can be determined. These functional relationships permit the estimation of the ration level for maximum growth rate, the maintenance ration level, and the optimum ration level (i.e., the ration level at which food conversion efficiency is maximized).

It is a widely held view that the most efficient approach to pond aquaculture involves a combination of fertilization and supplemental feeding. Because it is not presently possible to develop reliable production functions for either practice independent of the other, it is not possible to optimize production systems combining both practices. Although many fish culturists have operated such systems successfully, it is uncertain whether these systems approach maximum economic efficiency.

Purpose of this Book

In compiling this series of articles, the intent has been to survey the existing information about pond aquaculture with the ultimate goal of developing the production functions needed to maximize the efficiency of pond culture systems. The underlying question is: what is presently known and what needs to be known to accomplish this goal?

To present the information on pond aquaculture in a useful manner, this report has been organized into three main parts. In part one, the basic biological, chemical and physical principles governing the operation of pond culture systems are examined in a series of eight papers. The principles are examined in an integrated analysis in the first six papers. Chemical interactions are considered in the seventh paper and an additional perspective on fish-plankton interactions is presented in the eighth and final paper.

Pond culture practices including stocking, feeding, water quality, disease, competitors, pests, predators, and public health considerations are addressed in a series of six papers in part two.

Finally, part three is an examination of the modelling of aquaculture processes and systems. Two papers on the modelling of pond culture systems treat (1) hydromechanical and water quality responses of aquaculture systems, and (2) a survey of the mathematical models pertinent to fish production and tropical pond aquaculture.

It is hoped that a synthesis of these contributions will provide an assessment of where we are today in moving aquaculture towards a highly developed agricultural technology.

J.E. Lannan
R. Oneal Smitherman
George Tchobanoglous

LITERATURE CITED

Brett, J.R., J.E. Shelbourn, and C.T. Shoop. 1969. Growth rate and body composition in fingerling sockeye salmon, Oncorhynchus nerka, in relation to temperature and ration size. J. Fish. Res. Bd. Can. 26:2363-2394.

Hepher, B. 1978. Ecological aspects of warm-water fish pond management. Chapt. 18 In S.D. Gerking, (ed), Ecology of Freshwater Fish Production. John Wiley and Sons, New York. 520 pp.

ACKNOWLEDGEMENTS

The contributions of the several authors contributing to this report is acknowledged gratefully. Without their willingness to give unselfishly of their time and expertise to the preparation of the papers this report would not have been possible.

Special thanks are due to Dinah Pfoutz and Carman McBride, who assisted in the typing and preparation of this book.

The preparation of this book was supported in part by the U.S. Agency for International Development as part of the Title XII Collaborative Research Support Program in Pond Dynamics/Aquaculture under Specific Support Grant #AID/DSAN/G-0264, and by the participating U.S. universities.

PART 1

PRINCIPLES OF POND AQUACULTURE

BIOLOGICAL PRINCIPLES OF POND CULTURE: AN OVERVIEW

by

William Y.B. Chang

INTRODUCTION

The purpose of this and the following five papers is to report the state of knowledge on the principles of pond culture. Emphasis is on the ecological dynamics of factors related to maximizing fish (including finfish and shellfish, principally shrimp and prawns) production. Ponds used in aquaculture differ in origins, structure, function, and geographic location and exhibit marked differences in fish production. However, this production is essentially governed by interactions among the physical, chemical, and biological conditions of the water and the species of fish or fishes present.

POND CLASSIFICATION

Recognizing the limitations inherent in any classification of pond systems from a production standpoint, the following classification is adopted here for purposes of convenience.

1. warmwater low-intensity fish production ponds,
2. warmwater high-intensity fish production ponds,
3. coolwater ponds (generally at high elevation or high latitudes), and
4. brackish-water ponds (including mangrove swamps).

Although the principles of pond culture are similar for all of these four broad types, the quantity and quality of information available differ greatly among them. Among the types, there are physico-chemical and operational differences, including origin (natural vs. man-made), stocking density, kind and amount of fertilization, fallowing periods, supplemental feeding, and the fish species used either in monoculture or in various levels of polyculture.

STATE OF KNOWLEDGE

Available knowledge is greatest for warmwater ponds that are intensively managed, least for coolwater ponds, and needed research differs accordingly. Solutions to narrow but highly technical problems, mainly dealing with transfers between trophic levels, are most needed for intensively farmed ponds (Figure 2) and brackish-water culture systems (Figure 4). For low intensity production (Figure 1) and cool water ponds (Figure 3), which are less well studied, broader research topics, such as feasibility, selection of species to be cultured, and relationships within trophic levels are the most needed and promising areas of research.

Ponds in the tropics differ considerably from those in temperate regions. As the majority of research in aquatic ecology and physiology has been done in temperate waters, it is important here to focus on some of these differences. For example, lowland ponds near the equator show only slight seasonal variations in temperature compared to those in the temperate zone. Most tropical ponds experience stronger seasonal variation (wet vs. dry season) than temperate water bodies. Equatorial high-altitude, coolwater ponds (predominantly Andean) may show large diel changes in temperature that remain relatively constant over the year. The foregoing examples are only illustrative; much more detail can be given to precisely delimit (perhaps by biogeographical region) the average and range of physico-chemical conditions present in each pond type in relation to bioproduction. Among these conditions, examination of the role(s) in production of such previously obscure factors such as naturally occurring trace elements (e.g., cobalt, molybdenum, etc.) and dissolved secondary compounds of aquatic and terrestrial plant manure (tannins, saponins, and alkaloids) along with metabolites of aquatic organisms themselves may prove valuable for managing aquaculture systems. Such a perspective will be of unusual importance in the application of information or transfer of technology from one geographical area to another. Comparison of pond ecosystems with marked production differences in the same geographical area is an efficient way of exploring important parameters which affect production in them--here what is learned from the "good" may be used to alleviate problems in the "poor" and from the "poor" to avoid problems in the "good".

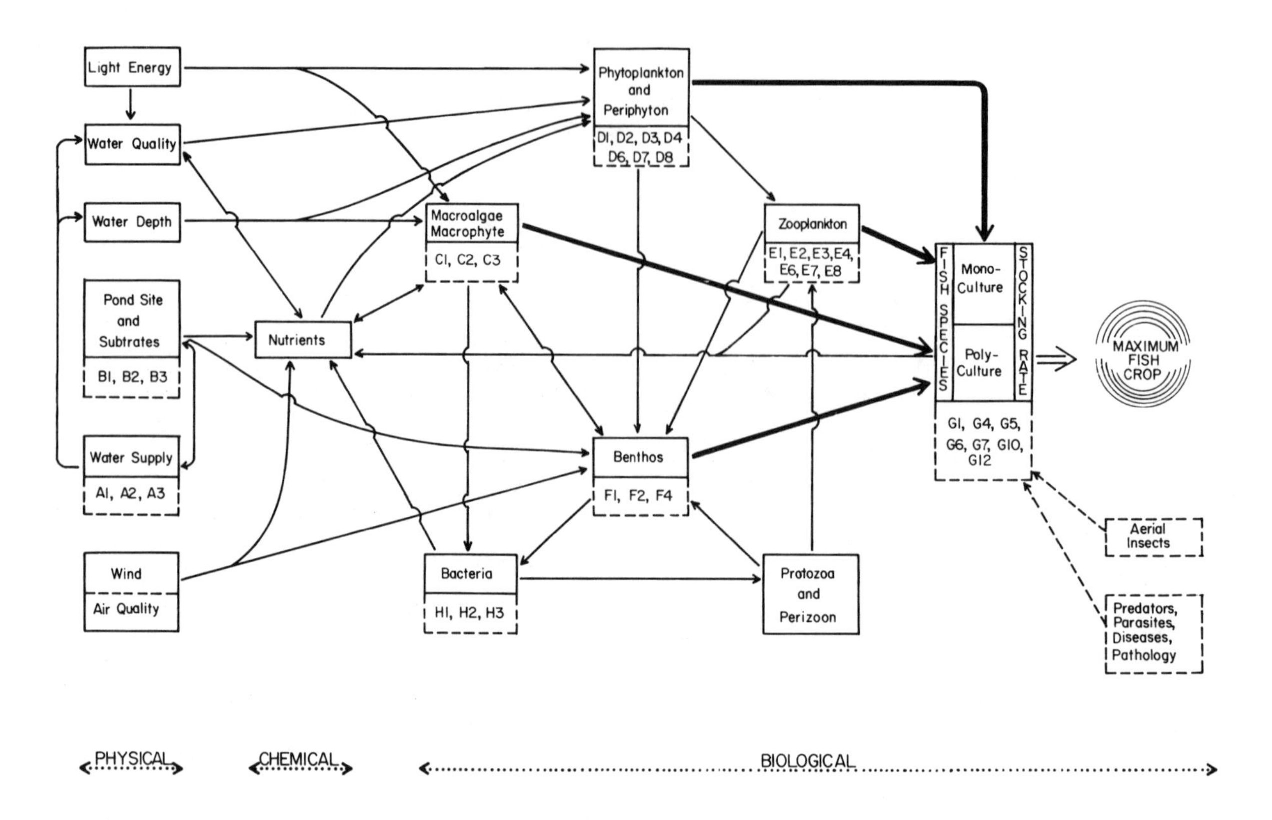

Figure 1.

Low-intensity tropical ponds.

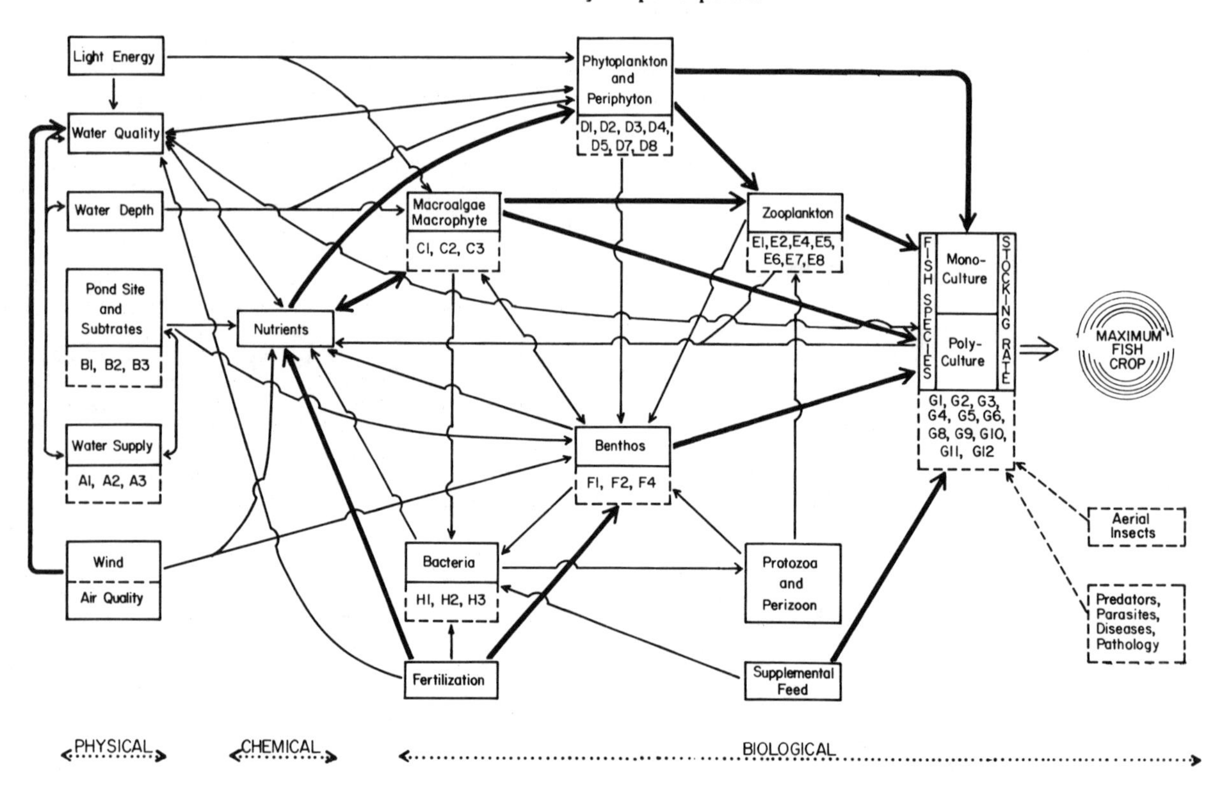

Figure 2.

High-intensity tropical ponds.

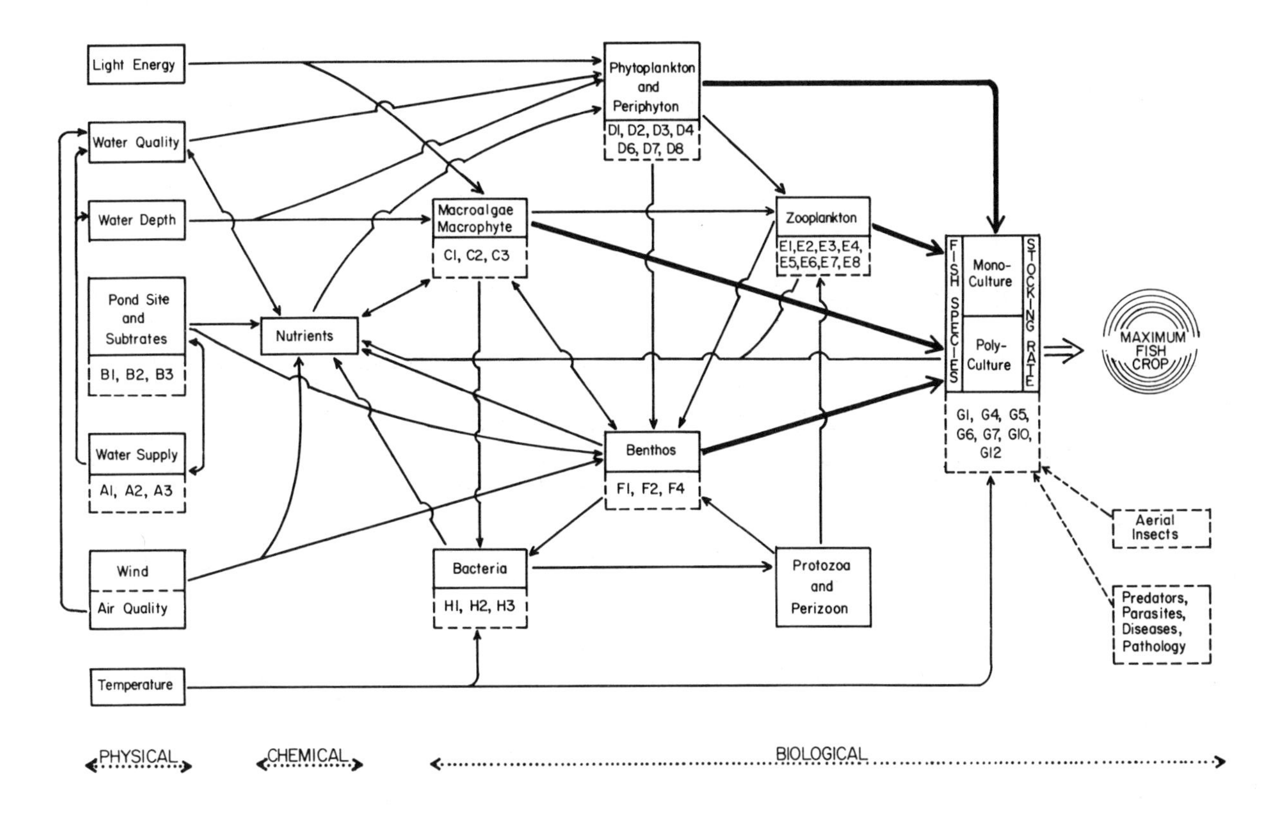

Figure 3.

Cool-water tropical ponds.

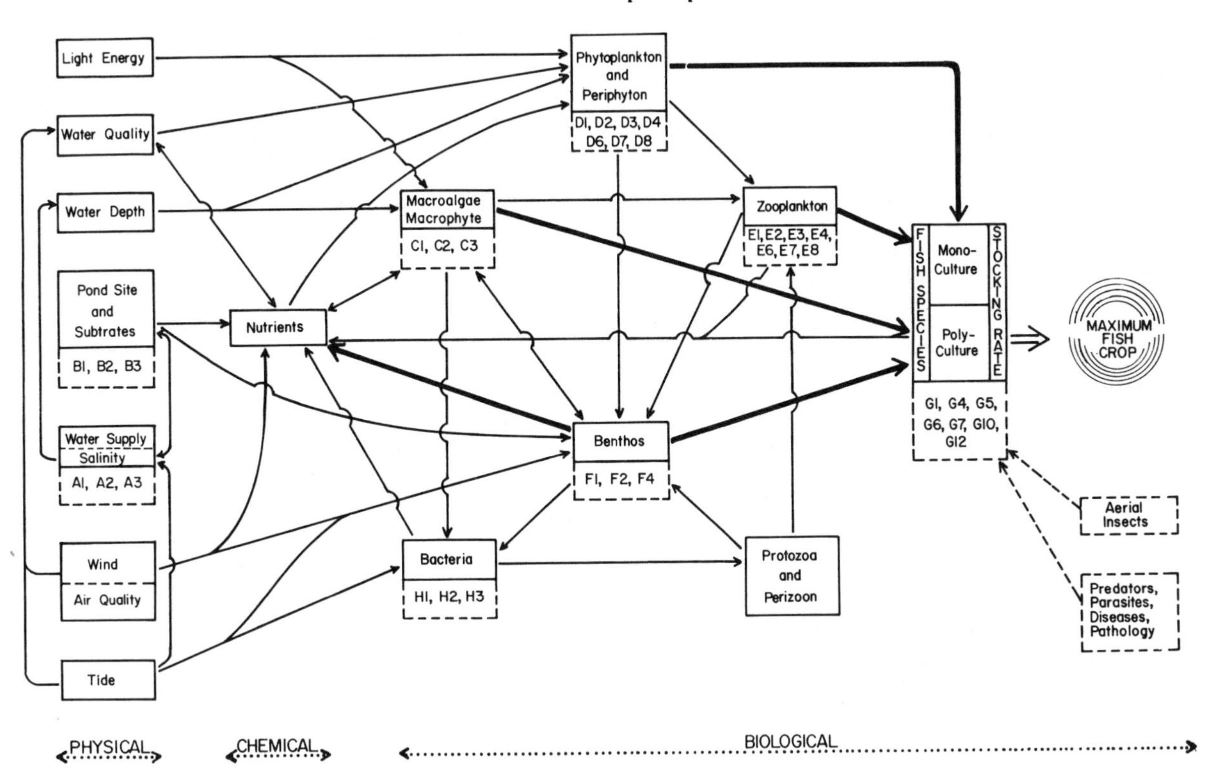

Figure 4.

Brackish-water ponds.

POND DYNAMICS

Four conceptual models (Figures 1-4) graphically display interactions among compartments in the four types of pond culture systems. These models also show something of the intricate nature of pond productivity, the relative importance of the factorial interaction of various components, and linkages of these components in a system. The relative productive importance of the interactions among components differs (as represented by the boldness of the lines connecting the compartments in the figures). These differences can be grouped under three major headings: physical, chemical, and biological. The selected research topics important to each compartment or trophic level are indicated in the lower part of the compartment. The emphasis, however, should be placed on the process(es) involved in the relationship and/or transfers between trophic levels.

Thus the five major dynamic areas discussed in the following papers deal with: (1) bacteria and nutrient cycling, (2) benthos and sediments, (3) phytoplankton and macrophytes, (4) zooplankton, and (5) fish. A brief overview of these topics is presented below.

Bacteria and Nutrient Cycling

Successful management of an aquaculture system depends on a constant supply of basic nutrients necessary for optimal growth of the cultured species. This pervading fact is particularly important in extensive culture systems where food sources for the propagated species are maintained internally, as are most of the essential nutrients such as nitrates, phosphates, and carbon-containing compounds. The constant supply of nutrients depends heavily on rapid recycling, which is one of the most important factors in maximizing production in the pond culture. Phosphate and nitrate play a significant role in the production of aquatic organisms, especially micro- and macro-plants, and rely on the biochemical process of recycling for their conversion into a form available to the organisms involved. Discussion of nutrient cycling processes is provided in this study. As nutrient cycling cannot be accomplished without the assistance of bacterial activity (for example, cycling of nitrogen depends on the activity of both nitrifying and denitrifying bacteria), a discussion of nutrient cycling is provided in connection with bacteria. The discussion is also extended to cover the roles of bacteria as removers of substances toxic (NH^+, NO_2) to the cultured species and as a food source for plankton, which in turn become direct food items for the cultured species, or indirectly so for the small animal organisms on which they feed.

Benthos and Sediment

Inasmuch as the supply of nutrients is also dependent on the fertility of the indigenous soil and/or on soft sediment, which provides living environments for bottom organisms and which can variously affect the productivity of an aquaculture system, a discussion of these issues is presented in a subsequent section. As might be expected, it shows that pure (inorganic) sand provides the least productive environment, and fine silts and muck the most productive. The annual productivity of some cultured species is directly proportional to the organic carbon content in soil. Under anaerobic or reduced conditions on the sediment surface, nutrients and some ions are released to the water, whereas the overlying water is kept at oxygen saturation in most aquaculture systems. The net flow of nutrients is from water to soil, with the soil acting as a sink for nutrients. Regeneration of these nutrients then comes primarily through biological interaction at the soil-water interface and through subsurface mixing such as bioturbation, in which the oligochaete worms and certain insect larvae act as conveyor belts to redeposit subsurface sediment at the soil-water interface.

Benthic algae and bacteria, which are directly controlled by the soil type, can form a watery (subaqueous) horizontal layer on the sediment surface. They often create chemical gradients by providing an increasingly rapid flux of nutrients from the soils, and also serve as food sources for species such as milkfish. A discussion of these topics is therefore provided in connection with the section on benthos and sediments.

Phytoplankton and Macrophytes

The growth of phytoplankton and aquatic macrophytes is known to be critical for augmenting fish production in pond culture, especially as most fish raised in warmwater ponds in less developed nations are dependent largely upon natural foods. The section on phytoplankton and macrophytes provides a synthesis of the relevant literature between these plants and fish yields. As managed fish ponds are often fertilized, and have nutrient concentrations which are generally elevated and primary productivity greater than the average natural water, the impacts of excessive fertilization are discussed. Included among these impacts arising from phytoplankton blooms are a lowered level of light penetration, a reduced compensation point which may depress productivity per unit area in the pond, reduced species diversity, and stimulation of unwanted species of algae such as certain blue-greens, which can limit light penetration, lower CO_2 concentration, and deplete nutrients in surface water. These conditions can lead to massive and disastrous die-offs from severe depletion of oxygen following decay of the dead algae. Preventing such blooms is of great concern to many intensive aquaculturists. Some success has been achieved by manipulation of nutrient levels where varying N/P ratios result in changes in the dominance between blue-green and other types of phytoplankton.

The occurrence of macrophytes is normally considered to be undesirable in most kinds of pond management because they compete with phytoplankton for nutrients and interfere with the fish harvest. However, there are occasions when aquatic macrophytes in fish ponds become beneficial as food sources for certain fishes or serve as expanded host substrates for attached and/or clambering invertebrates, which are food for some fish.

Zooplankton

In low and/or moderately managed systems, a good relationship has been found between secondary production and fish yield, but zooplankton populations have been measured infrequently in most aquaculture systems. Nonetheless their role in fish culture is important, and their abundance and composition are clearly affected by fish standing stocks and they can influence the fish yield. A strong correlation has been noted between food conversion rate (kg feed supplied:kg fish yield) and zooplankton standing stock at zooplankton densities of 0.1 to 1.1 mg dry weight/L, but this correlation appears to be weakest at the highest zooplankton densities. The supply of zooplankton in this instance is important in fish production, and is also important as a food for many species of fish fry. In many intensively managed systems zooplankton is grown to improve the survival of the early developmental stages of fish. For example, the rotifer Branchionus and the brine shrimp Artemia are commonly reared as food items for fish fry. In addition, it is to be noted that the maximum growth of fish fry depends not only on having the proper number and size of zooplankton present as food items, but also on having these zooplankton contain the appropriate nutrients. Other than the positive association between zooplankton and fish, some zooplankters can also act as predators on fish larvae. For example, cyclopoid predation on fish larvae occurs commonly and has been reported by a number of freshwater and marine researchers. Certain species of copepods are also fish parasites. They damage the fish either by disrupting tissue or through encouraging secondary infections. When this occurs, there can be extensive negative effects on fish production in a high density culture system.

Fish

The major goal of this investigation is to provide information on principles related to achieving optimal fish production, which is intimately affected by stocking density (and species mix in polyculture), mortality, and growth in individual weight. These parameters are then related through mechanisms of biotic interaction with physico-chemical environments. Growth relations with density are thought to be the most important areas for research on pond culture for fish production. The rationale for this focus is detailed in a subsequent section, where the ecological and physiological mechanisms governing managed and natural reproduction, growth, mortality, and recruitment to harvestability of fish are discussed extensively with respect to those species commonly cultured in tropical ponds. The effects on growth in individual weight of the cultured fish stock are organized into the following categories: (i) intra- and interspecific competition; (ii) effects of antagonism and facilitation; (iii) abiotic limiting factors; (iv) effects through efficiency of food conversion and assimilation; (v) nutrition and diet; (vi) endocrinology and genetics; and (vii) pathology and growth (growth influences due to dietary deficiency-related diseases or bacterial, viral, and parasitic diseases). The integration of these factors is examined through a bioenergetic model which illustrates the effects on fish production of various primary mechanisms and ecological principles. This model provides an in-depth overview of indirect components which interact with fish culture, offers explanations for high yield of polyspecific culture systems, and highlights impacts stemming from various aquaculture practices.

BIOLOGICAL PRINCIPLES OF POND CULTURE: BACTERIA AND NUTRIENT CYCLING

by

Russell Moll

An important aspect of successful aquaculture system management is a constant supply of basic nutrients necessary for optimal yield of the cultured fish (Edwards 1980). This fact is especially important in extensive culture systems where food sources for the culture species are typically maintained internally, as are most of the nutrients such as nitrogen, phosphorus, and carbon compounds. Adding manures to a pond or culturing tank as a method of ensuring a proper nutrient balance may also be used in intensive cultures. Irrespective of the method used, nutrient cycling is one of the most important factors in efficient aquaculturing.

Nutrient cycling involves a number of chemical and biological processes, but a major contributor is bacteria, which utilize substances in the water and sediments and make them available to other organisms in the fish pond. In fact, many of these substances can be harmful to other organisms in the system (Seymour 1980). It would be difficult to discuss bacterial activities in aquaculture without including nutrient cycling and, conversely, nutrient cycling could not be examined accurately without incorporating bacterial activity. Therefore, both subjects will be addressed in this section.

Many substances in the water and sediments of an aquaculture system are useful as nutrients. These substances include mineral salts, various carbon, nitrogen and phosphorus compounds (Fogg 1975). Of these, nitrogen and phosphorus compounds are usually found to have the greatest importance to primary production by aquatic organisms (Likens 1972). Excess phosphorus compounds do not have the same detrimental effects on an aquaculture system as do certain nitrogenous substances (e.g., NH^+, NO_2). This fact has supported studies on the removal of nitrogenous compounds or conversions to useful products. Nonetheless, levels of phosphorus compounds have a major effect on the amount of primary production in all aquatic systems (Rigler 1973). The ratio of P to C to N can cause changes in algal species composition as well as abundance (Schindler 1975) which in turn affect the entire food web.

Bacteria undergo a series of biochemical processes in aquatic environments which are beneficial in providing increased secondary production, if not essential to production (Kuznetsov 1970). Cycling of nitrogen through the biochemical activity of nitrifying and denitrifying bacteria accomplishes two significant ends: first, it removes substances toxic to the cultured species (e.g. NH^+, NO_2) and second, it produces useful nutrients which aid in increased production. The cycling of phosphorus, on the other hand, primarily involves mineralizing organic phosphorus compounds to inorganic forms available for algal uptake. Bacteria also decompose solid wastes produced by animals in fish ponds through the same mechanisms to recover nutrients as additions of organics (such as manures) to the culturing tank or pond.

Microorganisms are also useful as food sources for zooplankters which in turn become foodstuffs for fish (Schroeder 1978). Systems aimed toward more intensive production have successfully incorporated bacterial protein into feed pellets as a less expensive protein source than fish meal (Kaushik and Luquet 1980; Matty and Smith 1978; Atack et al. 1979).

A final aspect of bacterial presence in aquaculture systems is the pathogenic nature of certain bacterial species. Because this section primarily addresses bacterial nutrient cycling, only minor discussion will be made concerning pathogens in aquaculture. Pathogens seldom become a major setback in extensive culturing systems while their major damage is done in intensive culture systems.

NUTRIENT CYCLES

Nitrogen Cycle

The nitrogen cycle in aquaculture ponds is the most important biogeochemical cycle of these systems. Nitrogen is essential to sustaining high primary production; but nitrogen can also be toxic to many organisms in high concentrations. The correct balance of available nutrient and controlled wastes is necessary in any aquaculture system.

Bacterial metabolism plays an essential role in controlling unwanted nitrogen as well as maintaining available inorganic nutrient levels. The exact form the nitrogen cycle takes in any pond or lake depends on the location of the majority of nitrogen associated metabolism (Kuznetsov 1970). Denitrification occurs in anaerobic sediments and waters, while aerobic waters support nitrification.

In aquaculture ponds, high levels of fish and shellfish metabolism result in the production of various nitrogenous waste products. The major nitrogen-containing substance released by these organisms is ammonia (NH_3), which reacts with water to become the soluble ammonium ion NH_4^+. Either NH_3 or NH_4^+ is fairly toxic to aquatic organisms, as is nitrite ion (NO_2^-) and its aqueous counterpart, nitrous acid HNO_2. Through bacterial activity, ammonia and nitrate are metabolized to yield nitrate (NO_3^-), which is fairly safe for aquatic organisms and useful to phytoplankton as a nutrient source. The biochemical reactions which occur to produce nitrate are given below:

$$2\ NH_3 + 3.5\ O_2 \xrightarrow{\text{Nitrosomonas}} 2\ NO_2 + 3\ H_2O$$

$$NO_2 + .5\ O_2 \xrightarrow{\text{Nitrobacter}} NO_3$$

As can be seen from these two equations, the production of the end product, nitrate, is dependent on the presence of oxygen and the activity of Nitrobacter to metabolize nitrite ion. Colt and Tchobanoglous (1976) calculated that the doubling time for Nitrosomonas was thirteen hours at 30°C and fourteen hours for Nitrobacter. Therefore, in a closed or semi-closed recycling system, two problems develop in the normal nitrogen cycle. First, the bacterial flora cannot respond rapidly enough to prevent a buildup of toxic substances, especially nitrite, since ammonia must first be reduced to nitrite and then to nitrate. Second, oxygen can be rapidly depleted during nitrification and prevent further oxygenation of nitrogenous waste products.

Although Nitrosomonas and Nitrobacter are the most common genera of nitrifying bacteria, there are others. Some other nitrifiers include: Mycobacterium, Nocardia, Streptomyces, Micromonospora, Streptosporangium, Arthrobacter, Agrobacterium, Bacillus, Corynebacterium, and Pseudomonas (McCoy 1972). Several fungi have also been found to be nitrifiers.

Nitrification rates are greatly affected by pH values (Srna 1976). Higher pH values greatly increase the oxidation of ammonia to nitrite. For example, the rate constant for ammonium oxidation at pH 7.13 is about 1/3 of the rate at pH 7.78 in a multi-stage biological filter (Srna 1976), whereas oxidation of nitrite to nitrate occurs more efficiently and rapidly at lower pH values.

Ammonia may also be used directly by certain phytoplankton through the following equation:

$$\left.\begin{matrix} NH_3 \\ NO_3 \end{matrix}\right\} + H_2O + PO_4 + CO_2 + \text{Light} \xrightarrow{\text{Photosynthesis}} \text{Organic} + O_2$$

When oxygen becomes depleted in the water column, or organic matter reaches the sediments, denitrification takes place instead of nitrification.

$$2\ NO_3 + CH_2O \xrightarrow{\text{Denitrification}} N_2,\ N_2O$$

The process of denitrification takes place only in the absence of oxygen and/or in the presence of large amounts of organic nitrogen. Denitrification is conducted by a variety of denitrifying bacteria, and primarily between pH 5.5 and 8.0. Thus, as a pond or lake becomes anaerobic because of high metabolism levels, nitrogen is tied up either as organic nitrogen in the sediments, or is converted to free nitrogen (N_2). Eventually, these pools of nitrogen will be oxidized back to nitrate. But, the short-term management of the proper balance of different nitrogen forms is essential to maintaining an aquaculture system.

Phosphorus Cycle

In many natural freshwater ecosystems, available soluble phosphorus is the essential nutrient limiting primary production. But, once supplies of nutrients are elevated by anthropogenic or other inputs, phosphorus tends to lose its position as the limiting nutrient; in these cases other factors such as water clarity or dissolved oxygen levels become essential for continued algal growth (Lean 1973). Nonetheless, the cycling of phosphorus remains a key aspect in the maintainance of primary and secondary production in all aquatic environments. Proper elemental ratios of C to N to P must be sustained for protein synthesis. For these reasons, the phosphorus cycle is just as important to aquaculture pond dynamics as the nitrogen cycle, although the two are fundamentally different (Wetzel 1975).

Phosphorus in aquatic ecosystems occurs as phosphate in organic and inorganic compounds. Free orthophosphate is the only form of phosphorus believed to be taken up directly by phytoplankton (Rigler 1973) and thus represents a major link between organic and inorganic phosphorus cycling in lakes. In natural waters, orthophosphate occurs in ionic equilibrium, i.e.

$$\text{as } H_3PO_4 \xrightarrow{pK = 2.2} H_2PO_4^- \xrightarrow{pK = 7.2} HPO_4^{2-} \xrightarrow{pK = 12.3} PO_4^{3-}$$

with $H_2PO_4^-$ and HPO_4^{2-} being the predominant species over pH range of 5 to 9 (Stumm and Morgan 1970).

The cycling of phosphorus in small lakes was studied by Hutchinson and Bowen (1950) and re-evaluated by Kuznetsov (1970). The results show that most of the inorganic phosphorus is quickly utilized by phytoplankton and littoral aquatic macrophytes. When these plants die, the phosphorus is carried as organic matter (usually as a ferric phosphate complex) to the sediments. If the sediments become anaerobic along with the overlying water, phosphorus is released back into the water column. This release of phosphorus from the sediments is caused indirectly by bacterial metabolism. Anaerobic muds are normally rich in reduced sulfur compounds because of the activity of sulfur bacteria (Kuznetsov 1970). The reduced sulfur-rich muds quickly reduce ferric phosphate to soluble ferrous phosphate, which readily migrates from the sediments back into the water.

Thus, in small ponds and lakes with anaerobic bottom waters, phosphorus would primarily cycle through the sediments. Most small aquaculture ponds would be expected to have this type of phosphorus cycle. But, in larger or well-oxygenated ponds, phosphorus cycling is primarily carried out by a large variety of bacteria in the water column (Rigler 1973). The uptake and release of phosphorus by pelagic bacteria in either organic or inorganic form has been difficult to quantify. Bacterial activity is influenced by temperature, availability of substrate, oxygen concentration, pH, and chemical composition of the substrate (DePinto and Verhoff 1977; Foree and Barrow 1970; Kuznetsov 1970). Jones (1972) observed a relationship between senescent algae and bacterial biomass; a rapid increase in the amount of decaying algal cells in the water column, such as when a phytoplankton bloom declines, is usually followed by an increase in bacterial biomass.

Despite large variability in decomposition rates, some preliminary estimates of inorganic release by bacterial metabolism have been made. Field estimates of P turnover have been particularly inaccurate in that liberated P is rapidly re-used and rarely measured as available soluble P. Fuhs (1973) and Charlton (1975) estimated that water-column P was recycled from 10 to 34 times during one growing season. Fallon and Brock (1979, 1980) have estimated that 60-70 percent of the organic matter from decaying phytoplankton was decomposed in the water column, primarily in the epilimnion.

More exact rates of P release by bacteria are available from laboratory studies. Golterman (1973) found that 70-80 percent of the organic P released by algal autolysis was converted to inorganic forms within a few days. DePinto and Verhoff (1977) conducted a study of the regeneration of P and N in axenic algal cultures and bacteria-algae cultures. They found that the P released as inorganic soluble reactive phosphorus was very slow when axenic cultures were placed in the dark. On the other hand, the bacteria-algae mixtures liberated up to 75 percent of the total P pool after several weeks. The rate of regeneration of the soluble P was a function of the initial P composition of the algae; P-limited algae produced regeneration rates of 0.16-0.39 %g P/mg algae (dry weight)/day. DePinto and Verhoff (1977) stated that excess cellular P is stored in algae in an inorganic state. This would explain the much higher P regeneration rates from decaying algae growing in a P-rich environment. Therefore, bacteria mineralizing phosphorus-rich algal cells would likely be adapted to high levels of organic P while bacteria associated with P-limited algae would have slower regeneration rates and use more inorganic compounds for their P source.

BIOLOGICAL FILTRATION

The knowledge gained from the various nitrification and denitrification studies has been put to use in the development of specialized and efficient biological filtration units. A biological filter has a great advantage in the conversion of toxic nitrogenous substances to nutrients over natural microbial activity. The major advantage is that of concentration of functional species. Biological filtration units have immense surface areas of suitable substrates for growth and maintenance of nitrifying and denitrifying bacteria. Nitrifying bacteria require oxygen to carry out nitrification of ammonia and nitrite and biological filters offer the ability to fully aerate all surfaces where bacterial activity is maintained. Denitrifying bacteria require anoxic conditions to convert nitrate to nitrogen gas or N_2O. This can easily be accomplished by incorporating a second filter column to the system which is sealed from atmospheric conditions, thus maintaining an anaerobic environment. It should be noted here that all culturing systems do not need the added nitrification and denitrification capabilities of biological filters. Biological filters are basically used in intensive aquaculture systems where recirculating water is necessary due to short water supply.

NUTRIENT SOURCES

Many products, both natural and synthetic, can serve as nutrient sources or fertilizers for aquaculture systems. The most practical substances are products of natural sources which are readily available, inexpensive, and in most cases enhance ecological and aesthetic goals most effectively. These substances include animal wastes and other manures, composts, excreta from culture organisms, perished culture organisms, and some less likely sources such as petrochemicals and brewers' effluents.

In aquaculture, the greatest amount of research has been done regarding the use of various manures as the primary fertilizer for fish ponds (especially for cultures of _Tilapia_, common carp, silver carp, white amur (grass carp), and bighead carp). The manures which have been analyzed most for their efficiencies in producing useful foods for the fish are liquid cow manure, poultry manure, mustard oil cake, liquid swine manure, and human wastes.

The research and/or studies involving the response of aquaculture ponds to different nutrient sources have been primarily mechanistic in approach. The results have tended to show what type or amount of additives yield the best harvest from the ponds. In essence, the complexity of the various nutrient and degradation cycles are reduced to a simple "black box" approach. Various studies (Ghosh 1975; Behrends et al. 1980; Asare 1980; Wahby 1974) have identified efficient and successful results with different manures as nutrient additives. But, these studies do not recognize the interrelationships among the types of nutrient (i.e., which compounds), the necessary degradative process to change the form of the additive, and the size and activity of the bacterial flora required to conduct the nutrient degradation. Currently, the results from basic limnological studies of microbial nutrient cycling must be extrapolated to aquaculture ponds. Research should be initiated to determine if this extrapolation is valid, or if a different set of microbiological principles hold in aquaculture systems.

BACTERIA AS FOOD

The role of bacteria is not strictly limited to the recycling of organic nutrients; in aquaculture systems bacteria have also been used as a direct food source in the culture of plankton as well as useful protein sources for culture species (Atack et al. 1979). There are many aquatic organisms which devour bacteria quite actively - especially protozoans and zooplankton of the class Rotifera. Large populations of plankton can result from this process, which in turn serve as a viable food source for secondary production.

Some fish ingest and digest bacteria when bacteria are associated with detritus eaten by detritivores (e.g. _Tilapia_) or through filter feeding by use of gill rakers. The bacteria which are ingested by filter feeders (e.g. silver carp) are generally attached to some larger organism or piece of detritus which is caught by the gill rakers or actively devoured. Schroeder (1978) showed that a two- to three-fold increase in fish yield occurred in ponds receiving chemical fertilization and manure treatments over ponds receiving only chemical fertilization, even though primary productivity remained the same. Abundance of large "food organisms" in the water also could not account for the difference in production. Therefore, it was determined that the fish were utilizing the bacteria and protozoans attached to detritus and particles of straw from the manures as a source of high quality protein. Autotrophic production is limited by the amount of light reaching the cells which decreases as phytoplankton numbers increase to a saturation point. Heterotrophs are not limited by light transmittance and therefore can continue to produce as long as nutrients are present for their growth and suitable aeration is maintained.

A more recent use of bacteria as a food source is that of bacterial protein in pelleted feeds. Fish meal is typically the major source of protein in pelleted feeds but is becoming financially unfeasible as the costs of this protein source increase. Bacterial proteins are relatively inexpensive and have been shown to be of high quality, being utilized well by the fish species tested (Kaushik and Luquet 1980; Atack et al. 1979). Another study showed little financial incentive for using bacterial feeds, but did maintain the feasibility of bacterial protein utilization in pelleted fish feeds (Matty and Smith 1978).

PATHOGENS

As in any system where living organisms congregate in large numbers, aquaculture systems are afflicted by numerous pathogenic organisms and diseases. Since the advent of more intensive culturing, pathogens have become an increasing problem in the process of raising fish and shellfish. It has also been found that disease problems are more acute in cold water aquaculture than in warm water cultures (Sarig 1979). Even in extensive culturing ponds the problems of disease can be very detrimental to good fish yields. Although the subject of pathogens is an important one, its significance to bacterial nutrient cycling is not very great. Therefore, only a brief discussion will be made of this highly diverse and complex subject.

The major preventative factor against disease in aquaculture remains proper management and husbandry practices; for example, maintenance of nutrient loads without large excesses, balanced nutrition for cultured species, maintenance of proper pH, and consistent monitoring and adjustment for changes in water temperature, ammonia concentration, and dissolved oxygen (Fisher et al. 1978; Sarig 1979).

There are many pathogenic bacteria, but the most commonly encountered groups are listed here: 1) *Aeromonas* and *Pseudomonas*, 2) Corynebacteria, 3) Enterobacteria, 4) Hemophilus, 5) Mycobacteria and *Nocardia*, 6) Myxobacteria, 7) Streptomyces, and 8) *Vibrio* (Sarig 1979).

There are many treatments for diseases in fish and shellfish once the pathogen has become a problem. These include various antibiotics which are usually added to feeds, vaccines (usually injected), and several chemotherapeutics (added to the culturing water). Other techniques which aid in prevention of pathogenic bacterial infections include ultraviolet light disinfection (Brown and Russo 1979), ozonation, and use of diatomaceous earth filters to reduce bacterial inflow to the culturing tanks or pools (Illingworth et al. 1979).

The use of chemotherapeutics presents a great risk of destroying the native bacterial population. Use of these general antibiotics tends to indiscriminately destroy both pathogenic and native bacterial flora and their associated processes. Thus, although a pathogen may be effectively controlled, the natural degradative processes will be arrested. Without these degradative processes, nitrogenous wastes will build up in the aquaculture system and nutrient supplies will become tied up in organic matter. In essence, a broad application of antibiotic may lead to a cure that is detrimental to the functioning of the entire pond system.

If human wastes are used as fertilizers in culturing ponds it should be realized that fish may be vectors of human infectious diseases. This is not a serious problem as fish are not more suspect than cattle, sheep, and other organisms which are cultured by man, but one should not assume that fish cannot be carriers of human diseases. Further research is needed to determine the extent (if any) of human pathogen-carrying capacities of fish and other cultured organisms in aquaculture (Janssen 1970).

FUTURE RESEARCH

Aquaculture is a practice which strives to maximize fish and shellfish yields with minimum energy inputs. Despite the evolution of the most efficient practices, much research has yet to be conducted before a useful understanding of aquaculture pond principles is developed. The role of bacteria in aquaculture has been only minimally examined and further research can be conducted in many areas.

Biological filtration has become widely accepted as an inexpensive and effective means of converting toxic metabolic wastes to useful compounds in culturing systems. Srna (1976) showed that studies on the kinetics of nitrification can aid in developing more efficient and compact biological filters, yet further research could be conducted to increase efficiencies in biological filtration units incorporating different substrates in various combinations.

Bacteria have been used successfully as inexpensive and nutritious substitutes for fish meal in pelleted fish feeds (Kaushik and Luquet 1980; Atack et al. 1979; Matty and Smith 1978). Research could be conducted to determine the most efficient combinations of bacterial and other single-celled proteins to use and which species of bacteria provide the most nutritious sources of protein. As Schroeder (1978) observed, a large percentage of aquaculture production could not directly be accounted for in ponds receiving chemical and supplemental manure treatments, indicating that some source of nutrition was present that has as yet not been detected. It was determined that bacteria and protozoa provided the additional food to the cultured species. Further studies need to be conducted to verify this determination and to investigate how these organisms can be utilized more efficiently.

Attempts have been made to isolate various bacterial groups to be used as reference strains for early warning detection of pathogens (Egidius and Andersen 1977; Gilmour et al. 1976). Investigation in this area could be expanded to encompass other pathogenic strains to develop a more complete and accurate early warning system to protect the vast investments involved in aquacultural endeavors.

Further studies into the use of bacterial activity indicators could prove beneficial, especially in ponds fertilized by manures, as was shown by the use of cotton strips to measure cellulase activity in manured ponds (Schroeder 1978). This measurement would aid the user in determining maximum amounts of manure which could be added to maximize fish production.

LITERATURE CITED

Asare, S.O. 1980. Animal waste as a nitrogen source for Gracilaria tikvahiae and Neoagardhiella baileyi in culture. Aquaculture 21:87-91.

Atack, T.H., K. Jauncey, and A.J. Matty. 1979. The utilization of some single cell proteins by fingerling mirror carp (Cyprinus carpio). Aquaculture 18:337-348.

Behrends, L.L., J.J. Maddox, C.E. Madewell, and R.S. Pile. 1980. Comparison of two methods of using liquid swine manure as an organic fertilizer in the production of filter feeding fish. Aquaculture 20:147-153.

Brown, C., and D.J. Russo. 1979. Ultraviolet light disinfection of shellfish hatchery sea water. I. Elimination of five pathogenic bacteria. Aquaculture 17:17-23.

Charlton, M.N. 1975. Sedimentation: measurements in experimental enclosures. Verhein. Internat. Verein. Limnol. 19:267-272.

Colt, J., and G. Tchobanoglous. 1976. Evaluation of the short term toxicity of nitrogenous compounds to channel catfish Ictalurus punctatus. Aquaculture 8:209-224.

DePinto, J.V., and F.H. Verhoff. 1977. Nutrient regeneration from aerobic decomposition of green algae. Eng. Sci. Technol. 11:371-377.

Edwards, P. 1980. A review of recycling organic wastes into fish, with emphasis on the tropics. Aquaculture 21:261-279.

Egidius, E., and K. Andersen. 1977. Norwegian reference strains of Vibrio anguillarum. Aquaculture 10:215-219.

Fallon, R.D., and T.D. Brock. 1979. Decomposition of blue-green algal (cyanobacterial) blooms in Lake Mendota, Wisconsin. Appl. Environ. Microbiol. 37:820-830.

_______, and ________. 1980. Planktonic blue-green algae: production, sedimentation and decomposition in Lake Mendota, Wisconsin. Limnol. Oceanogr. 25:72-88.

Fisher, W.S., E.H. Nilson, J.F. Steenbergen, and D.V. Lightner. 1978. Microbial diseases of cultured lobsters: a review. Aquaculture 14:115-140.

Fogg, G.E. 1975. Algal Cultures and Phytoplankton Ecology. 2nd ed., Univ. of Wisconsin Press, Madison. 175 pp.

Foree, E.G., and R.L. Barrow. 1970. Algal growth and decomposition: effects on water quality, phase II. Nutrient regeneration, composition and decomposition of algae in batch culture. Office of Water Resources Research, Report No. 31. Kentucky Water Resources Institute.

Fuhs, G.W. 1973. Improved device for the collection of sedimenting matter. Limnol. Oceanogr. 18:989-993.

Ghosh, S.R. 1975. A study on the relative efficiency of organic manures and the effect of salinity on its mineralisation in brackishwater fish farm soil. Aquaculture 5:359-366.

Gilmour, A., M.C. Allan, and M.F. McCallum. 1976. The unsuitability of high pH media for the selection of marine Vibrio species. Aquaculture 7:81-87.

Golterman, H.L. 1973. Natural phosphate sources in relation to phosphate budgets: a contribution to the understanding of eutrophication. Water Res. 7:3-17.

Hutchinson, G.E., and V.T. Bowen. 1950. Limnological studies in Connecticut. 9. A quantitative radiochemical study of the phosphorus cycle in Linsley Pond. Ecology 31:194-203.

Illingworth, J., F.M. Patrick, P. Redfearn, and D.M. Rodley. 1979. Construction, operation and performance of a modified diatomaceous earth filter for shellfish hatcheries. Aquaculture 17:181-187.

Janssen, W.A. 1970. Fish as potential vectors of human bacterial diseases, pp. 284-290. In S.F. Snieszko (ed), A Symposium on Diseases of Fishes and Shellfishes. Special Publication No. 5, American Fisheries Society.

Jones, J.G. 1972. Studies on freshwater bacteria: association with algae and alkaline phosphatase activity. J. Ecol. 60:59-75.

Kaushik, S.J., and P. Luquet. 1980. Influence of bacterial protein incorporation and of sulphur amino acid supplementation to such diets on growth of rainbow trout, Salmo gairdnerii Richardson. Aquaculture 19:163-175.

Kuznetsov, S.I. 1970. The Microflora of Lakes and its Geochemical Activity. Univ. of Texas Press, Austin. 503 pp.

Lean, D.R.S. 1973. Phosphorus dynamics in lakewater. Science 179:678-680.

Likens, G.E. 1972. Eutrophication and aquatic ecosystems, pp. 3-13. In G.E. Likens (ed), Nutrients and Eutrophication: the Limiting-Nutrient Controversy. Amer. Soc. Limnol. Oceanogr. Spec. Symposia I.

Matty, A.J. and P. Smith. 1978. Evaluation of a yeast, a bacterium and an alga as a protein source for rainbow trout. I. Effect of protein level on growth, gross conversion efficiency and protein conversion efficiency. Aquaculture 14:235-246.

McCoy, E.F. 1972. Role of bacteria in the nitrogen cycle in lakes. USEPA Water Pollution Control Research Series No. 16010 EHR 03/72.

Rigler, F.H. 1973. A dynamic view of the phosphorus cycle in lakes, pp. 539-572. In E.J. Griffith et al. (eds), Environmental Phosphorus Handbook. Wiley.

Sarig, S. 1979. Fish diseases and their control in aquaculture, pp. 190-197. In T.V.R. Pillay and W.A. Dill, (eds), FAO Technical Conference on Aquaculture, Kyoto, Japan, 1976. Advances in Aquaculture, Fishing News Books, Ltd., England.

Schindler, D.W. 1975. Whole-lake eutrophication experiments with phosphorus, nitrogen and carbon. Proc. Int. Assoc. Theor. Appl. Limnol. 19:3221-3231.

Schroeder, G.L. 1978. Autotrophic and heterotrophic production of microorganisms in intensely manured fish ponds, and related yields. Aquaculture 14:303-325.

Seymour, E.A. 1980. The effects and control of algal blooms in fish ponds. Aquaculture 19:55-74.

Srna, R.F. 1976. A modular nitrification filter design based on a study of the kinetics of nitrification of marine bacteria. Univ. Del. Sea Grant Publ. DEL-SG-7-76.

Stumm, W., and J.J. Morgan. 1970. Aquatic Chemistry: An Introduction Emphasizing Chemical Equilibria in Natural Waters. John Wiley, New York.

Wahby, S.D. 1974. Fertilizing fish ponds. I. Chemistry of the waters. Aquaculture 3:245-259.

Wetzel, R.G. 1975. Limnology. W.B. Saunders Co., Philadelphia. 743 pp.

BIOLOGICAL PRINCIPLES OF POND CULTURE: SEDIMENT AND BENTHOS

by

David White

For most types of aquaculture systems, the relationships of sediments or "soils" and the benthic organisms occupying the soils are poorly understood. Further, the processes which occur at this level of freshwater and estuarian ecosystems have been examined in depth only within the past twenty years. Thus, benthic ecological theory is lacking in many potentially important soil-water processes. Much of the information in this review comes either from the general limnological literature or the aquaculture literature involving benthic feeding fishes. In practice, the relationship of soils and benthos to aquaculture should be applicable to and have a direct bearing on not only the culture of benthic feeders but also on other types of aquaculture in both cool and warm water, and in low and high intensity farming. In much of the aquaculture literature, soils and benthos are mentioned only in passing beyond pond site selection and manipulation. By example, Bardach et al. (1972) list fewer than a dozen citations directly related to the role of soils or benthos. This is somewhat surprising as the potential importance of soils in pond fertility has been known for some time (Pershall and Mortimer 1939; Mortimer 1949).

The origins and compositions of bottom soils are determined in part by the geological formations and soils of the basin, in part by autochthonous processes in the water column and the water-soil interface, and in part by allochthonous materials entering the system via runoff and anthropogenic sources. The benthic or bottom-dwelling plant and animal community structure is directly related to the composition of the soils and is modified by the physiochemical conditions in the overlying water (Ruttner 1952; Hynes 1970).

Considered here is the role of soils as it relates to the benthic community which thereby influences aquaculture practices. It is assumed here that the pond site is active. Selection of the site and site preparation will directly affect sediments through potential loading as a function of the basin hydrology (Bardach et al. 1972: Tapiador and Henderson 1977). The drainage patterns will influence a site through runoff of silts, human and animal wastes, agricultural fertilization, and toxins such as pesticides, herbicides, heavy metals, and toxic organics. Site dynamics must also provide for reasonable water circulation to prevent local areas of stagnation (Delmendo and Gedney 1976).

SEDIMENTS AND NUTRIENTS

The productivity of an aquaculture system is directly dependent on the fertility of the sedimentary soils, with the type and texture being the basic governing factors (Djajadiredja and Poernomo 1972). Pure sands provide the least productive (Schuster 1949; also see reviews in Golterman 1977). The annual yields of some cultured species are directly proportional to the organic carbon content (Liang and Huang 1972).

Under anaerobic or reduced conditions in the soil and overlying water (if the pond stratifies), nutrients and some ions are released to the water. This process is independent of biologically mediated exchange. The magnitude and direction of the exchange process is proportional to the amounts of nutrients available. In most systems where soils contain silts and organic muck, the net flow of nutrients and ions will be from the soil to water. In most aquaculture systems, the overlying water is kept at or near oxygen saturation (Bardach et al 1972) and the net flow of nutrients will be from water to soil; thus the soils act as a sink (Mortimer 1941, 1942, 1971; Fillos 1977). Regeneration of the nutrients in these systems primarily comes through biological interactions at the soil-water interface and subsurface mixing. Aquaculture practices can account for nutrient build-up in soils through various manipulations such as draining or flushing of pond sites and subsequent reworking of the soils. Figure 1 shows the typical annual pond manipulations in culture of milkfish (Chanos chanos, a brackish-water benthic algal feeder) through data compiled from Bardach et al. (1972); Delmendo and Gedney (1976); Pillay (1972); Smith et al. (1978); Spotte (1979); and Tapiador and Henderson (1977). Following harvest, ponds are drained. While still moist, the soil is tilled to a specific colloidal texture based on the Atterberg scale (Liang and Huang 1972) and tested for levels of organic matter. If needed, lime is added to adjust the pH. Both organic and inorganic fertilizers are added as required, and then allowed to settle for two to three weeks. The site is then flooded minimally (1-10 cm) to induce the development of benthic algal culture. In this type of system, it is important that light penetrate to the soil surface for maximum algae growth over the next one to six weeks. The benthic or epipelic algae grown in this manner (along with the associated detrital aggregate) is the primary food source for the milkfish. Once a sufficient culture has developed, the pond is flooded (35-100 cm) and restocked with fish of various sizes.

Harvest

drain pond

dry

till soil

lime (if required)

organic and inorganic fertilizers

flood to 10 cm

allow for algal growth

flood to 35-100 cm

stock fish

Figure 1

Pond manipulation for growth of benthic algae in milkfish (Chanos chanos) culture.

Species of cultured fish which directly feed on soils or the detrital components in the system (various carps, Tilapia) also rely on the quality and level of nutrients in the soil. For the Tilapia sarotherodon mossambicus, Bowen (1980) has shown that it is not the amount of detrital aggregate available but the levels of nonprotein amino acids and their composition in the detritus that will determine growth rates. The detrital aggregate process may have a direct relationship to the level of nutrients in the soil coupled with more complex organic compounds in the water column. However, it is not truly known at this time what factors produce the detrital aggregate or how the nonprotein detrital amino acids are related to fertilization and maintenance of aquaculture systems (Bowen 1979, 1980).

SEDIMENT AND ALGAL-BACTERIAL RELATIONSHIP

The soil type directly controls the fertility of the aquaculture system, particularly algal and bacterial production in what has been termed the subaqueous horizontal layer (Rosell and Arguelles 1936). A list of some of the more important algal taxa is given in Table 1. Epipelic algae, both diatoms and filamentous, create chemical gradients by their uptake providing a more rapid flux of nutrients from the soils (Lee 1970). This, in turn, stimulates planktonic algal growth and nutrient uptake in the water column (Golterman et al. 1969). Release of nutrients from sediments under aerobic conditions is dependent on the types and densities of algae, and it is thought that the interaction with bacteria may control the form of nutrients released (Porcella et al. 1970). The exact mechanisms involved are not well known. Porcella et al. (1970) further found that mats of filamentous algae (e.g., Oscillatoria), which are a primary food source for species such as milkfish, may promote anaerobic conditions through metabolic activities in still water. The mats themselves disrupt soil-water mixing which further results in the release of organics and nutrients from the soils. From what is known, epipelic production and community structure may be directly proportional to nutrient contents of the soil, particularly nitrogen levels (see Patrick 1977 for effects of specific ions and nutrients on algal growth).

Effective aquaculture practices often require the selective growth, protection, and replenishment of desirable epipelic algal species (Blanco 1972). Improper nutrient balance or overfertilization of the soil may accelerate nutrient release sufficient for both epipelic and planktonic algal blooms. Such blooms often are responsible for fish mortality and/or fish flesh taste deterioration. In brackish-water pond culture, nutrients from soil fertilization provide the basis for growth of "lab-lab," a complex of epipelic diatoms, green and blue-green algae, and the associated benthic fauna of protozoans, copepods, worms, insect larvae, etc., which together serve as a staple for benthic feeding fish (Gopalakrishnan 1972).

Bacteria, while usually not serving directly as a food source for cultured fish, function in releasing or rereleasing nutrients from soils, particularly in the detrital aggregate component, by assimilating various organic compounds (Neilson 1962). Decomposition and assimilation mechanisms are poorly understood but are known to be greatly enhanced by larger benthic invertebrates and fish which physically stir the soils (Hargrave 1970; Provini and Marchetti 1976). The mixing process primarily allows for greater contact with overlying water. This creates more effective bacterial activity as higher nutrient gradients are produced.

Table 1

EXAMPLES OF BENTHIC ALGAE AND MACROINVERTEBRATES ASSOCIATED WITH SEDIMENTS IN AQUACULTURE, PRIMARILY FRESHWATER

Diatoms	Benthic Invertebrates
Caloneis spp.	Tubificidae (Oligochaeta)
Diploneis spp.	*Tubifex* spp.
Fragilaria spp.	*Limnodrilus* spp.
Gyrosigma spp.	Gastropoda
Navicula spp.	Cerithidae spp.
Pinnullaria spp.	Physidae spp.
Nitzschia spp.	Lymnaeidae spp.
Surirella spp.	Insecta
	Chironomidae -
Filamentous algae	tribe Chronominae in
Oscillatoria spp.	particular
Cladophora spp.	

The presence of toxic substances and antibiotics which kill bacteria or limit their growth may alter rates and modes of decomposition and assimilation (Fillos 1977; Hayes and Phillips 1958) causing distinct changes in overall productivity of a pond system. The magnitude and direction of the changes, however, are not predictable.

SEDIMENT AND MACROBENTHOS RELATIONSHIPS

The larger benthic invertebrates (Table 1) enhance exchange between soils and water not only by metabolic processes but by directly feeding on soils and detritus which increases porosity, mixes soils, and alters particle size (Fisher et al. 1980). Such mechanical changes, termed bioturbation, are more rapid and complete with the oligochaete worms which act as conveyor belts to deposit subsurface sediments at the soil-water interface. Other organisms (Chironomidae larvae, gastropods, decapods, etc.) also act to alter soil profiles but to a lesser extent. Bioturbation increases exchange rates and solubility of nutrients naturally occurring and from fertilization. Under laboratory conditions, some oligochaete species may rework the entire top 5 cm every two weeks (Fisher et al. 1980). In natural conditions these rates are much lower but still highly significant (Krezoski et al. 1978).

Both oligochaetes and chironomid larvae are ubiquitous in fine sediments of virtually all aquatic systems and often become particularly abundant in aquaculture ponds (Bardach et al. 1972). Beyond physical mixing of soils, these taxa pump proportionally large quantities of water into the soil, increasing pore water exchange which ultimately results in increased release of nutrients. Holdren and Armstrong (1980) have reported a direct relationship between silica and phosphorus levels in overlying waters and the presence and abundance of particular chironomid species. Aquaculture systems which promote and establish growth and maintenance of the benthic community should more effectively utilize the nutrients in the system. This concept in aquaculture seems not to have been greatly explored.

In small ponds, mass emergences of adult chironomids may result in heavy losses of nutrients from the system, although again little is known of energy budgets of entire systems to predict this type of flux. This may be most important in tropical aquaculture where chironomid densities often are very high (Oldroyd 1964). Further information on nutrient exchange between soils and water as modified by the benthic community of temperate waters has been summarized by Nalepa and Quigley (1980).

Benthic macroinvertebrates can influence fish production in other manners. Several chironomid species feed directly on filamentous algae, as do various gastropods (Cerithidae). These species may compete directly with cultured fish that are also benthic feeders (Bardach et al. 1972; Liang and Huang 1972; Pillai 1972), but the amount of energy lost to the system in this manner is unknown (Pillay 1972). Several physical and chemical methods are used to reduce population of these species (Pillai 1972), but effects on beneficial species have not been shown nor have the effects on nutrient cycling. It has been further shown that certain benthic invertebrates may release growth substances which promote growth of algae (Mundie 1956).

BENTHIC INVERTEBRATES AS A FOOD SOURCE IN AQUACULTURE

Most aquaculture, even low intensity cool water systems, does not rely on natural or added benthic invertebrate populations as a major source of food for fish (Bardach et al. 1972). Seasonality of production and poor knowledge or unpredictability of life cycles have made this impractical. Even under the best of conditions, it is often difficult to establish and maintain the required biomass in either new or ongoing sites. Some elaborate methods of rearing benthic organisms have been attempted with varying success. Pillay (1972) noted one method of rearing larval chironomids in Israel. Shallow pans 1 m in diameter are filled with soil-Asmal, chicken manure, and fish meal. Adult chironomids are attracted by the odor and lay their eggs in the pans. About 200 g/m^2 can be obtained every 2 to 3 days. The success of this method depends on a large stock of wild adult flies being present. When one considers a standard conversion ratio of 10 gm larvae to 1 gm fish, a very large number of pans has to be worked each day to provide sufficient food for even modestly intense aquaculture.

SUMMARY

The greatest contribution of benthic organisms to aquaculture is their ability to modify sediments and enhance release of nutrients. Where algae serve directly to provide a food source for fish, nutrient transfer may be immediate. Benthic bacteria and the larger benthic invertebrates are most important in reworking soils and in facilitating release of nutrients from the soil by assimilating compounds and keeping the top few centimeters in constant movement. Production of fish will be directly proportional to the fertility and type of soils present.

LITERATURE CITED

Bardach, J.E., J.H. Ryther, and W.O. McLarney. 1972. Aquaculture. The farming and husbandry of freshwater and marine organisms. Wiley-Interscience, John Wiley & Sons, Inc., New York. 868 pp.

Blanco, G.J. 1972. Status and problems of coastal aquaculture in the Philippines, pp. 60-67. In T.V.R. Pillay (ed), Coastal Aquaculture in the Indo-Pacific Region. Fishing News Books, London.

Bowen, S.H. 1979. A nutritional constraint in detritivory by fishes: The stunted population of Sarotherodon mossambicus in Lake Sibaya, South Africa. Ecol. Monogr. 1979:17-31.

__________. 1980. Detrital nonprotein amino acids are the key to rapid growth of Tilapia in Lake Valencia, Venezuela. Science 107:1216-1218.

Delmendo, M.N., and R.H. Gedney. 1976. Laguna de Bay fish pen aquaculture development--Philippines. Paper presented, Seventh Annual Meeting World Mariculture Soc., 26-29 January 1969, San Diego, CA 8 pp.

Djajadiredja, R., and A. Poernomo. 1972. Requirements for successful fertilization to increase milkfish production, pp. 398-409. In T.V.R. Pillay (ed), Coastal Aquaculture in the Indo-Pacific Region. Fishing News Books, London.

Fillos, J. 1977. Effects of sediments on the quality of overlying water, pp. 266-271. In H.L. Golterman (ed), Interactions Between Sediments and Fresh Water. Dr. W. Junk B.V. Publishers, The Hague.

Fisher, J.B., W.J. Lick, P.L. McCall, and J.A. Robbins. 1980. Vertical mixing of lake sediments by tubificid oligochaetes. J. Geophysical Res. 85:3997-4006.

Golterman, J.L. (ed). 1977. Interactions Between Sediments and Fresh Water. Dr. W. Junk B.V. Publishers, The Hague.

__________, C.C. Bakels, and J. Jacobs-Mogelin. 1969. Availability of mud phosphates for the growth of algae. Verh. Internat. Verein. Limnol. 17:467-479.

Gopalakrishnan, V. 1972. Taxonomy and biology of tropical fin-fish for coastal aquaculture in the Indo-Pacific region, pp. 120-149. In T.V.R. Pillay (ed), Coastal Aquaculture in the Indo-Pacific Region. Fishing News Books, London.

Hargrave, B.T. 1970. The effect of a deposit-feeding amphipod on the metabolism of benthic microflora. Limnol. Oceanogr. 15:21-30.

Hayes, F.R., and J.E. Phillips. 1958. Lake water and sediment. IV. Radiophosphorus equilibrium with mud, plants, and bacteria under oxidized and reduced conditions. Limnol Oceanogr. 3:459-475.

Holdren, G.C., Jr., and D.E. Armstrong. 1980. Factors affecting phosphorus release from intact lake sediment cores. Environ. Sci. Tech. 14:79-87.

Hynes, H.B.N. 1970. The Ecology of Running Waters. Univ. of Toronto Press. 555 pp.

Krezoski, J.R., S.C. Mozley, and J.A.Robbins. 1978. Influence of benthic macroinvertebrates on mixing of profundal sediments in southeastern Lake Huron. Limnol. Oceanogr. 23:1011-1016.

Lee, G.F. 1970. Factors affecting the transfer of materials between water and sediments. Littr. Rev. No. 1, Water Resources Center, Univ. of Wisconsin, Madison, Wisc.

Liang, J., and C. Huang. 1972. Milkfish production in a newly reclaimed tidal land in Taiwan, pp. 417-428. In T.V.R. Pillay (ed), Coastal Aquaculture in the Indo-Pacific Region. Fishing News Books, London.

Mortimer, C.H. 1941. The exchange of dissolved substances between mud and water in lakes. J. Ecol. 29:280-329.

____________. 1942. The exchange of dissolved substances between mud and water in lakes. III and IV, Summary and References. J. Ecol. 30:147-201.

____________. 1949. Seasonal changes in chemical conditions near the mud surface in the two lakes of the English Lake District. Verh. Internat. Verin. Limnol. 10:353-356.

____________. 1971. Chemical exchanges between sediments and water in the Great Lakes; speculations on probable regulatory mechanisms. Limnol. Oceanogr. 16:387-404.

Mundie, J.H. 1956. The biology of flies associated with water supply. J. Inst. Publ. Hlth. Eng. 1956:178-193.

Nalepa, T.V., and M.A. Quigley. 1980. The abundance and biomass of the macro- and meiobenthos in southeastern Lake Michigan. NOAA Data Report ERL GLERL-17. 12 pp.

Neilson, C.O. 1962. Carbohydrates in soil and litter invertebrates. Oikos 13:200-215.

Oldroyd, H. 1964. The Natural History of Flies. W.W. Norton & Co., Inc., New York. 324 pp.

Patrick, R. 1977. Ecology of freshwater diatoms and diatom communities, pp. 284-332. In D. Werner (ed), The Biology of Diatoms. University of California Press, Berkeley.

Pershall, W.H., and C.H. Mortimer. 1939. Oxidation-reduction potentials in water-logged soils, natural waters and muds. J. Ecol. 22:483-501.

Pillai, T.G. 1972. Pests and predators in coastal aquaculture systems of the Indo-Pacific region, pp. 456-471. In T.V.R. Pillay (ed), Coastal Aquaculture in the Indo-Pacific Region. Fishing News Books, London.

Pillay, T.V.R. (ed). 1972. Coastal Aquaculture in the Indo-Pacific Region. Fishing News Books, London. 497 pp.

Porcella, D.B., J.S. Kumagai, A.M. Ase, and E.J. Middlebrooks. 1970. Biological effects on sediment-water nutrient interchange. J. Sanit. Engin. Div. 96:911-926.

Provini, A., and R. Marchetti. 1976. Oxygen uptake rate of river sediments and benthic fauna. Bull. Zool. 43:87-110.

Rosell, D.Z., and A.S. Arguelles. 1936. Soil types and growth of algae in bangos fish ponds. Philipp. J. Sci. 61(1):1-7.

Ruttner, F. 1952. Fundamentals of Limnology. University of Toronto Press. 307 pp.

Schuster, W.H. 1949. De viscultuur in de kustvijvers op Java. Dep. van Landbouw Visserij, Onderafdeling Binnenvisserij, (2). 143 pp.

Smith, I.F., F.C.Cas, B.P. Gibe, and L.M. Romillo. 1978. Preliminary analysis of the performance of the fry industry of the milkfish (Chanos chanos Forskal) in the Philippines. Aquaculture:14:119-219.

Spotte, S. 1979. Fish and Invertebrate Culture: Water Management in Closed Systems. 2nd ed., Wiley-Interscience, John Wiley & Sons, Inc., New York. 179 pp.

Tapiador, D.D., and H.F. Henderson. 1977. Freshwater fisheries and aquaculture in China. FAO Fisheries Technical Paper, No. 168 (FIR/T168). 84 pp.

BIOLOGICAL PRINCIPLES OF POND CULTURE: PHYTOPLANKTON AND MACROPHYTES

by

C. Kwei Lin

Freshwater fish culture has a long history, but an enormous expansion has taken place during the last few decades. The fishes most commonly used in pond culture in warm climates are various herbivorous species, such as carp (grass carp - *Ctenopharyngodon idellus*, silver carp - *Hypothalmichthys molitrix*, bighead carp - *Aristichthys nobilis*, and mud carp - *Cirrhina molitorella*), species of *Tilapia* and grey mullet (*Mugil cephalus*). The important species in tropical fish ponds are herbivores, plankton feeders, or omnivores that can thrive on detritus material. Generally, a mixture of two or more species having different feeding habits and stocked in the same pond result in greater gross yields.

Although some freshwater fish-farming is carried out in cold temperate climates in Europe and North America, those high-yield intensive cultures (catfish, trout, etc.) require expensive fishmeal. The acute need for animal protein in many underdeveloped tropical countries has turned them to supplies from pond cultured fish. In the tropics, the potential growth rates of fishes under the warmer conditions and extended active growing season provide the most productive fish culture systems (Hickling 1968; Horn and Pillay 1962; Chen 1976). Fish farming in ponds has a long history of empirical development and "green-thumb" expertise, but the scientific investigation and documentation fall relatively short.

PHYTOPLANKTON COMMUNITIES AND SPECIES COMPOSITION IN FISH PONDS

The growth of phytoplankton and aquatic macrophytes is the most critical aspect of fish production in pond culture. The phytoplankton growth and its ecological factors in fish ponds have concerned fish farmers the world over. For example, Lin (1970) stated that many Chinese carp farmers judge the water quality of fish ponds by their color - the degree of greenness reflects the abundance of phytoplankton. Planktonic algae are food for silver carp as well as for zooplankton, which is food for bighead carp. It also means abundance of dissolved oxygen in the pond produced by phytoplankton. Unfortunately, such expertise seldom provides precise information on species composition, abundance, and related water quality parameters. In recent years, however, many investigations on those aspects have been carried out and documented in scientific periodicals, and are particularly abundant in journals such as *Bamidgeh* (Israeli) and *Hydrobiologia*.

Species composition of phytoplankton communities in fish ponds is important because different taxa of planktonic algae present different diet values in various developmental stages of fish or zooplankton. In general, phytoplankton species occurring in fish ponds include members of Chlorophyta (green algae), Cyanophyta (blue-green algae), Chrysophyta (diatoms and golden brown algae), Euglenophyta (euglenoids), and Pyrrhophyta (dino-flagellates). Literature surveys show that the following genera are most commonly found in warmwater fish ponds: Blue-green algae (*Microcystis*, *Aphanozomenon*, *Anabaena*, *Oscillatoria*, and *Spirulina*); green algae (*Scenedesmus*, *Pandorina*, *Ankistrodesmus*, *Chodatella*, *Dictyosphaerium*, *Sphaerocystis*, *Coelastrum*, *Tetradron*, *Pediastrum*, *Staurastrum*, *Selenastrum*, *Oocystis*, *Closterium*, *Golenkinia*, *Kirchneriella*); Diatoms (*Cyclotella*, *Fragilaria*, *Synedra*, *Nitzschia*, and *Navicula*); Englenoids (*Phacus*, *Euglena*, and *Trachelomonas*); and Dinoflagellates (*Ceratium* and *Gymnodinium*) (Wiebe 1930; Sreenivasan 1964a, 1964b; George 1966; Vaas and Vaas-van Oven 1959; Zafar 1967; Boyd 1973; Seenayya 1972).

The phytoplankton species composition in fish ponds may vary from a few to a large number of species. Green and blue-green algae are usually most abundant in warmwater fish ponds. The species diversity concept, developed by Margalef (1958) for phytoplankton, is useful in describing the significance of species abundance in relation to the stability of the phytoplankton community. In general, higher species diversity indicates a greater stability in a given ecosystem (Odum 1971), because the fluctuations in abundance of individual species would have less influence on the entire community than the systems of lower diversity. In most fish ponds, nutrient enrichment by artificial fertilization is commonly practiced in order to increase the fish production (Chen 1976;

Hepher 1962; Chiou and Boyd 1974). As a result, the phytoplankton species diversity is lowered in the nutrient enriched system, which leads to a wide fluctuation of total population density. The individual species of phytoplankton in fertilized ponds often undergo rapid cycles of population bloom and massive die-off, and may cause oxygen super-saturation as well as severe oxygen depletion that results in fish kill (Swingle 1968; Boyd 1975). Excessive blooms of blue-green algae frequently occur in nutrient-rich water and become the most damaging situation in fish ponds. Blue-green algae produce geosmin, a compound with an earthy-musty flavor and odor, which is excreted into the water and absorbed by fish, giving them an off-flavor taste (Lovell and Sackey 1973).

The most notorious blooms are commonly formed by members of blue-greens - Anabaena, Aphanizomenon and Microcystis. Those blue-green algae possess buoyant gas vacuoles and accumulate at the surface of a pond during warm, calm weather. The development of blue-green algal scums limits light penetration, lowers CO_2 concentration, and depletes nutrients in surface water. Those conditions lead to massive die-offs and severe depletion of dissolved oxygen following the decay of the dead algae. The photo-oxidation process was also reported as an important factor causing the bloom to collapse (Abeliovich and Shilo 1972; Abeliovich et al. 1974).

After observing many fertilized ponds, Boyd (1979) noticed that the phytoplankton abundance and the species composition in adjacent ponds that receive the same treatment may differ greatly. The causes of these differences are unknown, and therefore no existing management procedure will consistently result in a particular type of phytoplankton community. Although blue-green algal blooms always occur in ponds containing high concentrations of phosphorus and nitrogen, not all eutrophic ponds undergo this event.

To prevent eventual collapse of blue-green algal blooms, destratification of the water column by mechanical means has been proven to be effective (Swingle 1968; Zarnecki 1967). The most fundamental way of preventing excessive blooms is through manipulating nutrient levels, keeping certain critical elements at a steady-state limiting situation. Experience with nutrient enrichment in a whole lake involving variation of the N/P ratio gives new insight into changing dominance between blue-green and other types of phytoplankton (Schindler 1977). Nutrient bioassay might provide a rapid, economical means to investigate the nutrient requirement for phytoplankton species composition and abundance desirable in fish ponds. The nutrient bioassay techniques have been widely used in predicting phytoplankton production in relation to eutrophication in natural waters. The systems analysis and numerical models that have been developed for eutrophication in natural waters may provide useful managerial procedures for fish ponds.

The food and feeding habits of freshwater fishes have received much attention (Alikunhi 1952, 1958; Singh 1958, 1961; Das and Moitra 1955; Gupta and Ahmad 1966; Vaas and Vaas-van Oven 1959). Philipose (1960) has reviewed the role of phytoplankton in inland fisheries. The algae commonly used are diatoms, some members of blue-green algae, and a few greens. The observations show that diatoms are readily digested in fish stomachs (Fish 1951; Evans 1960). Tay (1960-61) showed that Anabaenopsis was more readily digested than Anabaena. It is suggested that no further studies on food value of individual phytoplankton components rather than populations as a whole are needed. In spite of considerable work done on the utilization of algae by fishes, little work has been done so far on the effect of feeding selectivity by fishes on sizes and the nutritious value of algae.

PRODUCTIVITY

Since most fish raised in warmwater ponds in underdeveloped countries are dependent largely upon natural foods, the fish production is closely related to the levels of primary productivity of aquatic plants - phytoplankton and macrophytes. Boyd (1979) illustrated that in fertilized ponds stocked with sunfish (Lepomis spp.) and largemouth bass (Micropterus salmoides) phytoplankton represents the base of the food web which culminates in fish production. The production of sunfish is almost directly related to concentrations of particulate organic matter in the ponds (Smith and Swingle 1938; Swingle and Smith 1938). In those fish ponds, the phytoplankton biomass usually comprises most of the particulate organic materials. Maleck (1976) reported that fish yields increased in African and Indian lakes with increasing gross photosynthesis. Production of Tilapia aurea also depends greatly upon plankton density (Almazan and Boyd 1978). Sreenivasan (1964a,b) reported that primary productivity in three tropic Indian lakes was 0.46, 2.9, and 3.8 g C/m^3/day and the fish production was 5.3, 31.6, and 75 kg/hr respectively.

Since managed fish ponds are often fertilized, their nutrient concentrations are generally higher and phytoplankton productivity greater than the average natural waters. Boyd (1973) reported that chlorophyll concentrations in a series of fertilized ponds averaged about one order of magnitude greater than those in the unfertilized ones. In Israel the chlorophyll concentrations in fertilized ponds reach between 103 and 212 μ g/L, as stimulated by the nutrient released from the feed.

Gross phytoplankton productivity is closely related to the fertility of the ponds. Hall et al. (1970) reported that primary productivity was ten to fifteen times greater in fertilized ponds than in unfertilized controls. In Israeli ponds (Hepher 1962), the average productivity ranged from 3.3 to 6.4 g C/m^2/day and the conversion rate to fish production was 1.3-2.3 percent C. In most fish ponds, rates of gross productivity are highest near the surface water and decline rapidly with depth because the dense phytoplankton standing crop reduces light penetration. The compensation point at which the amount of oxygen produced by photosynthesis equals the amount consumed by respiration is extremely shallow. Boyd (1973) recorded that the depth of the compensation points at the Auburn experiment ponds were 0.4-0.75 m, and Hepher (1962) observed them usually at 0.4 m in Israeli ponds. It is concluded that nutrient enrichment of ponds increases phytoplankton production in the upper layers of water and decreases productivity in deeper water. Therefore, excessive fertilization may lower productivity per unit area over the pond.

MACROPHYTES AND BENTHIC ALGAE

Aquatic macrophytes which commonly occur in fish ponds include macroalgae, mosses, ferns, and flowering vascular plants. They appear in various growth forms - submersed, floating, or emergent. The biology of aquatic macrophytes has been a subject of much investigation and reviewed in great detail by Gessner (1959), Sculthorpe (1967), and Hutchinson (1975). Macrophytic algae include those species dwelling on the bottom, e.g., Chara and Nitella and species of filamentous algae which form mats floating on the surface, such as Spirogyra, Cladophora, Rhizoclonium, and Pithophora. Most common vascular plants are species of Eichhornia, Potamogeton, Ceratophyllum, Myriophyllum, Elodia, Najas, and Lemna. Ferns, such as Azolla and Salvinia, are common free floaters.

Macrophytes normally occur in ponds with relatively transparent waters. While floating macrophytes absorb their nutrients from the water column, the rooted ones can use the nutrients from both the water column and sediments (Boyd 1971; Denny 1972; Bristow 1975). Boyd (1979) reported that relatively unproductive fish ponds may support luxurious macrophyte growth due to their ability to use nutrients from the mud. The potential for macrophyte production is greater in hard water than the soft, acid water. In hard water the macrophytes often cause calcium and magnesium to precipitate in colloids.

Occurrence of macrophytes is normally undesirable in managed ponds because they compete for nutrients with phytoplankton, interfere with fish mobility and feeding, and interfere with fish harvest. The most common and serious problems in a pond result from extensive growth of floating macroalgae and higher plants, such as Eichhornia, Lemna, and Azolla because they prevent light penetration causing light limitation for phytoplankton growth. Dobbins and Boyd (1976) reported that macrophyte growth was an important factor in causing unpredictable plankton production, and thus presented a difficult management procedure in fertilizing the ponds. However, there are occasions where aquatic macrophytes in fish ponds become beneficial. Little (1979) gave an extensive review on the utilization and nutritional value of aquatic plants for fish culture throughout the world. The best example is grass carp culture ponds (Chen 1976), where the fingerlings feed on Lemna and mature fish on a large variety of aquatic and even land vegetation. In semi-managed ponds, the aquatic macrophytes can also be beneficial because they harbor a large variety of attached invertebrate fauna such as snails and gammarids which are sources of food for certain fish.

Although macrophyte growth increases with increasing nutrient concentrations, it is often inhibited by turbidity (Spence 1964; Boyd 1971). Smith and Swingle (1941) suggested that inorganic fertilization of ponds often triggers plankton blooms which shade the bottom and prevent macrophyte growth. The maximum depth at which macrophytes occur exhibits a linear relationship with Secchi disc transparency (Hutchinson 1975). Boyd (1975) found that growth of underwater macrophytes was inhibited at depths greater than twice the Secchi disc visibility in ponds in Auburn, Alabama.

Benthic algae have seldom been reported to play a major role as fish food in pond culture. Although microscopic diatoms have been reported as an important food source for detritus feeders such as Tilapia mossambica in tropical lakes (Bowen 1978, 1979), in most fish ponds the dense phytoplankton growth often prevents penetration of light to the bottom of the ponds. Therefore, benthic algae growth is light limited. To fully utilize benthic algae for the culture of Tilapia it may be desirable to design ponds shallow enough to promote the healthy growth of benthic algae. A benthic algal bed has been successfully used for milkfish culture in Taiwan (Ling 1966). There the good growth of benthic algae is responsible for the success of milkfish farming. The benthic algae usually consist of blue-green algae (Oscillatoria, Lynghya, Phormidium, Spirulina, and Micrococcus) and diatoms (Navicula, Mastogloia, Stauroneis, Amphora, and Nitzschia.

RESEARCH NEEDS

As a result of the brief literature review, the following aspects of phytoplankton ecology require further knowledge and research in freshwater fish culture ponds:

1. The environmental conditions governing species composition and succession, particularly control of blue-green algae blooms.
2. The desirable phytoplankton species composition, biomass, and productivity for sustaining an optimal yield of selected species of fish.
3. The relationship between nutrient ratio (N:P) and dominance of phytoplankton species.
4. The nutritional value of phytoplankton for herbivorous fish: protein, carbohydrate, and lipid content of phytoplankton.
5. Phytoplankton response to fertilization (organic and inorganic).
6. Nutrient assimilation capacity of a phytoplankton community, with an emphasis on its utilization of ammonia released by fish, and anaerobic decomposition of organic matter.

LITERATURE CITED

Abeliovich, A., and M. Shilo. 1972. Photooxidative death in blue-green algae. J. Bacteriol. 111:682-689.

____________, D. Kellenberg, M. Shilo. 1974. Effect of photooxidative conditions on levels of superoxide dismutase in Anacystis nidulans. Photochem. Photobiol. 19:379-382.

Alikunhi, K.H. 1952. On the food of young carp fry. J. Zool. Soc. India 4:77-84.

____________. 1958. Observations on the feeding habits of young carp fry. Indian J. Fish. 5:95-106.

Almazan, G., and C.E. Boyd. 1978. An evaluation of Secchi disc visibility for estimating pond density in fish ponds. Hydrobiologia 65:601-608.

Bowen, S.H. 1978. Benthic diatom distribution and grazing by Sarotherodon mossambicus in Lake Sibaya, South Africa. Freshwater Biology 8:449-453.

____________. 1979. A nutritional constraint in detritivory by fishes: the stunted population of Sarotherodon mossambicus in Lake Sibaya, South Africa. Ecological Monogr. 49:17-31.

Boyd, C.E. 1971. The limnological role of aquatic macrophytes and their relationship to reservoir management, pp. 153-166. In G.E. Hall (ed), Reservoir Fish. Limnol. Amer. Fish . Soc., Spec. Publ. No. 8.

____________. 1973. Summer algal communities and primary productivity in fish ponds. Hydrobiologia 41:357-390.

____________. 1975. Competition For Light by Aquatic Plants in Fish Ponds. Auburn Univ. Agr. Expt. Sta. Circ. 215. 19 pp.

____________. 1979. Water Quality in Warmwater Fish Ponds. Auburn Univ. 369 pp.

Bristow, J.M. 1975. The structure and function of roots in aquatic vascular plants, pp. 221-236. In J.G. Torrey and D.T. Clarkson (eds), The Development and Function of Roots. Academic Press, New York.

Chen, T.P. 1976. Aquaculture Practices in Taiwan. Page Bros. (Norwich) Ltd. 161 pp.

Chiou, C. and C.E. Boyd. 1974. The utilization of phosphorus from muds by the plankter, Scenedesmus demorphus, and the significance of these findings to the practice of pond fertilization. Hydrobiologia 45:345-355.

Das, S.M., and S.K. Moitra. 1955. Feeding habits of freshwater fishes of Uttar Pradesh. Curr. Sci. 24:417-418.

Denny, P. 1972. Sites of nutrient absorption in aquatic macrophytes. J. Ecol. 60:819-829.

Dobbins, D.A., and C.E. Boyd. 1976. Phosphorus and potassium fertilization in sunfish ponds. Trans. Amer. Fish Soc. 105:536-540.

Evans, J.H. 1960. Algological studies. In Annual Rept. East African Freshwater Fish. Og. Jinga, Uganda.

Fish, G.R. 1951. Digestion in Tilapia esculenta. Nature 167:900-901.

George, M.G. 1966. Comparative plankton ecology of five fish tanks in Delhi, India. Hydrobiol. 27:81-108.

Gessner, F. 1959. Hydrobotanik. Die Physiologischen. Grundlagen der Pflanzenverbreibing in Wasser. II. Stoffhaushalt, Berlin. VEB Dentscher Verlage der Wissenschafter. 710 pp.

Gupta, H.B., and M.R. Ahmad. 1966. Studies on the effect of feeding some freshwater fishes with Scenedesmus obliquus (Turpin) Knetzing. Hydrobiologia 28:42-48.

Hall, D.J., W.E. Cooper, and E.E. Werner. 1970. An experimental approach to the production dynamics and structure of freshwater animal communities. Limnol. Oceanogr. 15:839-928.

Hepher, B. 1962. Ten years of research in fish pond fertilization in Israel. I. The effect of fertilization on fish yields. Bamidgeh 14:29-38.

Hickling, C.F. 1968. The Farming of Fish. Pergamon, Oxford.

Horn, S.L., and T.V.R. Pillay. 1962. Handbook on Fish Culture in the Indo-Pacific Region. FAO, United Nations, Rome.

Hutchinson, G.E. 1975. A Treatise on Limnology. Vol. IV. Limnological Botany. John Wiley and Sons, New York. 660 pp.

Lin, S.Y. 1970. Fish pond fertilization and the principle of water conditioning. China Fisheries Monthly 209.

Ling, S.W. 1966. Feeds and feeding of warm-water fishes in Asia and the Far East. Proc. FAO.

Little, E.C.S. 1979. Handbook of Utilization of Aquatic Plants; A Review of World Literature. FAO Fish. Tech. Pap. Rome. 187 pp.

Lovell, R.T., and L.A. Sackey. 1973. Absorption by channel catfish of earthy-musty flavor compounds synthesized by cultures of blue-green algae. Trans. Amer. Fish. Soc. 102:774-777.

Maleck, J.H. 1976. Primary productivity and fish yields in tropical lakes. Trans. Amer. Fish. Soc. 105:575-580.

Margalef, R. 1958. Temporary succession and spatial heterogeneity in phytoplankton, pp. 323-349. In A.A. Buzzati-Traverso (ed), Perspectives in Marine Biology. Univ. Calif. Press, Berkeley.

Odum, H.T. 1971. Fundamentals of Ecology. 3rd. ed., H.B. Saunders Co., Philadelphia. 574 pp.

Philipose, M.T. 1960. Freshwater phytoplankton of inland fisheries, pp. 272-291. In Proc. Symp. Algology. I.C.A.R. and U.N.E.S.C.O., New Delhi, India.

Schindler, D.W. 1977. Evolution of phosphorus limitation in lakes. Science 195:260-262.

Sculthorpe, C.D. 1967. The Biology of Aquatic Vascular Plants. St. Martin's Press, New York. 610 pp.

Seenayya, G. 1972. Ecological studies in the plankton of certain freshwater ponds of Hyderabad-India. Phytoplankton 2. Hydrobiologia 39:247-271.

Singh, V.P. 1958. Algal food of some local fishes. Indian J. Agr. Sci. 28:403-408.

____________. 1961. An investigation into the composition of the algal food of Gadusia chapra (Ham). Proc. Zool. Soc. (Calcutta) 14:53-60.

Smith, E.V., and H.S. Swingle. 1938. The relationship between plankton production and fish production in ponds. Trans. Amer. Fish Soc. 68:309-315.

____________, and __________. 1941. The use of fertilizer for controlling several submerged aquatic plants in ponds. Trans. Amer. Fish. Soc. 71:94-101.

Spence, D.H.N. 1964. The macrophytic vegetation of freshwater lochs, swamps, and associated fens, pp. 326-425. In J.H. Burnett (ed), The Vegetation of Scotland. Oliver and Boyd, London.

Sreenivasan, A. 1964a. Limnological studies and fish yield in three upland lakes of Madras State, India. Limnol. Oceanogr. 9:564-575.

____________. 1964b. The limnology, primary production, and fish production in a tropical pond. Limnol. Oceanogr. 9:391-396.

Swingle, H.S. 1968. Fish kills caused by phytoplankton blooms and their prevention. Proc. World Symp. on Warm Water Pond Fish Culture. FAO, United Nations. Fish. Rep. 44:407-411.

__________, and E.V. Smith. 1938. Fertilizers for increasing the natural food for fish in ponds. Trans. Amer. Fish. Soc. 68:126-135.

Tay, G. 1960-61. Annual Report Tropical Fish Culture Research Inst. Malacca. 24-26 pp.

Vaas, K.F., and A. Vaas-van Oven. 1959. Studies on the production and utilization of natural food in Indonesian carp ponds. Hydrobiol. 12:308-392.

Wiebe, A.H. 1930. Investigations on plankton production in fish ponds. Bull. U.S. Bureau Fish. 46:136-176.

Zarnecki, S. 1967. Algae and fish relationships, pp. 459-477. In D.F. Jackson (ed), Algae, Man and the Environment. Syracuse Univ. Press.

Zafar, A.R. 1967. On the ecology of algae in certain fish ponds of Hyderabad, India. Hydrobiol. 30:96-112.

BIOLOGICAL PRINCIPLES OF POND CULTURE: ZOOPLANKTON

by

Marlene Evans

STATE OF THE ART OF ZOOPLANKTON IN AQUACULTURE SYSTEMS

The role of zooplankton in aquaculture systems has been investigated to various levels of precision. The crudest is simply a faunal list of zooplankton inhabiting waters that are used or have the potential for being used to rear fish. In general, most of these studies are not at the level of sophistication that characterizes oceanographic and limnological studies conducted in Europe and North America.

Survey Studies

Survey studies have been conducted to determine the suitability of a water body for the introduction of a particular fish species. For the most part, these studies have been conducted in large bodies of water such as lakes, reservoirs, and rivers. Given the size of these systems, little or no management of the water body was intended as part of the stocking program. The major decision regarding the suitability of a water body for the introduction of a fish species has been its physical and chemical characteristics. However, most such studies have determined the major floral and faunal components.

In India, several studies have investigated the hydrobiology of various waters which have the potential for fish culture. Ganapati and Chacko (1951) studied one lake, two reservoirs, and four rivers in Madras State, India (elevation approximately 7,000 ft.). Kodaikanal Lake supported a rich plankton population. English carp (*Carassius carassius*) and baril (*Barilus gatensis*) were successfully introduced into the lake and formed part of a fishery. In addition to these fish, the fishery was based upon minnows, a few Mahseer (*Barbus tor*), and murrels (*Ophicephalus gachua*). Conversely, Pambar Reservoir and Berijam Reservoir were poor in plankton. English carp and rainbow trout had been introduced into these reservoirs although without success. The reason for this failure was not investigated but may have been related to low plankton production. While rainbow trout previously had been successfully introduced into certain rivers, they were decimated by overfishing and heavy otter predation. The authors concluded that the rivers investigated in their study and the three lakes could be used to culture food and game fish. A total of five genera of rotifers, two genera of cladocerans (*Daphnia*, *Ceriodaphnia*), and two genera of copepods (*Microcyclops*, *Eucyclops*) were observed in the seven bodies of water. Zooplankton occurred more commonly in Kodaikanal lake and Berijam Reservoir, although not enough information was provided to interpret the ecological significance of these differences.

Chacko and Ganapati (1952) conducted a hydrobiological survey in the Surula River (elevation 4,850-6,000 feet) in India and concluded that the river was suitable for the introduction of rainbow trout. Zooplankton included three genera of protozoans (*Holophyra*, *Lacrymaria*, *Ulvella*), six genera of cladocerans (*Acroperus*, *Alonella*, *Anuraea*, *Ceriodaphnia*, *Diaphanosoma*, and *Simosa*), one ostracod (*Stenocypris*), and three rotifers (*Rotifer*, *Rattulus*, *Pedalion*); copepods apparently were not collected.

Huq and Sirajul Islam (1977) conducted a survey of manmade ponds in a village in Bangladesh to determine their productive capacity, current use, and reasons for underutilization for aquaculture. The study was an opinion survey of the villagers and provided little data on the physical, chemical, and biological properties of the water. However, the study is noteworthy because, although ponds were located within a relatively small distance of each other, they apparently differed in their abilities to support fish growth. About 61 percent of the ponds supported good fish growth while about 12 percent of the ponds did not.

In Africa, Munro (1966) conducted a limnological survey of Lake McIlwaine, comparing this Rhodesian lake with other lakes in Central South Africa. Lake McIlwaine is utilized as a sport fishery and also supports a gill net fishery. The lake is mildly polluted by sewage and has an abundant plankton population. The author reports ten species of rotifers, six species of cladocerans, and three species of copepods in addition to a littoral cladoceran and ostracod, a *Chaoborus* species, and a species of the coelenterate *Limnocnida*. No estimates were made of zooplankton standing stocks although some seasonal percent composition data were presented.

The state of the art for surveying water bodies for their potential use in low-intensity aquaculture systems is not well developed. The present approach consists of surveying a water body with respect to its physical and chemical characteristics, with relatively little attention addressed to the indigenous biota. However, for fish to be successfully introduced into a stream, river, or reservoir, the biota must be present in sufficient quantities and types to satisfy the nutritional requirements of the fish throughout its life. Zooplankton-consuming fish must be presented with a sufficient array of forms to ensure its survival, healthy growth, and successful reproduction.

Survey studies should contain detailed information on the composition, abundance, and seasonality of zooplankton. All too often, zooplankton data are reported simply as species lists based on single date collections. Survey studies also should provide some estimate of secondary production so that the value of a body of water for fish production can be estimated. Similarly, production estimates are necessary to estimate appropriate stocking programs.

Many studies have been conducted in developing regions of the world, providing information on the ecology of tropical water bodies. For example, Lewis (1978) conducted an excellent study of the dynamics and succession of phytoplankton in a tropical lake in the Philippines. Zooplankton grazing activity (seven species of rotifers, three species of cladocerans, and two species of copepods) was estimated and found to remove only a small percentage (< 7 percent) of the daily primary production. Systems such as these may be particularly suitable for the introduction of phytophagous fish. Matsumura-Tundisi and Tundisi (1976) studied Broa reservoir in Brazil. Populations were more stable in the cold dry season than in the warm rainy season. The reasons for this were not investigated. Rotifers were the most abundant zooplankter (77.6 percent), consisting of fifteen species. Cladocerans consisted of three major species and accounted for 13.6 of the zooplankton. Copepods were the least abundant (8.7 percent) taxonomic group, consisting of six species.

Zooplankton As Indicator Species

A number of researchers have observed that zooplankton composition varies as a function of the physical, chemical, and biological characteristics of the water. Because of this, zooplankton have often been used as indicators of certain trophic conditions of water bodies. Arora (1966) studied rotifers as indicators of the trophic natures of two lakes, a tank, and an oxidation pond in Nagpur, India. Gorewara Lake was considered mildly polluted, while Jumma Tank was more strongly polluted with sewage and industrial effluent. Thirty-six species of rotifers were observed in this study. Considering the two lakes and the tank, the highest density (1,991/L) of rotifers was observed in Jumma Tank while the lowest (<9/L) was observed in Gorewara Lake. Populations were lowest in July and August, possibly due to the turbidity of the water during these monsoon months. *Brachionus falcatus*, *B. forficula*, *B. quadrilidentatus*, and *Tetramastix opoliensis* were characteristic of the clear waters of Gorewara Lake (maximum rotifer density 245/L) while *Filinia terminalis*, *Epiphanes macrurus*, *Asplanchna intermedia*, *Pedalia intermedia*, *B. angularis*, and *B. calyciflorus* were characteristic of the polluted waters of the aeration pond. Rotifer densities in this pond attained a maximum of 14,600/L.

Zooplankton Composition as Affected by Planktivorous Fish

Zooplankton composition is affected by the presence of size-selective planktivorous fish which remove the larger components of the zooplankton. Classic studies were conducted in fish ponds by Czechoslovakian researchers in the 1960s. Straskraba (1967) showed that, in the absence of fish predators, the zooplankton community in the littoral region of fish ponds was dominated by the large *Simocephalus vetulus* and *Daphnia pulex*. However, when predatory fish were present, these zooplankton were replaced by the smaller *Chydorus sphaericus* and other euplanktonic cladocerans. Postolkova (1967) observed that the standing crop of zooplankton was greater inside a fenced portion of the pond than outside: fish predators were excluded by these enclosures and thus were unable to reduce the standing stock of zooplankton. In an earlier study, Straskraba (1963) determined that phytoplankton production was lower in the littoral region than in the open region of the pond due to macrophyte shading. Phytoplankton production was not sufficient to support the littoral zooplankton community. The author concluded that the growth of macrophytes should be discouraged in fish ponds.

Zooplankton and Fish Stocking

Zooplankton composition in a water body also is a function of the rate of fish stocking. Marvan et al. (1978) described a number of south Bohemian ponds stocked with various numbers and sizes of fish. Ponds with low fish stocks (700/ha) were dominated by the cladocerans Daphnia pulicaria and D. longispina. Polytrophic ponds with high concentrations of fish (about 20,000/ha with small fish and fry) were dominated by the smaller Bosmina longirostris, Ceriodaphnia affinis, Diaphanasoma brachyurum, cyclopoida, and rotifers in addition to some D. longispina and D. pulicaria.

While it is well-known that planktivorous fish affect zooplankton composition in a water body, many questions have yet to be answered. Predation rates of fish on zooplankton have not been well qualified and are of particular relevance to aquaculture systems. Strategies by which zooplankton avoid predation have been investigated in a number of natural systems and include morphological (increased transparency) and behavioral (vertical migration) adaptation. However, more work must be done in aquaculture systems. How do zooplankton populations compensate for grazing? Laboratory experiments and certain field studies indicate that, under such stresses, zooplankton fecundity increases. However, the birth rate can increase only to a certain level beyond which losses due to predation cannot be compensated for and the population crashes. Grazing may have a greater effect on copepods which tend to have somewhat longer developmental times than cladocerans and rotifers; longer development time to the adult increases the probability that development will not be complete before predation occurs. In addition, as copepods reproduce sexually, the population must remain dense enough for males and females to locate one another. Conversely, cladocerans and rotifers reproduce parthogenetically and probably can withstand greater grazing pressures. There is the need for more aquaculture research at the theoretical level on zooplankton-planktivorous fish interactions, particularly with regard to predation avoidance mechanisms and the ability of the zooplankton population to withstand heavy grazing pressure.

Zooplankton and Fish Yield

Several researchers have suggested that there is a strong correlation between primary productivity, zooplankton productivity, and fish yield. This has been confirmed experimentally for phytoplankton-fish production relationships by a number of researchers including Hall et al. (1970) and Goodyear et al. (1972).

Das and Upadhyay (1979) studied the qualitative and quantitative fluctuations of plankton in two Kuman lakes in India. Nainital Lake (altitude 1,938 m) was highly eutrophic and polluted: plankton densities ranged from 0.9 to 1.8 mL/m^3 and formed the major food for the larvae, fry, and fingerlings of the Mahseer fish population. Oligotrophic Bhimtal Lake (elevation 1,376 m) with low plankton populations (0.08-4.4 mL/m^3) apparently did not support significant commercial fisheries. In both lakes, population maxima occurred in spring (March) and after the monsoon (August).

Sreenivasan (1964) conducted a limnological study and estimated fish yield in three upland lakes in Madras State, India. Kodaikanal Lake had low primary production averaging 0.46 gC/m^2/day. Zooplankton (50-120 ind./L) were dominated by copepods, cladocerans, and the rotifers Brachionus sps. and Eubranchipus. Fish production (Carassius carassius and Cyprinus carpio) was also low (5.3 kg/ha). In comparison, Yercaud Lake had a higher average primary production rate (2.9 gC/m^2/day) and zooplankton were more abundant (80-600 ind./L). Copepods, cladocerans (Aneuraea), and rotifers (Brachionus, Eubranchipus, Asplanchna) were the dominant taxa. Fish yield (C. carpio and some Tilapia mossambica) was high (31.6 kg/ha). Ooty Lake, polluted by sewage input, had the highest average primary production rate of 3.8 gC/m^2/day and abundant (42-12,280/L) zooplankton populations. Dominant zooplankton included copepods, cladocerans (Aneuraea, Daphnia), and rotifers (Monostyla, Triarthra, Eubranchipus, and Brachionus). Fish yield was highest, exceeding 75 kg/ha.

Gophen and Landau (1977) studied the trophic interaction between zooplankton and the sardine Mirogrex terraesanctae in Lake Kinneret, Israel. They found that the biomass and productivity of cladocerans was higher (30-170 percent) in 1973 and the first half of 1974 than the average for 1969-1972. This increase in food resources apparently contributed to an increase in the sardine population in 1974-1975. Juvenile sardines grazed heavily upon the zooplankton, reducing both the cladoceran population and the adult female copepod population. Females compensated for this decrease in numbers by increased fecundity.

Schroeder (1973) investigated factors affecting feed conversion ratios (kg feed supplied:kg fish yield) in fish ponds in Israel, and found that there was a strong correlation between feed conversion ratio and zooplankton standing stock at zooplankton densities of 0.1 to 1.1 mg dry weight/L. Within this range, natural feed was the major factor governing the efficiency with which supplemental food was used by fish. At higher zooplankton densities, the feed conversion ratio remained unchanged (approximately 1.8) suggesting that there may have been excess natural food in the pond. Fish were stocked at densities of 1,000-5,000 kg/ha. Zooplankton were dominated by rotifers, *Cyclops*, and the cladoceran *Moina*. Schroeder also determined that zooplankton standing stocks in fish ponds were affected by flushing rates, with population density decreasing with increased flushing.

In a later study, Schroeder (1978) showed that ponds fertilized with cow and chicken manure, and with nitrogen and phosphorus fertilizers, could produce 15-30 kg/fish/ha/day (common carp, *Tilapia*, silver carp). Fish obtained half their food supply from organisms greater than 37 mm in size and the remaining half by directly consuming smaller particles. These smaller particles consisted of bacteria and protozoans. The addition of inorganic fertilizers alone increased primary productivity, while the addition of manures along with those fertilizers did not increase primary productivity of plankton standing stocks. The highest fish yields (32/kg/ha/day) were obtained when fertilizers and manures were added, while yields (10-15 kg/ha/day) were lower when only fertilizers were added. Schroeder provided some information on plankton composition and zooplankton production rates in his experimental ponds. He stated that young water had low densities of zooplankton which were dominated by rotifers and were followed by a pulse of *Moina*. Old water contained copepods (mainly *Cyclops*). In addition, he conducted enclosure experiments and observed at the end of a week that zooplankton retained in fine net enclosures were about five times more abundant than zooplankton in ambient waters. This suggests that fish predation on the zooplankton was intense.

These aquaculture studies confirm what previously has been demonstrated in other freshwater systems: there is a good correlation between primary production, zooplankton standing stock, and fish yield. However, many areas require further investigation. For example, what are zooplankton production rates in pond aquaculture systems? How do these rates vary with primary production and allochthonous inputs? What is the transfer efficiency between various trophic levels? What are the major energetic pathways? What are the seasonal dynamics of primary and secondary production and how do these relate to fish growth and yield?

Zooplankton in Polyculture Systems

Many aquaculture ponds are used to rear more than one species of fish. However, the success of such a polyculture system depends upon the appropriate combination of noncompeting species. For example, two phytophagous fish species would be a poor choice in a polyculture system while a phytophagous fish may be successfully reared with a particulate-feeding fish. Consequently, there has been a great deal of research investigating fish feeding habits in polyculture systems, including the role of zooplankton in fish diet. In addition, there has been some work on the effect of various combinations of fish species and standing stocks on the zooplankton community. Fish can affect zooplankton either directly through grazing or indirectly by modifying the ecological balance of the pond. For example, a fish species may through various mechanisms contribute to the growth of an algal form which is a good food source for one species of zooplankton but not another.

Grygierek (1973) investigated the influence of phytophagous fish and the common carp on zooplankton populations in several experimental ponds in Poland. He found that increased carp density resulted in an increase in the abundance of zooplankton, more rapid seasonal changes in numbers, and a shift in species dominance from the large *Daphnia longispina* to the smaller *Bosmina longirostris* and *Ceriodaphnia quadrangularis*. In pond systems with the same number of carp fingerlings and phytophagous fish (common carp, grass carp, silver carp, bighead carp), the total number of zooplankton was reduced. However, the larger *D. longispina* and *Moina rectirostris* were relatively more abundant, although the actual composition was dependent upon the common carp-phytophagous carp combinations. Grygierek proposed that two factors were important in affecting zooplankton composition with the ponds. One factor related to differences in the size-selective feeding behavior of the various carp species. Grass carp were highly size-selective in their feeding while the common carp was slightly less selective. Silver carp and bighead carp fingerlings fed on zooplankton but were not as highly size-selective.

The second important factor in Grygierek's study was the effect of fish composition on the microbial and phytoplankton communities. Both communities tended to be more abundant in ponds containing phytophagous fish and common carp than in ponds containing only common carp. In ponds with bighead carp, blue-greens were abundant, while diatoms were abundant in ponds containing only common carp. Differences in zooplankton composition between those ponds may have been related, in part, to differences in the microbial and phytoplankton communities which provided food resources to these invertebrate grazers.

Spataru (1977) studied the gut contents of silver carp and trophic relations to other fish species in a polyculture system. Four types of ponds were studied, consisting of a two-factor combination of fertilizer (fluid manure, fluid manure + sorghum) and treatment (storage pond, pond allowed to dry out following harvest). The poorest growth was obtained in ponds which were fertilized with manure only and allowed to dry out at the end of the summer. Best growth was obtained in ponds which were fertilized with manure and sorghum and which maintained a planktonic community throughout the year. In these ponds, there was excellent growth of the algae Scendedesmus, the preferred food of the silver carp, and sufficient benthic and detrital matter was present for Tilapia. In addition, common carp grew well on sorghum. This study suggests the importance of maintaining an ecologically balanced community even with supplemental feed. However, the nature of this ecological balance, including the zooplankton community, was not investigated.

An additional interesting observation reported by Spataru (1977) is that, while silver carp consumed zooplankton such as Rotaria and Brachionus, these rotifers passed through the gut unharmed. No additional information was provided on zooplankton composition and abundance in these ponds.

Hiatt (1944) investigated feeding habits of mullets, milkfish, and ten pounders in Hawaiian ponds. Mullets and milkfish are herbivores, feeding on similar phytoplankton. Zooplankton apparently were not consumed, possibly because of their larger size. Ten pounders consumed shrimp and mosquitofish (a secondary consumer according to Goodyear et al. 1972). Hiatt suggested that fish yield in polyculture systems could be improved by adding ten pounders while competition between milkfish and mullet probably reduced yield.

Cremer and Smitherman (1980) studied the food habits of silver carp and bighead carp in ponds in Alabama. Silver carp consumed primarily phytoplankton and grew successfully in cages. There was no difference in silver carp growth between ponds receiving and not receiving artificial feed. Conversely, bighead carp consumed large particles, including large quantities of zooplankton and detritus. Zooplankton food items included copepods, cladocerans, ostracods, rotifers, mites, amphipods, and chironomids. Bighead carp grew poorly in cages and responded well to artificial feeding.

More questions require further investigations in order to understand the dynamics of polyculture systems. Competition between phytophagous fish and herbivorous zooplankton has not been investigated and may affect fish yield. The effect of several fish species on zooplankton composition and abundance have been examined only superficially. For example, a planktivorous fish will consume only zooplankton which are of the appropriate size and palatable. Such predation will affect the competitive balance between zooplankton species, but in ways that are not well understood. For example, will the successful competitor be of a form that is a food source for another fish in the pond or will it be a form that competes with fish? Alternately, the form may be beneficial in other ways, either by serving as an energy source for invertebrate predators which are consumed by fish or by altering the phytoplankton community to forms that are more suitable to the phytophagous fish.

Zooplankton and Pond Balance

Several studies have indicated the importance of maintaining a balanced fish population in pond ecosystems. As stated previously, Spataru (1977) observed that fish yield was higher in ponds which did not dry out over winter and which could maintain a natural assemblage of organisms, while growth was poorer in ponds which do dry out resulting in the disruption of the natural succession. In ponds where intense algal blooms develop, the production of large amounts of ammonia from decaying algae can produce diseased fish (Seymour 1980). By properly managing ponds so that algal growth is not excessive and so that grazing rather than nutrient depletion causes a reduction to phytoplankton standing stocks, such occurrences may be avoided.

Laventer et al. (1968) studied a series of fish ponds in Israel to see how fish production could be increased. Pond water level decreased over the winter and increased between January and March with the spring rains. Algal standing stocks (primarily diatoms and greens) increased at the latter period. This was then followed by an increase in crustacean abundance. Laventer stated that this was the best time to add carp to the ponds although carp did not feed on plankton. Since chironomid larvae, the main food of the carp, did not appear in the pond until April, it is not clear what organisms were consumed by the carp at this time. Nevertheless, the particular balance which was established in the pond in mid-spring (abundant crustacean populations) was ideal for carp growth.

Overview of Plankton Ecology and Fish Yield in Aquaculture Systems

The literature review to this point has discussed the composition and production of primary, secondary, and tertiary trophic levels of organisms in relation to fish production. In low- and moderately-managed systems, there is a good relation between primary production and fish yield. The zooplankton abundance and composition have been measured less frequently in these systems, and they are clearly affected both by primary production and by fish standing stocks.

Primary production in pond systems can be increased by the addition of chemical fertilizers to a point where increased addition of chemical fertilizers does not result in increased algal growth. This may be due to limiting of other nutrients or to self shading.

Two mechanisms exist to further increase fish yield. Where ponds are stocked with fish that feed on large organisms such as zooplankton and benthos, diet can be supplemented by the addition of pellets, grains, offal, and other food items. Conversely, in ponds which contain filter-feeding fish which can handle particles only a few tens of microns in size, increased yield can be accomplished by adding manures so that a large microbial community is established.

The role of zooplankton in pond systems has not been well-quantified and will vary as a function of allochthonous inputs and the abundance and composition of fish. Relatively little work has been done to estimate zooplankton production in various aquaculture systems. Korinek (1966) estimated production of adult female *Daphnia pulicaria* in a carp pond in South Bohemia while Gophen and Landau (1977) estimated zooplankton production in Lake Kinneret. Techniques for estimating zooplankton production are well developed and production estimates have been made for a variety of waters (Comita 1972; Burgis 1971; Hall et al. 1970). Although procedures for estimating zooplankton production are relatively simple, they do involve more intensive sampling than primary production estimates. This probably is the major reason that both zooplankton and benthos production have been estimated less commonly than primary production.

Relatively few studies have addressed other aspects of zooplankton biology (growth, respiration, excretion, fecundity) in aquaculture systems. Gophen (1976a,b) investigated the physiology of *Mesocyclops leuckarti* and *Ceriodaphnia reticulata* in Lake Kinneret. O'Brien and Vinyard (1978) studied polymorphism in *Daphnia carinata* in two south Indian ponds, along with the effects of invertebrate predators. These research studies have provided valuable contributions to the understanding of zooplankton and indicate further areas of basic zooplankton research in low intensity pond and lake aquaculture systems.

Zooplankton as First Food for Fish Fry

The role of zooplankton in intensive pond systems is not well understood, particularly for adult fish. However, in such systems where large numbers of fry are reared in nursery ponds, the importance of zooplankton as a first food for many species of fish is well recognized. Many intensively managed systems grow zooplankton to improve the survival of the early developmental stages of fish. The rotifer *Brachionus* and the brine shrimp *Artemia* are the most commonly reared food items (Stickney 1979) although marine researchers have experimented with brackish-water harpacticoids (Kahan 1979). Since such copepods are benthic, their usefulness in pond systems would appear to be limited to bottom-feeding larvae.

Some researchers have investigated the biochemical composition of zooplankton, although much more work is needed in this area. Scott and Baynes (1978) showed that the biochemical composition of the rotifer *Brachionus plicotalis* was affected by starvation. Although rotifers did not decrease in size, they did decrease in weight under low-food conditions. More importantly, there was a reduction in lipid, protein, and carbohydrate levels. Thus, in the early stages of a fish's life when larval growth is rapid, it is vital not only that the proper number and size of zooplankton be present as food items, but that these zooplankton contain the appropriate nutrients. Farkas et al. (1977) demonstrated that the carp *Cyprinus carpio* can develop nutrient deficiencies when their diet lacks essential fatty acids.

Methods for Improving Zooplankton Harvest

Paulsen (1977) developed an apparatus for collection of invertebrates from ponds. It essentially consists of pumping water from the pond through a series of filters ranging in size from 75 to 1,600 μ m for use in collecting food items for fish. Since the device did sort the invertebrates by size, the smaller fractions could be used to feed fish fry while the larger fractions could be used to feed fish which consume benthos and/or other large particulates. The system has potential use in ponds with high flushing rates which tend to deplete their plankton population (Schroeder 1973).

Green and Merrick (1980) developed a covered pond system for improving the survival of fry. Although gold and silver perch fry will consume pellet feeds, growth is better when they are fed plankton. However, in some systems, invertebrate predation on plankton decimates the population and leads to poorer fry growth in these nursery ponds. Green and Merrick developed their system to exclude invertebrate predators (dragonflies) which entered the pond and preyed heavily upon the plankton. An added benefit to their system was that diel temperature fluctuations were reduced and pond waters did not reach the temperature extremes observed in more open systems. A final advantage not mentioned by the authors was the probable reduction in evaporation rates. Lowering of pond levels has been a problem in a number of Israeli studies (Spataru 1977).

Invertebrate Predators on Zooplankton

Zooplankton are subject to predation not only by fish and dragonflies but also by a variety of invertebrates living in pond systems. Clady and Ulrickson (1968) reported that hydra consumed large numbers of *Daphnia pulex* in a tank system. Because the cladoceran population was abundant, the hydra increased in numbers. After the zooplankton population was decimated, the hydra attacked young bluegill fry.

Zooplankton as Predators on Fish Larvae

Some zooplankton prey upon fish larvae. Sukhanova (1968) demonstrated that silver carp larvae were attacked by the cyclopoid *Acanthocyclops vernalis* in Indian ponds. Cyclopoid predation occurs commonly and has been reported by a number of fresh water (Davis 1959; Hartig et al. 1980) and marine (Lillelund and Lasker 1971) researchers.

Zooplankton as Parasites

Certain species of copepods are fish parasites, notably *Ergasilus*, *Argulus*, and *Lernaea*. These parasites damage the fish either directly by disrupting tissue or secondarily through infection (Stickney 1979; Khalifa and Post 1976). Problems are particularly severe in storage ponds where the high density of fish promotes the rapid dispersal of the parasite (Lahav et al. 1964).

CONCLUDING REMARKS

This report reviews the known role of zooplankton in various aquaculture systems. It suggests areas of future research needs. In some aquaculture systems, where an adequate descriptive data base exists, relatively sophisticated studies can be conducted to investigate processes and test hypotheses. Conversely, in areas where aquaculture programs are in the early stages of development, more basic research on the limnology of pond ecosystems must first be conducted.

The literature review focuses on studies conducted outside North America and Europe although some such studies are discussed when comparable studies have not been conducted in less developed regions of the world. This review includes a range of water bodies which support or have the potential for supporting a freshwater fishery: lakes, reservoirs, and ponds which have been affected to varying degrees by human activity (primarily eutrophication). Apart from North American and European studies, most of the published extensive aquaculture research has been conducted in India and Israel. Both countries have well-developed aquaculture programs (Pillay 1979) and publish much of their research results in English in the western literature. Conversely, although China has an extensive aquaculture program (Pillay 1979), much of its research has not appeared in the western literature. Japan also supports a highly developed aquaculture program, and current research is published in English in the western literature. The Japanese aquaculture projects are highly intensive, involving large capital investments and high maintenance costs. Because this report is primarily concerned with cultivation systems which do not necessitate large investments and maintenance costs (extensive systems), a review of the Japanese literature was not included.

This review provides an overview of aquaculture systems in various parts of the world. Much of the aquaculture research has not been abstracted, and computer searches generated a relatively small number of citations. Most of the literature was located through more traditional means of literature search. Some articles have been published in journals and reports which are not held by the University of Michigan library and were not located by the time of this writing. As a consequence, the bibliography consists of two parts; one part contains articles which were read and included in this article, while the other part contains a list of additional reading.

LITERATURE CITED

Arora, H.C. 1966. Rotifera as indicators of trophic nature of environments. *Hydrobiol.* 27:146-159.

Burgis, M.J. 1971. The ecology and production of copepods, particularly *Hyalinus thermocyclops*, in the tropical Lake George, Uganda. *Freshwater Biology* 1:169-192.

Chacko, P,I., and S.V. Ganapati. 1952. Hydrobiological survey of the Suruli River in the Highwavys, Madurai District, India, to determine its suitability for the introduction of the rainbow trout. Archiv. Hydrobiol. 46:128-141.

Clady, M.D., and G.U. Ulrickson. 1968. Mortality of recently hatched bluegill fry as a result of hydra. Prog. Fish-Cult. 30(1):3940.

Comita, G.W. 1972. The seasonal zooplankton cycles, production, and transformation of energy in Severson Lake, Minnesota. Arch. Hydrobiol. 70:14-66.

Cremer, M.C., and R.O. Smitherman. 1980. Food habits and growth of silver and bighead carp in cages and ponds. Aquaculture 20:57-64.

Das, S.M. and J.C. Upadhyay. 1979. Studies on qualitative and quantitative fluctuations of plankton in two Kumaon lakes, Nainital and Bhimtal (India). Acta Hydrobiol. 21(1):9-17.

Davis, C.C. 1959. Damage to fish fry by cyclopoid copepods. Ohio J. Sci. 59:101-102.

Farkas, T., I. Scengeri, F. Majoros, and J. Olah. 1977. Metabolism of fatty acids in fish. I. Development of essential fatty acid deficiency in the carp, Cyprinus carpio Linnaeus (1758). Aquaculture 11:147-157.

Ganapati, S.V., and P.I. Chacko. 1951. A hydrobiological survey of the waters of the Upper Palnis with a view to fish culture. Archiv. Hydrobiol. 45:543-556.

Goodyear, C.P., C.E. Boyd, and R.J. Beyers. 1972. Relationships between primary productivity and mosquitofish (Gambusia affinis) production in large microcosms. Limnol. Oceanogr. 17(3):445-450.

Gophen, M. 1976a. Temperature dependence of food intake, ammonia excretion and respiration in Ceriodaphnia reticulata (Jurine) (Lake Kinneret, Israel). Freshwater Biol. 6:451-455.

____________. 1976b. Temperature effect on lifespan, metabolism, and development time of Mesocyclops leuckarti (Claus). Oecologia 25:271-277.

____________, and R. Landau. 1977. Trophic interactions between zooplankton and sardine Mirogrex terraesanctae populations in Lake Kinneret, Israel. Oikos 29:166-174.

Green, L.C., and J.R. Merrick. 1980. Tropical freshwater fish culture: a covered pond improves plankton and fry production. Aquaculture 2:389-394.

Grygierek, E. 1973. The influence of phytophagous fish on pond zooplankton. Aquaculture 2:197-208.

Hall, D.J., W.E. Cooper, and E.E. Werner. 1970. An experimental approach to the production dynamics and structure of freshwater animal communities. Limnol. Oceanogr. 15:839-928.

Hartig, J.G., D.J. Jude, and M.S. Evans. 1980. Cyclopoid predation on Lake Michigan fish larvae. Abstr. 23 Conference on Great Lakes Res., Internat. Assoc. Great Lakes Res.

Hiatt, R.W. 1944. Food-chains and the food cycle in Hawaiian fish ponds. Part I. The food and feeding habits of mullet (Mugil cephalus), milkfish (Chanos chanos), and the ten pounder (Elops machnata). Trans. Amer. Fish. Soc. 74:250-261.

Huq, M.F., and A.K.M. Sirajul Islam. 1977. A survey of man-made water areas in a village in Bangladesh to determine their productive capacity, current use and reasons for underutilization. Aquaculture 12:75-88.

Kahan, D. 1979. Vegetables as food for marine harpacticoid copepods. Aquaculture 16:345-350.

Khalifa, K.A., and G. Post. 1976. Histopathological effect of Lernaea cyprinacea (a copepod parasite) on fish. Prog. Fish-Cult. 38(2):110-113.

Korinek, V. 1966. The production of adult females of Daphnia pulicaria Forbes in a carp pond estimated by a direct method. Verh. Internat. Verein. Limnol. 16:386-391.

Lahav, M., S. Sarig, and M. Shilo. 1964. The eradication of Lernaea in storage ponds of carps through destruction of the copepodida stage by Dipterex. Bamidgeh 16(3):87-94.

Laventer, C., Y. Dagan, and D. Mires. 1968. Biological observations in fish ponds in the Na'Aman region, 1964-65. Bamidgeh (20)1:16-30.

Lewis, W.M., Jr. 1978. Dynamics and succession of the phytoplankton in a tropical lake: Lake Lanao, Philippines. J. Ecol. 66:849-880.

Lillelund, K., and R. Lasker. 1971. Laboratory studies of predation by marine copepod on fish larvae. Fish. Bull. 69:655-667.

Marvan, P., J. Komarek, H. Ettl, and J. Komarkova. 1978. Structure and functioning of algal communities in fish ponds. Structural elements. Principal populations of algae. Spatial distribution, pp. 295-313. In D. Dykyjova and J. Kvet (eds), Pond Littoral Ecosystems; Structure and Functioning. Ecol. Series 28. Springer Verlag, New York.

Matsumura-Tundisi, T., and J.G. Tundisi. 1976. Plankton studies in a lacustrine environment. Oecologia 25:165-270.

Munro, J.L. 1966. A limnological survey of Lake McIlwaine, Rhodesia. Hydrobiol. 28:281-308.

O'Brien, W.J., and G.L. Vinyard. 1978. Polymorphism and predation: The effect of invertebrate predation on the distribution of two varieties of Daphnia carinata in South India ponds. Limnol. Oceanogr. 23:452-460.

Paulsen, C.L. 1977. Apparatus for collection and separation of invertebrate organisms from a pond, for use as fish food. Prog. Fish Cult. 39(2):101-102.

Pillay, T.V.R., (ed). 1979. Advances in Aquaculture. Fishing News Bks. Ltd., London. 651 pp.

Postolkova, M. 1967. Comparison of the zooplankton amount and primary production of the fenced and unfenced littoral regions of Smyslov pond. Rozpravy Ceskoslovenske Acad. Ved. 77(11):63-80.

Schroeder, G.L. 1973. Factors affecting feed conversion ratio in fish ponds. Bamidgeh 25(4):104-113.

____________. 1978. Autotrophic and heterotrophic production of microorganisms in intensely manured fish ponds, and related fish yields. Aquaculture 14:303-325.

Scott, A.P., and S.M. Baynes. 1978. Effect of algal diet and temperature on the biochemical composition of the rotifer, Brachionus plicatilis. Aquaculture 14:247-260.

Seymour, E.A. 1980. The effects and control of algal blooms in fish ponds. Aquaculture 11:137-146.

Spataru, P. 1977. Gut contents of silver carp Hypophthalmichthys molitrix (Val.) and some trophic relations to other fish species in a polyculture system. Aquaculture 11:137-146.

Sreenivasan, A. 1964. Limnological studies and fish yield in three upland lakes of Madras State, India. Limnol. Oceanogr. 9:564-575.

Stickney, R.R. 1979. Principles of Warmwater Aquaculture. J. Wiley and Sons, New York.

Straskraba, M. 1963. Share of the littoral region in the productivity of two fish ponds in southern Bohemia. Rozpravy Ceskoslovenske Acad. Ved. 73(13):3-391.

____________. 1967. Quantitative study on the littoral zooplankton of the Poltruba backwater with an attempt to disclose the effect on fish. Rzopravy Ceskoslovenske Acad. Ved. 77(1):7-34.

Sukhanova, Y.R. 1968. The role of cyclops (Acanthocyclops vernalis Fisch.) in the survival of silver carp (Hypopthalmichthys) larvae. J. Ichthyology 8:467-468.

ADDITIONAL BIBLIOGRAPHY ON ZOOPLANKTON

Alihunhi, K.H., and S. Banerjee. 1955. On the mortality of carp fry in nursery ponds and the role of plankton in their survival and growth. India J. Fish. 2:257-313.

Allanson, B.R., and J. Gieskes. 1961. Investigations into the ecology of polluted inland waters in the Transvaal, Part 2. Hydrobiol. 18:77-94.

Allen, E.J., and E.W. Nelson. 1910. On the artificial culture of marine plankton organisms. J. Mar. Biol. Ass. U.K. 8:421-474.

Alverson, D.L., and M.J. Carney. 1975. A graphic review of the growth and decay of population cohorts. J. Cons. Int. Explor. Mer. 36:133-143.

Archibald, F.C. 1975. Experimental observations on the effects of predation by goldfish (Carassius auratus) on the zooplankton of a small saline lake. J. Fish Res. Board Can. 32:1589-1594.

Bhowmick, R.M., G.V. Kowtol, R.J. Jana, and S.D. Gupta. 1977. Experiments on second spawning of major Indian carps in the same season by hypophysation. Aquaculture 12:149-155.

Bryan, P.G., and B.B. Madraisau. 1977. Larval rearing of Seganus lineatus (Pisces:Siganidae) from hatching through metamorphosis. Aquaculture 10:243-252.

Clark, A.A., and J.C.H. Carter. 1974. Population dynamics of cladocerans in Sunfish Lake, Ontario. Can. J. Zool. 52:1235-1242.

Cummins, W.K., and W.J.C. Wuycheck. 1971. Caloric equivalents for investigations in ecological energetics. Mitt. Int. Ver. Limnol. 18:101-106.

Das, S.M., and V.K. Srivastva. 1959. Studies of freshwater plankton, qualitative composition, and seasonal fluctuations in plankton components. Proc. Nat. Acad. Sci. (India) 19:174-189.

____________, and ___________. 1966. Quantitative studies of freshwater plankton. Plankton of a fish tank in Lucknow, India. Proc. Nat. Ac. Sci. (India) 36:85-92.

Das, S.M., S.C. Choudhary, N. Ahmad, S. Akhtar, S. Peer, and M. Rashid. 1970. Studies on organic production in high altitude lakes of India. Part 2. The zooplankton, phytoplankton, and pedon of high altitude Kashmir lakes, Kounsarnag and Alpather. Kashmir Sci. 7:119-132.

Das, S.M., H. Daftari, H. Singh, S. Akhtar, S. Choudhary, and N. Ahmad. 1969. Studies on organic production in high altitude lakes of India. Part 1. The general ecology and zooplankton of Kashmir lakes. Kashmir Sci. 6:12-22.

Fostier, A., G. Barnabe, and R. Billard. 1980. Conclusion: Etat actual des connaissances dans le domaine de la pisciculture en etangs et perspectives. pp. 429-434. In R. Billard (ed), La Pisciculture en Etang. I.N.R.A. Pub. Par.

Ganapati, S.V. 1940. The ecology of a temple tank containing a permanent bloom of Microcystis aeuruginosa (Kutz) Henfr. J. Bombay Nat. Hist. Soc. 42:65-67.

____________. 1957. Limnological studies of two upland waters in Madras State. Arch. Hydrobiol. 56:36-61.

Gophen, M. 1972. Zooplankton distributions in Lake Kinneret (Israel). Israel J. Zool. 21:17-27.

____________, B. Cavari, and T. Berman. 1974. Zooplankton feeding on differentially labelled algae and bacteria. Nature 247:393-394.

Goren, M., L. Fishelson, and E.Trewavas. 1973. The cyprinid fishes of Acanthabrama Heckel and related genera. Bull. Br. Mus. (Nat. Hist.) Zoology 24:291-315.

Green, J. 1967. The distribution and variation of Daphnia lumholtzii (Crustacea:Cladocera) in relation to fish predation in Lake Abert, East Africa. J. Zool. 151:181-197.

Harding, D. 1961. Limnological trends in Lake Kariba. Nature, Lond. 191:119-120.

Hepher, B. 1962. Primary production in fish ponds and its application to fertilizer experiments. Limnol. Oceanogr. 7:131-136.

Ikeda, T. 1973. On the criteria to select copepod species for mass culture. Bull. Plankton Soc. Japan 20:41-48.

Jana, B.B. 1973. Seasonal periodicity of plankton in a freshwater pond in West Bengal, India. Int. Revue ges. Hydrobiol. 58:127-143.

Jones, A. 1973. Observations on the growth of turob larvae, Scopthalmus maximum reared in the laboratory. Aquaculture 2:149-155.

Kitajima, C. 1973. Experimental trials on mass culture of copepods. Bull. Plankton Soc. Japan 20:54-60.

Mandal, B.K., and S.K. Moitra. 1975. Seasonal variations of benthos and bottom soil edaphic factors in a freshwater fish pond at Burdwan, West Bengal. Tropical Ecology 16:43-48.

McConnell, W.J. 1963. Primary productivity and fish harvest in a small desert impoundment. Trans. Amer. Fish. Soc. 92:1-12.

Navaneethakrishan, P., and R.G. Michael. 1971. Egg production and growth in Daphnia carinata King. Proc. Indian Acad. Sci. Sec. B. 53:117-223.

Neilsen, L.A., and J.B. Reynolds. 1975. Freshwater shrimp. Natural food for pond fishes. Farm Pond Harvest 9:8, 9, and 24.

Rothbard, S. 1976. Experiments in mass culture of the marine copepod Tibriopus japonicus (Mori) on a bed of crushed sea weed Ulva petrusa (Kjelman). Bamidgeh 4:80-105.

Smyly, W.J.P. 1970. Observations on the rate of development, longevity, and fecundity of Acanthocyclops viridis, in relation to the type of prey. Crust. 18:21-36.

__________, and V.G. Collins. 1975. The influence of microbial food sources and aeration on the growth of Ceriodaphnia quadrangulas (O.F. Muller) (Crustacea:Cladocera) under experimental conditions. Freshwater Biol. 5:251-256.

Snow, N.B. 1972. The effect of season and animal size on the caloric content of Daphnia pulicaria Forbes. Limnol. Oceanogr. 17:909-913.

Stross, R.G., J.C. Ness, and A.D. Hasler. 1961. Turnover time and production of planktonic Crustacea in limed and reference portion of a bog lake. Ecol. 42:237-245.

Tang, Y.A. 1970. Evaluation of the balance between fishes and available fish foods in multiple species fish culture ponds in Taiwan. Trans. Amer. Fish. Soc. 4:708-718.

Theilacker, G.H., and M.F. McMaster. 1971. Mass cultivation of the rotifer Brachionus plicatilis and its evaluation as a food for larval anchovies. Mar. Biol. 10:183-188.

BIOLOGICAL PRINCIPLES OF POND CULTURE: FISH

by

James Diana and David Ottey

Production of fish biomass in pond culture systems is regulated by three parameters: stock density, mortality, and growth in individual weight (Backiel and LeCren 1967, 1978; Chapman 1971). These parameters are related through mechanisms of biotic interactions and their physico-chemical environments (Fry 1947, 1971; Kerr 1980; Werner 1980), as previously shown in this document. For many cultured fish species, pond management for maximization of biomass production and standing stocks remains limited by knowledge of the ecological principles of the systems. The mechanisms relating stocking density, mortality, and growth are therefore key considerations of this study.

The relationship of growth to density, including definition of specific ecological mechanisms, may well be the most effective area for research on pond culture. Reasons for emphasis on density-dependent mechanisms of growth rather than on mortality or density-independent processes include: 1) A vast number of studies has dealt with nutrition and dietary effects on growth. 2) Stock densities have more generally defined and stronger effects on growth than on mortality in culture systems (Backiel and LeCren 1978). The plasticity of individual growth in exploited populations in fish is well-documented and indicates the importance of density in control of growth. 3) Within a cohort, growth and mortality are more commonly controlled by density-dependent rather than density-independent processes (Larkin 1978). This may be particularly germane as intensively cultured pond systems can be controlled for stocking rates, physico-chemical factors, and food or nutrient inputs. 4) Evaluation and quantification of mechanisms of density effects on growth will provide characters for genetic stock manipulation explicit to culture situations and species. Genetic control of growth in fish is well established, but selection of growth-related traits requires physiological-bioenergetic analyses (Weatherley 1976). 5) Many empirical and mechanistic (bioenergetic and optimum foraging) models exist on growth's relationship to density and ration (e.g., Ware 1975, 1978; Paloheimo and Dickie 1965, 1966; Kerr 1971a, 1971b; Sperber et al. 1977). Additionally, the balanced energy equation for metabolism and growth (Winberg 1956) provides a powerful paradigm for evaluation and analysis of factors affecting growth (Warren and Davis 1967; Webb 1978).

Unfortunately, the analysis of mechanisms and development of appropriate models for species and situations of tropical pond culture are yet in their infancy. The major purposes of this synopsis are to examine recent research efforts in pond culture principles, and to provide schemes for describing and examining density effects on growth.

GUIDELINES FOR STUDY

Recent (1976-1980) aquaculture and fish biology literature was reviewed while emphasizing ecological and physiological mechanisms governing reproduction, growth, mortality, and recruitment to harvest of fishes commonly cultured in tropical ponds. The reference search has thus been autecological in perspective, although some of the most important processes, such as interspecific competition and facilitation of trophic efficiency, are synecological in nature. The multidisciplinary approach taken by this CRSP research group required an interactive, integrated scheme for relating principles of the systems, but the extensive tropical pond aquaculture literature warranted an approach which categorized information available for a restricted set of individual species (Table 1).

The synopsis also was restricted to considering only those environmental factors which typically do not degrade culture conditions to levels detrimental to the raising of quality food fishes for human consumption. Although chemicals are frequently used to reduce levels of competing or deleterious organisms in pond culture (Hickling 1962), we have not considered topics such as environmental contamination and toxicology. Studies dealing solely with intermediary metabolism were also excluded, and research on catabolic-anabolic processes was included only if there was a direct relation to growth, recruitment, or mortality.

A scheme for categorizing research on principles of spawning, egg development, and growth and mortality of juvenile and adult fish is presented in Table 2 and 3. The classes are not mutually exclusive, nor are culture system dynamics defined by them in a noninteractive fashion, but in practice

Table 1

LIST OF FISH SPECIES COMMONLY CULTIVATED IN TROPICAL PONDS

Finfish species considered in the reviewed literature for the years 1976-1980. The list was compiled after consideration of Hickling (1962); Hora and Pillay (1962); Bardach and Ryther (1968); Bardach, Ryther, and McLarney (1972); Nelson (1976); and Brown (1977). The systematic status varies considerably for some taxa.

Family	Species
Anabantidae	Anabas testudineus
Anguillidae	Anguilla japonica
Belontiidae	Trichogaster pectoralis
	Trichogaster trichopterus
Chanidae	Chanos chanos
Channidae	Ophicephalus gachua
	Ophicephalus marulius
	Ophicephalus punctatus
	Ophicephalus striatus
Cichlidae	Etroplus suratensis
	Tilapia aurea
	Tilapia andersonni
	Tilapia galilaea
	Tilapia macrochir
	Tilapia melanopleura
	Tilapia mossambica
	Tilapia nigra
	Tilapia nilotica
	Tilapia rendalli
	Tilapia sparmanni
	Tilapia zilli
Clariidae	Clarius batrachus
	Clarius gariepinus
	Clarius macrocephala
	Clarius magur
Cyprinidae	Arstichthys nobilis
	Barbus carnaticus
	Barbus hexagonalis
	Carassius auratus
	Carassius carassius
	Catla catla
	Cirrhina cirrhosa
	Cirrhina militorella
	Cirrhina mrigala
	Cirrhina reba
	Ctenopharyngodon idella
	Cyprinus carpio
	Hypophthalmichthys harmandi
	Hypophthalmichthys molitrix
	Labeo bata
	Labeo calbasu
	Labeo collaris
	Labeo fimbriatus
	Labeo kontius
	Labeo rohita
	Labeobarbus tambroides
	Megalobrama bramula
	Mylopharyngodon aethiops
	Mylopharyngodon piceus
	Osteochilus hasselti
	Osteochilus thomassi
	Parabramis pekinensis
	Puntius belinka
	Puntius gonionotus
	Puntius japonicus
	Puntius javanicus
	Puntius orphoides
	Puntius schwanefeldi
	Squaliobarbus curiculus
	Tinca tinca
	Thynnichthys sandkhol
Eleotridae	Oxyeleotris marmoratus
Elopidae	Elops saurus Bleeker
	Megalops cyprinoides
Helostomidae	Helostoma temmincki
Heteropheustidae	Heteropheustes fossilis
Latidae	Lates calcarifer
Mugilidae	Mugil cephalus
	Mugil corsula
	Mugil dussumieri
	Mugil tade
Osphronemidae	Osphronemus goramy
Pangasiidae	Pangasius larnuaudi
	Pangasius micronemus
	Pangasius sanitwongsei
	Pangasius sutchi
Salmonidae	Salmo gairdneri
	Salmo trutta
Some generic synonymies:	Mugil = Rhinomugil, Tilapia = Sarotherodon, Puntius = Barbus in some classifications of species in the above list.

most research efforts could be categorized into one or several main areas.

RESULTS OF THE LITERATURE SEARCH

The literature examined for 1976-1980 was primarily confined to the aquaculture and fish biology journals published in English, French, and German (which were cataloged in Cvancara 1976-1980) due to considerations of time, uniformity and extent of the search, and the language biases of the authors. The results of the search for the years 1976-1980 are presented in Tables 2 and 3.

Comparison of the totals for the categories of pond culture principles in Tables 2 and 3 suggests that little change in the main directions of published research on dynamics of tropical fish culture occurred during the last half of the past decade. Even among the restricted set of species studied during 1976-1980, many areas appeared to have received quite minor effort. In some instances, research prior to 1976 may have been intense or concentrated in currently neglected areas. For example, the preceding decade saw much research on induced spawning of the mullet, Mugil cephalus (Kuo et al. 1973, 1974; Liao et al. 1971; Shehadeh and Ellis 1970; Shehadeh et al. 1973a, 1973b; Tang 1964;

Table 2

SUMMARY OF KEY INFORMATION CONTAINED IN ALL AQUACULTURE AND FISH BIOLOGY LITERATURE EXAMINED FOR THE YEARS 1976–78 FOR FISH SPECIES COMMONLY CULTIVATED IN TROPICAL PONDS.[a] NUMBER OF REPORTS DEALING WITH TOPIC.

	A. japonica	C. chanos	O. punctatus	O. striatus	E. suratensis	T. aurea	T. galilaea	T. rendalli	T. mossambica	T. nilotica	T. zillii	C. batrachus	C. macrocephala	C. auratus	C. catla	C. mrigala	C. idella	C. carpio	H. molitrix	L. bata	L. calbasu	L. rohita	T. tinca	H. fossilis	M. cephalus	T. trichopterus	S. gairdneri	S. trutta	Total
Natural maturation/ spawning				1	1				3					5			1	2		1				3		2	4		23
Fecundity relations																								1				1	
Fecundity management through stock																													0
Spawning habitat manipulation		1																											1
Spawning induction/ prevention									2	1	2	1	1	5	1	1	7	4	2					10			2		39
Embryonic/larvel development																													0
Antagonism-facilitation									1						1	1		1	1			1					3		9
Abiotic environmental limitation														1				2			1			2	1		7	1	15
Ingestion/feeding pattern				1		1	1				1				1		1							1	2		6		15
Locomotor activity				2										4				1						1					8
Evacuation time/egestion/excretion																									1		2	1	4
Empirical food conversion efficiency/ growth and nutrition	7		1	1		1	1	3	3	1	3	2		2	1	2	5	14				2	2	2	2		32	8	95

[a]Government publications and unpublished symposia, conferences, and dissertations were not included. Information pertaining to diet-related diseases and metabolic rate and scope were not included.

Table 3

SUMMARY OF KEY INFORMATION CONTAINED IN ALL AQUACULTURE AND FISH BIOLOGY LITERATURE EXAMINED FOR THE YEARS 1978–80 FOR FISH SPECIES COMMONLY CULTIVATED IN TROPICAL PONDS.[a] NUMBER OF REPORTS DEALING WITH EACH TOPIC.

	A. japonica	*C. chanos*	*O. punctatus*	*O. striatus*	*E. suratensis*	*T. aurea*	*T. galilaea*	*T. rendalli*	*T. mossambica*	*T. nilotica*	*T. zillii*	*C. batrachus*	*C. macrocephala*	*C. auratus*	*C. catla*	*C. mrigala*	*C. idella*	*C. carpio*	*H. molitrix*	*L. bata*	*L. calbasu*	*L. rohita*	*T. tinca*	*H. fossilis*	*M. cephalus*	*T. trichopterus*	*S. gairdneri*	*S. trutta*	Total
Natural maturation/ spawning		1	3	2					1	1	3	2		3				2						2			5	1	26
Fecundity relations			1							1				1															3
Fecundity management through stock																													0
Spawning habitat manipulation																	1												1
Spawning induction/ prevention		2	1			3			1	2		1		5			3	4	1					2			9		34
Embryonic/larval development																													0
Antagonism-facilitation						1												1											2
Abiotic environmental limitation			1			1						2	2	1			2	7	1			1		1			12	4	35
Ingestion/feeding pattern										1							1							1	1		1	1	6
Locomotor activity														1										1				1	3
Evacuation time/egestion/excretion											2																3		5
Empirical food conversion efficiency/ growth and nutrition	1	1	1	1		1		1	1	2	2			4		2	3	18	1				1	3			27		71

[a]Government publications and unpublished symposia, conferences, and dissertations were not included. Information pertaining to diet-related diseases and metabolic rate and scope were not included.

Yang and Kimm 1962; Yashouv 1969), which led to the successful practices currently used. Certainly, a large volume of research effort was neglected with the exclusion of government aquaculture and fishery publications, but there is little reason to believe that the directions and concentrations of research topics differed radically in this realm from that reported in the surveyed literature.

The little-studied principles appear related to factors which influence fish growth indirectly (Weatherley 1976). Nevertheless, such mechanisms may be of paramount importance in production dynamics and in prediction of yield. Weatherley (1966, 1972) stressed the importance to fish population dynamics of such ecological processes. The relatively well-researched categories in Tables 2 and 3, other than spawning and spawning induction, were ingestion rates and feeding patterns in 1976-1978, and the categories of abiotic environmental limitation and food conversion efficiency and nutrition during both periods. Ethological factors, such as locomotor activity and antagonism-facilitation interaction, in general were poorly researched. Such factors can significantly alter directions of energy flow through populations. For example, the effect of increased locomotor activity on metabolic scope can drastically reduce energy available for growth (Webb 1978).

One major area of research virtually ignored in this synopsis was that of larval development of cultured tropical fishes. This is particularly problematic because the production of viable fish fry is of utmost importance to intensive aquacultural projects. However, the current techniques of fry rearing are more realistically considered a management practice, rather than a biological principle, and were not a major part of this synopsis.

ECOLOGICAL PERSPECTIVE

The ecological processes relating stock density, growth, and mortality may be approached through Fry's (1947, 1971) paradigm of limiting controlling, masking, lethal, and directive factors, or through the Hutchinsonian niche (Whittaker and Levin 1975), in which somatic and protein growth are fitness values which are maximized in pond culture. The Fry perspective is primarily concerned with physico-chemical impacts on species populations and is thus fundamentally autecological, whereas niche considerations ultimately are synecological, as both intra-and interspecific impacts are at the foundations. The synthesis of these two perspecives into a sound practical and theoretical basis for fisheries management awaits the formulation and resolution of theories and models on intra- and interspecific interactions in metabolic terms (Werner 1980). Webb (1978) has provided an examination of the metabolic bases of ecological processes, although the relation of conspecific density or species interactions to mechanisms affecting growth was not stressed.

Although the empirical effects of population density on growth are generally established for fish (Beverton and Holt 1957; Backiel and LeCren 1967, 1978), an understanding of the various behavioral and physiological mechanisms controlling density-dependent responses has scarcely been realized.

Genetic selection of stocks for optimal growth under high stocking rates could be more effectively pursued if behavioral bases of metabolic costs were known. This genetic selection for rapid growth requires physiological and bioenergetic analyses of growth (Weatherley 1976). Assimilative capacity and growth potential must account for effects of population density in relation to frequency and size of food rations, due to conflicting factors of social facilitation and behavioral dominance (social inhibition) (Weatherley 1976). The energetics of activity need examination, because if caloric costs of stressed and routine activity are high, decrements to growth from net (physico-logically useful) energy may be significant (Noakes 1978). The proportion of net energy allocated to activity under stressed, routine, and basal metabolism differs substantially (Weatherley 1976; Webb 1978). Some metabolic features, such as specific dynamic action (SDA), which are indirectly related to stock density, may not be reduced through genetic selection, In such instances, densities in polycultured communities may be manipulated to optimize overall food utilization.

A bioenergetic approach for evaluating effects on growth and production arises from the balanced energy equation of Ivlev (1939) and Winberg (1956).

$$Q_G = pQr - Q_M \quad (1)$$

where:

Q_G = growth (anabolism)

P = proportion of food energy consumed which is assimilated,

Q_R = food consumed, and

Q_M = metabolism (catabolism)

These quantities are properly considered as rates in time. It should be noted that Q_G contains both somatic and gonadal components, and that Q_M may contain energy devoted to reproductive behavior. In its expanded form, the equation may be expressed (after Webb 1978):

$$Q_R - (Q_F + Q_N) = Q_S + Q_L + Q_{SDA} + Q_G \quad (2)$$

where:

Q_F = faecal loss,

Q_N = excretory or non-faecal loss,

Q_S = standard metabolism,

Q_L = locomotor (activity) metabolic cost, and

Q_{SDA} = apparent specific dynamic action.

Thus,

$$Q_M = Q_S + Q_L + Q_{SDA} \quad (3)$$

A further useful concept is that of metabolic scope, equal to:

$$Q_{m_{max}} - Q_S \quad (4)$$

An outline of the primary mechanisms of density effects on growth is presented in Table 4. Mechanisms leading to negative changes in Q_G in culture systems would be classified as intra- and interspecific competition (Pianka 1978), whereas various positive growth rate effects could be considered as protocooperation, commensalism, or mutualism. Protocooperation refers to both intra- and interspecific mutualism (obligate reciprocol positive) and commensalism (positive-neutral) impacts refer to interspecific interactions. Direct competitive interactions are classed as interference (production of toxins and agonistic or territorial behaviors), but less direct effects are exploitation (those arising through reduction of unheld resource levels). The competitive mechanisms of immune hypersensitivity, metabolite loading, and hormonal antagonisms are chemical interference interaction, while the negative density-dependent growth impacts from activity, food-habitat shifts, and social feeding antagonisms are primarily physical interferences.

Although acute immune hypersensitivity responses leading to disease and death are often observed (Smith 1977; Henderson-Arzapalo et al. 1980), chronic cutaneous anaphylactic reactions likely increase Q_L and Q_S, and decrease Q_R. Metabolite growth interferences have been summarized by Webb (1978). He indicates that Q_S increases in metabolism, through general excitability rates, appear to be the dominant growth-reducing effect of metabolites. Effects on Q_{SDA} involving costs per unit ration (through Michaelis-Menten kinetics) appear unevaluated. Hyperexcitability in fishes exposed to increased ammonia levels is well documented (Wuhrman and Woker 1948: Fromm and Gillette 1968; Olson and Fromm 1971). Growth reductions have been observed in rainbow trout and chinook salmon, respectively, following chronic dosings at levels from 0.005-0.015 mg/L un-ionized ammonia (Smith 1972; Burrows 1964).

Hormone effects, which like the former categories are observed as conditioning or crowding factors, may increase (Allee et al. 1940) or decrease (Pfuderer et al. 1974; Yu and Perlmutter 1970) growth rates, through Q_S, Q_L, and Q_R. The relative importance of these components, as well as the general importance of hormones to growth, remains largely speculative.

Activity (Q_L) effects on growth are produced through several density-controlled mechanisms. Increased efficiency of locomotion due to mucus reduction of drag (Breder 1976), and hydrodynamic considerations of vortex production (Weihs 1973; Breder 1965) are possibly advantageous aspects of schooling (Krebs 1976). Schooling positively contributes to Q_G in some species through declines in general excitability or "calming" effects (Parker 1973; Schlaifer 1938, 1939). Heightened general excitation levels may lead to increased metabolic rates in territorial species or in habitats where territoriality is enhanced (Yamagishi 1962; Li and Brocksen 1977). These general excitation effects due to increased density are similar to effects produced by metabolite buildup; probably both mechanisms often occur together with territorial species in intensively cultured pond systems with negligible water turnover. Spontaneous activities and increased responses to stimuli have an ultimate effect of increasing Q_M beyond Q_S. Elevation of routine metabolism may be a major diversion from growth as routine metabolic costs may reach one-third to one-half of metabolic scope (Webb 1978). Additionally, some species may exhibit high levels of agonistic activity as well as higher metabolic levels due to decreased locomotor efficiency in more rigorous microhabitats. Both demonstrably reduce Q_G (Magnuson 1962; Eaton and Farley 1974; Li and Brocksen 1977).

The definition of sociality used in Table 4 is an expansion of Noake's (1978) definition as behavior directly related to actual or potential encounters between conspecifics to include interspecific interactions involving communication. Interspecific agonistic behaviors and dominance relationships associated with feeding may impact particularly strongly on ration when man introduces exotic

Table 4

FISH POPULATION INTERACTIONS - MECHANISMS OF DENSITY EFFECTS ON GROWTH[a]

Primary mechanisms	Category of ecological interaction, losses (-), or gains (+) to growth rate (QG)	Principal rates affected and direction in balanced energy equation	Description	Examples
Immune hypersensitivity antigen-anitibody	Intraspecific competition - Interspecific competition -,-	$+Q_L$, $+Q_S$, $-Q_R$	Antigen anaphylactic responses	Henderson-Arzapalo et al., 1980; Smith 1977
Metabolite	Intraspecific competition - Intraspecific competition -,-	$+Q_L$, $+Q_S$, $-Q_R$	Sublethal NH_3, CO_2 effects	Burrows, 1964; Kawamoto, 1958, 1961; Kawamoto et al., 1957; Smith 1972; Yashouv, 1958
Hormone	Intraspecific competition - Intraspecific protocooperation +	$\pm Q_S$, $\pm Q_L$, $\pm Q_R$	Conditioning crowding factors	Allee et al. 1940; Pfuderer et al. 1974; Timms 1975; Yu and Perlmutter 1970
Activity	Intraspecific competition - Interspecific competition -,- Intraspecific protocooperation Intraspecific mutualism +	$\pm Q_L$	Schooling effects on hydrodynamics, non-chemical density effects on locomotion, agonistic and dominance interactions	Breder 1965; Li and Brocksen 1977; Marr 1963; Parker 1973; Schlaifer 1938, 1939; Weihs 1973; Yamagishi 1962, 1964
Social feeding facilitation/ antagonism	Intraspecific competition - Intraspecific mutualism + Intraspecific protocooperation + Interspecific competition -,- Interspecific protocooperation +,+	$\pm Q_R$, $\pm Q_{SDA}$, $\pm Q_N$, $\pm Q_F$	Schooling, density effects on diet, feeding rates	Ivlev 1961, Nikol'skii 1955; LeCren 1965; Kawanabe 1969; Brown 1946
Non-social feeding facilitation/ antagonism	Intraspecific competition - Intraspecific protocooperation + Interspecific protocooperation +,+ Interspecific commensalism +,0 Interspecific competition -,-	$\pm Q_R$, $\pm Q_{SDA}$, $\pm Q_N$, $\pm Q_F$	External environmental food processing effects on food conversion and trophic transfer efficiencies	Kilgen and Smitherman 1971; Yashouv 1966
Food-habitat shifts	Intraspecific competition - Interspecific competition -,-	$-Q_R$, $+Q_{SDA}$, $+Q_N$, $+Q_F$	Habitat-niche expansion and packing, occupation of sub-optimal marginal habitats	Nilsson 1963, 1967; Svardson 1976; Werner and Hall 1976, 1977

[a]Category of ecological interaction after Pianka, 1978. The balanced energy equation is the expanded version, Eq. (2), wherein Q_R = food energy consumed, Q_S = standard metabolism, Q_L = activity metabolism, Q_{SDA} = apparent specific dynamic action, A_N = excretory loss of non=faecal loss, Q_F = faecal loss.

species to communities (Nilsson 1967). Further influences of sociality on feeding broadly overlap with chemical communication effects of feeding enhancement included under hormonal mechanisms. Reduction in Q_G through decreased foraging activity ($-Q_R$) due to time spent in defense of territories (e.g., LeCren 1965) frequently will coincide with increases in level of antagonistic activity mentioned previously. Landless (1976) found that confined rainbow trout showed differences in operant behavior to obtain food associated with territorial-dominance reactions. Physical interference from dominants that decreased average Q_R where food was not abundant has also been shown for Salmo trutta (Brown 1946) and Oryzias latipes (Magnuson 1962).

Social interactions also can increase feeding rates and influence diet composition. The advantages of gregarious feeding, such as enhanced foraging efficiencies via increased probability of locating food (Keenleyside 1955), may be more important to predatory species, or in systems where habitat heterogeneity is high and food is patchily distributed. However, plankton-eating fish may feed more intensively when in schools than when dispersed (Nikol'skii 1955). Social structure changes observed by Kawanabe (1969) in ayu, Plecoglossus altivelis, enhanced exploitation of food resources under changing availability. Learning rate in operation of self-feeders in rainbow trout is increased by presence of conspecifics (Adron et al. 1973). Noakes (1978) commented that fish kept in groups feed more readily and adapt to dietary changes more rapidly than do isolated individuals.

Total fish production in pond systems can be increased dramatically through polyspecific culture (Jhingran 1976). This phenomenon has long been known, particularly in the cases of Chinese and Indian carps (Bardach et al. 1972). Species which operate on different trophic levels, or forage differently, can use single or diverse food resources more efficiently and thus provide higher levels of total pond fish production (Hickling 1962; Yashouv 1958, 1966). Behavioral and physiological description and modelling of the energetics of food processing are uncommon in the literature. Foraging efficiency (Q_R), as well as ingested food conversion efficiency (Q_{SDA}, Q_N, Q_F), are rates affected in the balanced energy equation. Food and habitat shifts are most frequently observed with exotic species introductions by man (Nilsson 1967). Segregation of habitat is probably the most important means of niche separation in freshwater fishes (Werner et al. 1977). Such shifts due to density effects may operate through a variety of interference and exploitation mechanisms (Nilsson 1967). The proximate causes of interference are similar to those present in social feeding antagonisms. The main difference lies in the displacement of less dominant individuals to habitats where suboptimal conditions depress feeding rates ($-Q_R$) and alter diets (effects on Q_{SDA}, Q_N, Q_F). The presence of a competitor should not alter the food consumption spectrum within a habitat, but may influence the amount of time spent foraging in that habitat relative to other habitats (Werner and Hall 1977). Depression of Q_G, therefore, is considered independently of costs of ambient agonistic interactions in this mechanism. It may be noted that when ration effects associated with habitat shifts occur, qualitative diet changes (e.g., away from carnivory or large food-item sizes) may actually decrease Q_{SDA} per unit ration (by reducing amount of protein in the diet), although total apparent SDA may be increased. Also, total pond species production may also be augmented through more efficient filling of available habitats, especially in polyculture. Analytic research on food-habitat shifts is rare, especially for communities cultured in tropical pond systems.

FUTURE RESEARCH NEEDS

Based on the literature search of selected publications and the evaluation of ecological principles presented in this review, recommendations for future research can be made. Considerations such as parameter requirements for specific models, or the dynamics of a particular culture system, certainly could alter these recommendations. However, the greatest area of potential innovation in the biological principles regulating tropical pond culture systems is in the understanding of genetic enhancement of fitness and of density effect on fish quality and production. The following areas are of particular interest.

Few of the commonly cultured tropical species have been evaluated for genetic selection of superior growth, disease resistance, or hardiness (Amend 1976; Shepherd 1978), although this research has been of great value to other culture systems (Donaldson and Olson 1957; Moav et al. 1976; Moav and Wohlfarth 1976; Gjedrem 1976).

Interactions between genetics, habitat structure, and bioenergetics in culture situations have seldom been examined (Bams and Simpson 1976), and may be very important in optimizing growth and yield of cultured fish.

Parasite and disease effects on growth rate have been little studied (Webb 1978), and the relationship between density (or competition) and parasite-disease effects on growth is unknown.

Polyculture community dynamics are not well understood, in spite of the prevalence of polyculture systems.

Density effects on food processing efficiency, occurring as a result of social interactions, metabolite levels, and locomotor effects, have received little attention for most species listed in Table 1.

The effects of habitat structure on optimal stocking density are virtually unknown in field situations (Bams and Simpson 1976; Leon 1975).

LITERATURE CITED

Adron, J.W., P.T. Grant, and C.B. Cowley. 1973. A system for the quantitative study of the learning capacity of rainbow trout and its application for the study of food preference and behavior. J. Fish. Biol. 5:625-636.

Allee, W.C., A.J. Finkel, and W.H. Hiskins. 1940. The growth of goldfish in homotypically conditioned water: A population study in mass physiology. J. Exp. Zool. 84:417-443.

Amend, D.F. 1976. Prevention and control of viral diseases of salmonids. J. Fish. Res. Board Can. 33:1059-1066.

Backiel, T., and E.D. LeCren. 1967. Some density relationships for fish population parameters. In S.D. Gerking (ed), Ecology of Freshwater Fish Production. Blackwell Scientific Publications, Oxford, England.

___________, and ___________. 1978. Some density relationships for fish population parameters, pp. 279-302. In S.D. Gerking (ed), Ecology of Freshwater Fish Production. Blackwell Scientific Publications, Oxford, England.

Bams, R.A., and K.S. Simpson. 1976. Substrate incubators workshop--1976. Report on the current state-of-the-art. Fish. Mar. Serv. Res. Dev. Tech. Rep. 689. 67 pp.

Bardach, J.E., and J.H. Ryther. 1968. The Status and Potential of Aquaculture, Particularly Fish Culture. Vol. 2. Part III. Fish Culture. American Institute of Biological Sciences, May 1968, 218 pp.

___________, ___________, and W.O. McLarney. 1972. Aquaculture. The Farming and Husbandry of Freshwater and Marine Organisms. Wiley-Interscience, New York. 868 pp.

Beverton, R.J., and S.J. Holt. 1957. On the dynamics of exploited fish populations. Fishery Invest., Lond. (II) 19:1-533.

Breder, C.M., Jr. 1965. Vortices and fish schools. Zoologica 50:97-114.

___________. 1976. Fish schools as operational structures. Fish. Bull. 74:471-502.

Brown, E.E. 1977. World Fish Farming: Cultivation and Economics. Avi Publ. Co., Westport, Conn. 397 pp.

Brown, M.E. 1946. The growth of brown trout (Salmo trutta L.) I. Factors influencing the growth of trout fry. J. Exp. Biol. 22:118-129.

Burrows, R.E. 1964. Effects of accumulated excretory products on hatchery-reared salmonids. Fish. Wildl. Serv., Bur. Sport Fish. Wildl. (U.S.) Res. Rep. 66:1-12.

Chapman, D.W. 1971. Production, pp. 199-214. In W.E. Richter (ed), Methods for Assessment of Fish Production in Fresh Waters. IBP Handbook No. 3, Blackwell Scientific Publications, Oxford.

Cvancara, V.A. 1976-1980. Current References in Fish Research. Vols. I-V. Dept. of Biology, Univ. Wisc., Eau Claire, WI

Donaldson, L.R., and P.R. Olson. 1957. Development of rainbow trout brood stock by selective breeding. Trans. Am. Fish. Soc. 85:93-101.

Eaton, R.C., and R.D. Farley. 1974. Growth and the reduction of depensation of zebrafish, Brachydanio rerio, reared in the laboratory. Copeia 1974:204-209.

Fromm, P.O., and J.R. Gillette. 1968. Effect of ambient ammonia on blood ammonia and nitrogen excretion of rainbow trout (Salmo gairdneri). Comp. Biochem. Physiol. 26:887-896.

Fry, F.E.J. 1947. Effects of the environment on animal activity. Univ. Toronto Std. Biol. Ser. 55. Publ. Ont. Fish. Res. Lab. 68:1-62.

___________. 1971. The effect of environmental factors on the physiology of fish, pp. 1-98. In W.S. Hoar and D.J. Randal (eds), Fish Physiology, Vol. 6. Academic Press, New York.

Gjedrem, T. 1976. Possibilities for genetic improvements in salmonids. J. Fish. Res. Board Can. 33:1094-1099.

Henderson-Arzapalo, A., R.R. Stickney, and D.H. Lewis. 1980. Immune hypersensitivity in intensively cultured Tilapia species. Trans. Am. Fish. Soc. 109:244-247.

Hickling, C.F. 1962. Fish Culture. Faber and Faber, London. 287 pp.

Hora, S.L., and T.V.R. Pillay. 1962. Handbook on Fish Culture in the Indo-Pacific Region. FAO Fisheries Biology Technical Paper No. 14, Rome, 1962. 204 pp.

Ivlev, I. 1939. Energy balance in the carp. Zool. Zh. 18:303-318.

__________. 1961. Experimental Ecology of Feeding of Fishes. New Haven, Yale Univ. Press, 302 pp.

Jhingran, V.G. 1976. Systems of polyculture of fishes in the inland waters of India. J. Fish. Res. Board Can. 33:905-910.

Kawamoto, N.Y. 1958. Influences of ammonia nitrogen excreted by fishes on their growth in the culture ponds. Rep. Faculty of Fish., Prefectural Univ. of Nie 3(1):104-121.

__________. 1961. The influence of excretory substances of fishes on their own growth. U.S. Bur. Sp. Fish. Wildl. Prog. Fish-Cult. 23:70-75.

__________, Y. Inovye, and S. Nakanishi. 1957. Studies on effect by the pond areas and densities of fish in the water upon the growth rate of the carp (Cyprinus carpio L.). Rept. of Faculty of Fish. Prefectural Univ. of Nie, Vol. 2, No. 3, July 30, 1957.

Kawanabe, H. 1969. The significance of social structure in production of the "Ayu", Plecoglossus altivelis. pp. 243-251. In T.G. Northcote (ed), Symposium on Salmon and Trout in Streams. H.R. Macmillan Lectures in Fisheries, Univ. British Columbia, Vancouver, B.C.

Keenleyside, M.H.A. 1955. Some aspects of the schooling behaviour of fish. Behaviour 8:183-248.

Kerr, S.R. 1971a. Prediction of fish growth efficiency in nature. J. Fish. Res. Board Can. 28:809-814.

__________. 1971b. A simulation model of lake trout growth. J. Fish. Res. Board Can. 28:815-819.

__________. 1980. Niche theory in fisheries ecology. Trans. Am. Fish. Soc. 109:254-257.

Kilgen, R.H., and R.O. Smitherman. 1971. Food habits of the white amur stocked in ponds alone and in combination with other species. Prog. Fish-Cult. 33:123-127.

Krebs, J. 1976. Fish schooling. Nature 264:701.

Kuo, C.-m., Z.H. Shehadeh, and C.E. Nash. 1973. Induced spawning of captive grey mullet (Mugil cephalus L.) females by injection of human chorionic gonadotropin (HCG). Aquaculture 1:429-432.

__________, C.E. Nash, and Z.H. Shehadeh. 1974. The effects of temperature and photoperiod on ovarian development in captive grey mullet (Mugil cephalus L.). Aquaculture 3:25-43.

Landless, P.J. 1976. Demand-feeding behaviour of rainbow trout. Aquaculture 7:11-25.

Larkin, P.A. 1978. Fisheries management--an essay for ecologists. Ann. Rev. Ecol. Syst. 9:57-73.

LeCren, E.D. 1965. Some factors regulating the size of populations of freshwater fish. Mitt. Verein. Theor. Angew. Limnol. 13:88-105.

Leon, K.A. 1975. Improved growth and survival of juvenile Atlantic salmon (Salmo salar) hatched in drums packed with a labyrinthine plastic substrate. Prog. Fish-Cult. 37:158-163.

Li, H.W., and R.W. Brocksen. 1977. Approaches to the analysis of energetic costs of intraspecific competition for space by rainbow trout (Salmo gairdneri). J. Fish. Biol. 11:329-341.

Liao, I.C., Y.J. Lu, T.L. Huang, and M.C. Lin. 1971. Experiments on induced breeding of the grey mullet, Mugil cephalus Linnaeus. Aquaculture 1:15-34.

Magnuson, J.J. 1962. An analysis of aggressive behaviour, growth, and competition for food and space in medaka (Oryzias latipes (Pisces, Cyprinodontidae)). Can. J. Zool. 40:313-363.

Marr, D.H.A. 1963. The influence of surface contour on the behavior of trout alevins, Salmo trutta (Linnaeus). Animal Behavior 11:412.

Moav, R., T. Brody, G. Wohlfarth, and G. Hulata. 1976. A proposal for the continuous production of F1 hybrids between the European and Chinese races of the common carp in traditional fish farms of Southeast Asia. FAO Technical Conference Aquaculture, Kyoto. Rome: FAO.

__________, and G. Wohlfarth. 1976. Two-way selection for growth rate in the common carp (Cyprinus carpio L.). Genetics 82:83-101.

Nelson, J.S. 1976. Fishes of the World. John Wiley & Sons, New York. 416 pp.

Nikol'skii, G.V. 1955. O biologicheskom znachenii stai u ryb. Trudy Sovesheh. ikhtiol. Kom. 5:104-107.

Nilsson, N.A. 1963. Interaction between trout and char in Scandinavia. Trans. Am. Fish. Soc. 92:276-285.

__________. 1967. Interactive segregation between fish species, pp. 295-313. In S.D. Gerking (ed), The Biological Basis of Freshwater Fish Production. John Wiley & Sons, Inc., New York.

Noakes, D.L.G. 1978. Social behaviour as it influences fish production, pp. 360-382. In S.D. Gerking (ed), The Ecology of Freshwater Fish Production. Blackwell Scientific Publications, Oxford.

Olson, K.R., and P.O. Fromm. 1971. Excretion of urea by two teleosts exposed to different concentrations of ambient ammonia. Comp. Biochem. Physiol. 40a:999-1007.

Paloheimo, J.E., and L.M. Dickie. 1965. Food and growth of fishes. I. A growth curve derived from experimental data. J. Fish. Res. Board Can. 22:521-542.

__________, and __________. 1966. Food and growth of fishes. III. Relations among food, body size, and growth efficiency. J. Fish. Res. Board Can. 23:1209-1248.

Parker, F.R., Jr. 1973. Reduced metabolic rates in fishes as a result of induced schooling. Trans. Am. Fish. Soc. 102:125-131.

Pfuderer, P., P. Williams, and A.A. Francis. 1974. Partial purification of the crowding factor from Carassium auratus and Cyprinus carpio. J. Exp. Zool. 187:375-382.

Pianka, E.R. 1978. Evolutionary Ecology. Harper and Row, New York. 397 pp.

Schlaifer, A. 1938. Studies in mass physiology: effect of numbers upon the oxygen consumption and locomotor activity of Carassius auratus. Physiol. Zool. 11:408-424.

__________. 1939. An analysis of the effect of numbers upon the oxygen consumption of Carassius auratus. Physiol. Zool. 12:381-392.

Shehadeh, Z.H., and J.N. Ellis. 1970. Induced spawning of the striped mullet Mugil cephalus L. J. Fish. Biol. 2:355-360.

__________, C.-m. Kuo, and K.K. Milisen. 1973a. Induced spawning of grey mullet Mugil cephalus L. with fractionated salmon pituitary extract. J. Fish. Biol. 5:471-478.

__________, W.D. Madden, and T.P. Dohl. 1973b. The effect of exogenous hormone treatment on spermiation and vitellogenesis in the grey mullet, Mugil cephalus L. J. Fish. Biol. 5:479-487.

Shepherd, C.J. 1978. Aquaculture: Some current problems and the way ahead. Proc. Roy. Soc. Edinburgh 76B:215-222.

Smith, A.C. 1977. Reactions of fish red blood cells with mucus and sera from other fish(es). California Fish and Game 63:52-57.

Smith, C.E. 1972. Effects of metabolic products on the quality of rainbow trout. Amer. Fishes and U.S. Trout News 17(3).

Sperber, O., J. Fromm, and P. Sparre. 1977. A method to estimate the growth rate of fishes, as a function of temperature and feeding level, applied to rainbow trout. Meddr. Danm. Fisk.-og Havunders. N.S. 7:275-317.

Svardson, G. 1976. Interspecific population dominance in fish communities of Scandinavian lakes. Inst. Freshw. Res. Drottningholm Rep. 55:147-171.

Tang, Y.A. 1964. Induced spawning of striped mullet by hormone injection. Jap. J. Ichthyol. 12:23-28.

Timms, A.M. 1975. Intraspecific communication in goldfish. J. Fish. Biol. 7:377-389.

Ware, D.M. 1975. Growth, metabolism, and optimal swimming speed of a pelagic fish. J. Fish. Res. Board Can. 32:33-41.

__________. 1978. Bioenergetics of pelagic fish: theoretical change in swimming speed and ration with body size. J. Fish. Res. Board Can. 35:220-228.

Warren, C.E., and G.E. Davis. 1967. Laboratory studies on the feeding bioenergetics, and growth of fish, pp. 175-214. In S.D. Gerking (ed), Ecology of Freshwater Fish Production. Blackwell Scientific Publications, Oxford, England.

Weatherley, A.H. 1966. The ecology of fish growth. Nature 212:1321-1324.

__________. 1972. Growth and Ecology of Fish Populations. Academic Press, Inc., London. 294 pp.

__________. 1976. Factors affecting maximization of fish growth. J. Fish. Res. Board Can. 33:1046-1058.

Webb, P.W. 1978. Partitioning of energy into metabolism and growth. In S.D. Gerking (ed), Ecology of Freshwater Fish Production. Blackwell Scientific Publications, Oxford, England.

Weihs, D. 1973. Hydromechanics of fish schooling. Nature 241:290-291.

Werner, E.E. 1980. Reply to: Niche theory in fisheries ecology. Trans. Am. Fish. Soc. 109:257-260.

__________, and D.J. Hall. 1976. Niche shifts in sunfishes: experimental evidence and significance. Science 191:404-406.

__________, and __________. 1977. Competition and habitat shift in two sunfishes (Centrarchidae). Ecology 58:869-876.

__________, __________, D.R. Laughlin, D.J. Wagner, L.A. Wilsmann, and F.C. Funk. 1977. Habitat partitioning in a freshwater fish community. J. Fish Res. Board Can. 34:360-370.

Whittaker, R.H., and S.A. Levin (eds). 1975. Niche: Theory and Application. Benchmark Papers in Ecology, Vol. 3. Dowden, Hutchinson and Ross, Stroudsburg, PA 448 pp.

Winberg, G.G. 1956. Rate of metabolism and food requirements of fishes. Belorussian State Univ. Minsk. Fish. Res. Bd. Canada Trans. Ser. No. 194. 1960.

Wuhrman, Von K., and W.H. Woker. 1948. Beitrage zur toxikologie der fische. H. Experimentelle untersuchungen uber die ammoniak and blausaureurevergiftung. Schweiz Z. Hydrol. 11:210.

Yamagishi, H. 1962. Growth relation in some small experimental populations of rainbow trout fry, Salmo gairdneri Richardson with special reference to social relations among individuals. Jap. J. Ecol. 12:45-53.

__________. 1964. An experimental study on the effect of aggressiveness to the variability of growth in the juvenile rainbow trout, Salmo gairdneri Richardson. Jap. J. Ecol. 14:228-232.

Yang, W.T., and U.B. Kimm. 1962. A preliminary report on the artificial culture of grey mullet in Korea. Indo-Pacific Fish. Coun. 9:62-70.

Yashouv, A. 1958. The excreta of carp as a growth-limiting factor. Bamidgeh 10:90-95.

__________. 1966. Mixed fish culture; an ecological approach to increase pond production. FAO World Symp. Warm-water Pond Fish. Cult., Rome, May 18-25, 1966.

__________. 1969. Preliminary report on induced spawning of M. cephalus (L.) reared in captivity in freshwater ponds. Bamidgeh 21:19-24.

Yu, M.L., and A. Perlmutter. 1970. Growth inhibiting factors in zebrafish (Brachydanio rerio) and the blue gourami (Trichogaster trichopterus). Growth 34:153-175.

ADDITIONAL BIBLIOGRAPHY ON GENETIC AND COMPETITION FACTORS GOVERNING GROWTH

Allee, W.C., and E.S. Bowen. 1932. Studies on animal aggregation: mass protection against colloidal silver among goldfishes. J. Exp. Zool. 61:185-207.

__________, B. Greenberg, G.M. Rosenthal, and P. Frank. 1948. Some effects of social organization on growth in the green sunfish, Lepomis cyanellus. J. Exp. Zool. 108:1-20.

Brockway, D.R. 1950. Metabolic products and their effects. Prog. Fish-Cult. 12:127-129.

Calaprice, J.R. 1976. Mariculture-ecological and genetic aspects of production. J. Fish. Res. Board Can. 33:1068-1087.

Chaudhuri, H., R.d. Chakrabarty, N.G.S. Rao, K. Janakiram, D.K. Chatterjee, and S. Jena. 1974. Record fish production with intensive culture of Indian and exotic carps. Curr. Sci. 43:303-304.

Chen, F.Y. 1965. The living-space effect and its economic implications. Appendix III, in Rep. Trop. Fish. Cult. Res. Inst. Malacca.

___________, and G.A. Prowse. 1964. The effect of living space on the growth rate of fish. Ichthyologica 3:11-20.

Davies, P.M.C. 1966. The energy relations of Carassius auratus L. II. The effect of food, crowding, and darkness on heat production. Comp. Biochem. Physiol. 17:893-995.

Forselius, S. 1957. Studies of anabantid fishes. I-III. Zool. Bidrag. Uppsala 32:93-598.

Frey, D.F., and R.J. Miller. 1972. The establishment of dominance relationships in the blue gourami, Trichogaster trichopterus (Pallus). Behaviour 42:8-62.

Fryer, G. 1959. The trophic interrelationships and ecology of some littoral communities of Lake Nyasa with special reference to the fishes, and a discussion of the evolution of a group of rock-frequenting Cichlidae. Proc. Zool. Soc. Lond. 132:153-281.

Gall, G.A.E. 1975. Genetics of reproduction in domesticated rainbow trout. J. Anim. Sci. 40:19-28.

___________. 1976. Possibilities for genetic improvements in salmonids. J. Fish. Res. Board Can. 33:1094-1099.

Hepher, B. 1978. Ecological aspects of warm-water fishpond management, pp. 447-468. In S.D. Gerking (ed), Ecology of Freshwater Fish Production. Blackwell Scientific Publications, Oxford, England.

Hiatt, R.W. 1944. Food chains and the food cycle in Hawaiian fish ponds. II. Biotic interaction. Trans. Am. Fish. Soc. 4:262-280.

Hines, R., G. Wohlfarth, R. Moav, and G. Hulata. 1974. Genetic differences in susceptibility to two diseases among strains of the common carp. Aquaculture 3:187-197.

Hiscox, J.I., and R.W. Brocksen. 1973. Effects of a parasitic gut nematode on consumption and growth in juvenile rainbow trout (Salmo gairdneri). J. Fish. Res. Board Can. 30:443-450.

Hodgkiss, I.J., and H.S.H. Man. 1977. Stock density and mortality assessment of Sarotherodon mossambicus (Cichlidae) in Plover Cover Reservoir, Hong Kong. Eng. Biol. Fish. 1:171.

Ihssen, P. 1976. Selective breeding and hybridization in fisheries management. J. Fish. Res. Board Can. 33:316-321.

Ingham, L., and C. Arme. 1973. Intestinal helminths in rainbow trout, Salmo gairdneri (Richardson): Absence of effect on nutrient absorption and fish growth. J. Fish. Biol. 5:309-313.

Kesteven, G.L., and R.R. Ingpen. 1966. The representation of relations, including those of sociality, in biotic systems. Proc. Ecol. Soc. Aust. 1:79-83.

Kripitschnikov, V. 1945. Viability, rate of growth, and morphology of carps of different genotypes as affected by rearing conditions. C.R. Aca. Sci. U.S.S.R., Vol. 47.

Kuronuma, K., and K. Nakamura. 1957. Weed control in farm pond and experiment by stocking grass carp. Proc. Indo-Pacific Fish. Count. 7:35-42.

MacArthur, R.H., and R. Levins. 1964. Competition, habitat selection, and character displacement in a patchy environment. Proc. Nat. Acad. Sci. 51:1207-1210.

Martin, R.G. 1975. Sexual and aggressive behavior, density, and social structure in a natural population of mosquito fish, Gambusia holbrooki. Copeia 1975:445-454.

Mayer, F.D.L., and R.M. Karmer. 1973. Effects of hatchery water reuse on rainbow trout metabolism. Prog. Fish-Cult. 35:9-10.

Miller, R.B. 1958. The role of competition in the mortality of hatchery trout. J. Fish. Res. Board Can. 15:27-45.

Miller, R.S. 1967. Pattern and process in competition. Adv. Ecol. Res. 4:1-74.

Moav, R., and G.W. Wohlfarth. 1966. Genetic improvements of yield in carp. FAO World Symp. Warm-water Pond Fish Cult., Rome, May 18-25, 1966.

___________, G. Hulata, and G. Wohlfarth. 1975. Genetic differences between the Chinese and European races of the common carp. I. Analysis of genotype environment interactions for growth rate. Heredity 34:323-340.

Roales, R.R. 1981. The effect of growth inhibiting factors on the total lipid content of the zebra fish, Brachydamo verio (Hamilton-Buchanan). J. Fish. Biol. 18:723-728.

Rose, S.M., and F.C. Rose. 1965. The control of growth and reproduction in freshwater organisms by specific products. Mitt. Internat. Verein. Limnol. 13:21-35.

Silliman, R.P. 1972. Effect of crowding on relation between exploitation and yield in Tilapia macrocephala. Fish. Bull. 70:693-698.

Sinha, V.R.P., and M.V. Gupta. 1975. On the growth of grass carp Ctenopharyngodon idella Val. in composite fish culture at Kalyani, West Bengal (India). Aquaculture 5:283-290.

Varghese, T.J., and B. Shantharam. 1979. Preliminary studies on the relative growth rates of 3 Indian major carp hybrids. Proc. Indian Acad. Sci. Sec. B, Part 1 88:209-216.

Walter, E. 1934. Grundlagender allgemeinen fischeilichen Produktionslehre. Handle. Binnenfisch. Mitteleur 4:480-662.

Wolny, P. 1962. The influence of increasing the density of stocked fish populations on the growth and survival of carp fry (Polish. Eng. Rus. Summ.) Roezn. Nauk. Roin. 81B(2):171-188.

A STATE-OF-THE-ART OVERVIEW OF AQUATIC FERTILITY WITH SPECIAL REFERENCE TO CONTROL EXERTED BY CHEMICAL AND PHYSICAL FACTORS

by

Darrel L. King and Donald L. Garling

INTRODUCTION

Efforts to produce high-quality animal protein for human consumption through aquaculture have often been directed towards intensive fish production using formulated feeds. Although many fish can efficiently convert artificial feeds to flesh, fish generally require two to four times greater dietary protein than warm-blooded domesticated animals (Mertz 1972). Fish diets rely heavily on high-quality protein meals which often are suitable for direct addition to the human diet. The conversion of edible protein sources to fish flesh in lesser developed countries is suspect if the goal is to increase available dietary protein for humans.

Channel catfish can be used as an example of an intensively cultured fish to demonstrate the efficiency of feed protein converted to edible fish protein. Catfish are harvested for human consumption after 18-22 months of production at an average weight of about 1.25 lbs. At a food conversion rate of 1.5:1, each fish has been fed about 1.875 lbs. of feed. Since commercial catfish feeds contain about 36 percent protein, each fish has consumed about 0.675 lbs. of protein. Only about 0.5625 lbs. of edible fish flesh is available from each fish produced (1.25 lbs. x 45 percent dressout to fillets) which contain about 0.107 lbs. of protein (0.5625 lbs. x 19 percent protein wet weight). Consequently, the percent of protein fed which is recovered is about 15.85 percent [(0.107 lbs. protein in flesh/0.675 lbs. protein fed) x 100].

Broiler chickens are as efficient as channel catfish in converting dietary protein to edible flesh protein. A broiler is ready for market after seven to eight weeks at a weight of approximately 4.12 lbs. Each chicken has been fed about 8.07 lbs of feed containing 1.74 lbs of protein (2.54 lbs. of starter at 23 percent protein and 5.53 lbs of finisher at 21 percent protein or 0.58 and 1.16 pound of protein, respectively) with a feed conversion of 1.96:1. The finished broiler provides about 0.29 lbs. of protein available for consumption (4.12 lbs. live weight x 66 percent dressout x 60 percent edible meat x 18 percent protein wet weight). Consequently, the percent protein fed which is recovered is about 16.67 percent [(0.29 lbs. protein in flesh/1.74 lbs. protein fed) x 100].

Despite similar protein conversion ratios, chickens would appear to be more desirable converters to high-grade protein because of their seven to eight week turnaround time, as opposed to the 18-22 months required for catfish. However, production of both chickens and catfish in a feedlot manner results in significant loss of edible protein.

Production of certain fishes in lesser developed countries has been limited by the availability of appropriate feed items in culture ponds throughout the rearing period. Supplemental feeding has often been suggested to augment pond fertilization practices. As an example, the production of *Chanos chanos* in developing countries has been limited by available periphyton in culture ponds near the end of production (Bardach et al. 1972). The protein requirements for maximum growth of *C. chanos* fry has been established at 40 percent of the diet (Lim et al. 1979). If protein requirements of adults are similar to fry and *C. chanos* is as efficient as channel catfish in converting protein fed to edible protein, the net result still would be a loss in edible protein during final production feeding.

The character of the nutrients in the animal wastes, particularly nitrogen resulting from degraded protein, again favors chickens. Chicken wastes are a concentrated nutrient-rich fertilizer which can be stored and applied when and where needed to enhance either terrestrial or aquatic fertility. The wastes from the fish release nutrients in a continuous and diffuse manner to the aquatic system yielding little opportunity to control aquatic productivity by adjusting nutrient availability.

The loss of edible protein with either fish or chickens fed protein-rich foodstuffs can lead to reductions in the protein level of the human diet in lesser developed countries in that the increased market cost of the higher grade protein is often beyond the means of a significant portion of the population. If the decision is made to upgrade protein on a large feedlot scale, the advantage would appear to lie with chickens in that the turn-around time is shorter and the waste nutrients are in a concentrated form allowing better potential for enhancing primary productivity. Overall, it appears that feedlot aquaculture should be avoided in lesser developed countries.

BASIN CHARACTERISTICS

As was shown in the preceding discussion, the probability of significantly increasing protein yield to the people of third world countries from aquaculture systems directly dependent on external food sources for the fish is, at best, extremely limited. The primary objective of aquaculture in such countries is to increase production of protein in the human diet, but the challenge is to maximize protein yield from the base supplied by aquatic photosynthesis without relying on protein- and energy-rich external foodstuffs.

Any manipulation of third world aquatic ecosystems to increase protein production must be accomplished, in most cases, without benefit of the energy- and resource-demanding technologies common to more industrialized societies. In such impoverished areas, enhancement of protein yield must be accomplished by simple manipulations aimed at approaching the maximum potential yield within the limits prescribed by the physical, chemical, and biological interactions peculiar to each system.

Ponds, like soups, differ greatly in nutritive content and the fertility of each pond, to a large extent, is dependent on where the water has been before it entered the pond. The remarkable solvent properties of water allow it to transport a multitude of both natural and man-added inorganic and organic materials in the dissolved state. The erosive character of water also allows the transport of a great variety of particulate materials. The combination of dissolved and particulate materials present in water interact with physical parameters such as temperature and light to dictate limits to both type and amount of aquatic photosynthesis. As such, the fertility of freshwater ponds reflect the drainage basin in which they lie. Thus, local geology and climate, together with terrestrial vegetation and type and amount of human use and perturbation of the land in the drainage basin to each pond, dictates the base physical and chemical character of the water in that particular pond. The interplay between the myriad variables involved guarantees variation in fertility from pond to pond and from region to region.

The various physical, chemical, and biochemical weathering reactions responsible for formation of soils from the parent rock yield significant cation release. With the abundance of water in humid climates, those released cations are washed from the soil leaving less fertile soils dominated by iron and aluminum silicates and oxy-hydroxides. The end result of a long history of such weathering in a humid climate is a basin which yields runoff water with low concentrations of dissolved minerals.

In more arid climates, the decreased water throughput associated with decreased precipitation and increased evaporation is not sufficient to wash out the cations released by weathering. The result is a basin dominated by more complex cation-rich clays from which the runoff water contains elevated concentrations of dissolved minerals.

In general, water from well-weathered drainage basins in humid climates or from areas of igneous rock with low solubility will contain reduced carbonate-bicarbonate alkalinity and total hardness. Waters from calcareous or historically more arid drainages will contain much higher alkalinity and hardness. An example of this is seen in data presented by Kempe (1979) which indicates an alkalinity averaging about 2.2 meq/l in the Elbe River compared with an alkalinity of only about 0.4 meq/l in the Amazon River.

ALKALINITY

In freshwater ponds and lakes, the bicarbonate-carbonate alkalinity plays a key role in dictating the potential productivity such systems can exhibit. The carbonate-bicarbonate alkalinity system serves simultaneously as the only significant buffer controlling the pH of the water and the only significant reserve source of carbon dioxide available for support of aquatic photosynthesis.

Carbon dioxide gas dissolves in water to form carbonic acid. Increased acidity associated with carbon dioxide gain by rainwater as it percolates through the organic-rich surface layers of terrestrial soils accelerates weathering of parent materials, yielding increased bicarbonate alkalinity in the water.

The two dissociations of carbonic acid shown in Equations 1 and 2 play a primary role in pH buffering in freshwater systems.

$$CO_2 + HOH \quad HCO_3^- + H^+ \qquad (1)$$

$$H^+ \; CO_3^= \quad HCO_3^- \qquad (2)$$

Combination of the first and second dissociations of carbonic acid to yield Equation 3 indicates the total reaction of this primary buffer system.

$$2HCO_3^- \quad CO_2 + CO_3^= + HOH \qquad (3)$$

The first and second dissociations of carbonic acid are both characterized by dissociation constants as shown in Equations 4 and 5.

$$K_1 = \frac{[HCO_3^-]\,[H^+]}{[CO_2]} \qquad (4)$$

$$K_2 = \frac{[CO_3^=]\,[H^+]}{[HCO_3^-]} \qquad (5)$$

Solving each of these equations for HCO_3 yields the equality shown in Equation 6.

$$[HCO_3^-] = \frac{K_1\,[CO_2]}{[H^+]} = \frac{[CO_3^=][H^+]}{K_2} \qquad (6)$$

Rearrangement of Equation 6 yields Equation 7, which, in an oversimplified fashion, indicates the interdependency of pH, the available carbon dioxide level, and the alkalinity system in freshwater.

$$[H^+]^2 = K_1K_2\,\frac{[CO_2]}{[CO_3^=]} \qquad (7)$$

The first and second dissociations of carbonic acid together with the solubility of carbonate salts, atmospheric carbon dioxide, and extraction and return of carbon dioxide by aquatic photosynthesis and respiration shown in Equation 8 determine the pH of freshwaters.

Atmospheric CO_2

$$Ca^{++} + 2HCO_3^- \quad CO_2 + Ca^{++} + CO_3^= + HOH \qquad (8)$$

Photosynthesis Respiration Ksp

Biota $CaCO_{3(s)}$

In the absence of nitrogen, phosphorus, and other nutrients required for aquatic photosynthesis, the pH of the water would be controlled by atmospheric carbon dioxide and the alkalinity of the water. Representative equilibrium pH values are presented in Figure 1 for an atmospheric carbon dioxide concentration of 340 ppm, a temperature of 20°C, and alkalinities of 5 to 300 mg $CaCO_3$/l. The total inorganic carbon present in the water under these conditions also is presented in Figure 1.

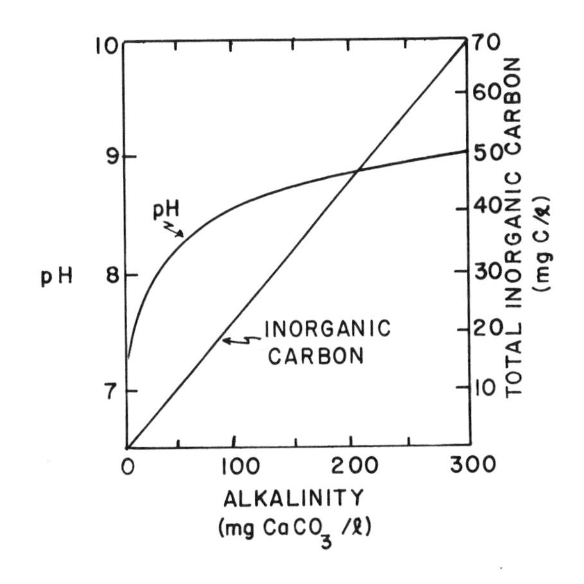

Figure 1.

Total inorganic carbon content and pH of freshwater at various carbonate-bicarbonate alkalinities at equilibrium with an atmospheric carbon dioxide level of 340 ppm at 20°C.

Addition of nutrients which allow photosynthetic uptake of carbon dioxide by aquatic plants at a rate faster than it can be supplied by the atmosphere or respiratory sources leads to an increased pH and a concomitant reduction in the carbon dioxide concentration of the water as shown in Figure 2:

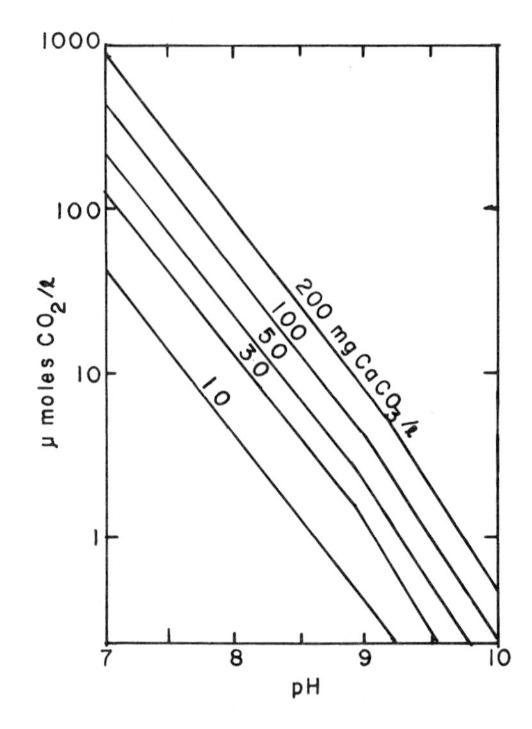

Figure 2.

Carbon dioxide content of freshwater as a function of pH and carbonate-bicarbonate alkalinity of the water.

In order for atmospheric carbon dioxide to enter water, the carbon dioxide concentration of the water must be reduced to a level less than that at atmospheric equilibrium (13.2 μ moles CO_2/l for an atmospheric content of 340 ppm and a water temperature of 20°C). In a pond, this reduction is accomplished by the photosynthetic fixation of carbon dioxide by aquatic plants at a rate greater than the rate of supply by respiration. However, the carbonate-bicarbonate alkalinity of water determines the amount of carbon dioxide which must be fixed before atmospheric recarbonation becomes a significant source of photosynthetic carbon.

The amount of photosynthetic carbon fixation necessary to reduce the free carbon dioxide concentration of the water (CO_2) from 13.2 μ moles/l to 1.32 μ moles/l (100 percent to 10 percent saturation at 20°C with an atmospheric content of 340 ppm carbon dioxide) is shown for a wide range of alkalinities in Figure 3. Clearly, a much greater amount of photosynthesis must occur at high alkalinities before atmospheric recarbonation reaches the rate found at low alkalinities after only a small amount of photosynthetic carbon fixation.

The amount of phosphorus required for algae to accomplish the amount of carbon fixation indicated in Figure 3, assuming a carbon-to-phosphorus atomic ratio of the algae of 100:1, also is given in Figure 3. If no phosphorus is available, there will be no algal photosynthesis and consequently, no diffusion of atmospheric carbon dioxide into the water. As indicated in Figure 3, the amount of phosphorus necessary to allow reduction of the CO_2 concentration from 13.2 to 1.32 μ moles CO_2/l is dependent upon the alkalinity of the water. It should be noted that at an alkalinity of 5 mg $CaCO_3$/l, the required amount of phosphorus is only 4.3 μ g P^3/l, while at an alkalinity of 300 mg $CaCO_3$/l, the required amount of phosphorus is 291.7 μ g P/l. Thus, the amount of phosphorus a pond can tolerate before exhibiting reduced carbon dioxide concentrations is a function of the alkalinity of the pond water which, in turn, is a function of the drainage basin in which the pond lies. Likewise, the amount of algal production allowed prior to establishment of low levels of carbon dioxide in the water also is a function of the alkalinity, increasing with increased alkalinity. As the carbon dioxide concentration of a water is reduced, the specific net carbon fixation rate for algae (King and Novak 1974; King 1980b) and aquatic plants (Liehr 1978; Craig 1978) is reduced. In general, the specific net carbon fixation rate of algae and submerged aquatic plants can be fit to the Michaelis-Menton model if corrections are made for the threshold carbon dioxide concentration required to initiate net photosynthesis as indicated in Equation 9.

$$\mu = \mu max \frac{C-C_q}{(K_c-C_q) + (C-C_q)} \qquad (9)$$

where: μ = specific net carbon fixation rate (Time $^{-1}$)

μ max = maximum specific net carbon fixation rate (Time $^{-1}$)

C = existing carbon dioxide concentration (μ moles CO_2/l)

K_c = carbon dioxide concentration at which μ = 0.5 μ max (μ moles CO_2/l)

C_q = threshold carbon dioxide concentration required to initiate net carbon fixation (μ moles CO_2/l)

At least for the alga <u>Chlorella vulgaris</u> the terms μ max, K_c and C_q are all functions of available light intensity (King 1980b) indicating marked interaction between the existing concentration of carbon dioxide and light intensity as shown in Equation 10.

$$\mu = f_1 light \frac{C - f_2 \text{ light}}{f_3 \text{ light} - f_2 \text{light} + (C - f_2 \text{ light})} \quad (10)$$

The threshold carbon dioxide concentration required to initiate net photosynthetic carbon fixation varies by orders of magnitude between different algae at any given light intensity (King 1980b). The blue-green algae <u>Anacystus nidulans</u> exhibits the lowest C_q of all algae evaluated to date (King 1980b). In nutrient enriched water, photosynthetic uptake of carbon dioxide by aquatic plants often exceeds the rate of recarbonation by atmospheric and respiratory carbon dioxide yielding decreasing carbon dioxide concentrations with time. This results in a decreased net specific carbon fixation rate by the plants but also in an increase in the net specific sinking rate of green algae (King 1980b). The result of this accelerated sinking is the loss of the green algae at carbon dioxide concentrations in two of the three orders of magnitude higher than the threshold concentrations to which they are physiologically capable.

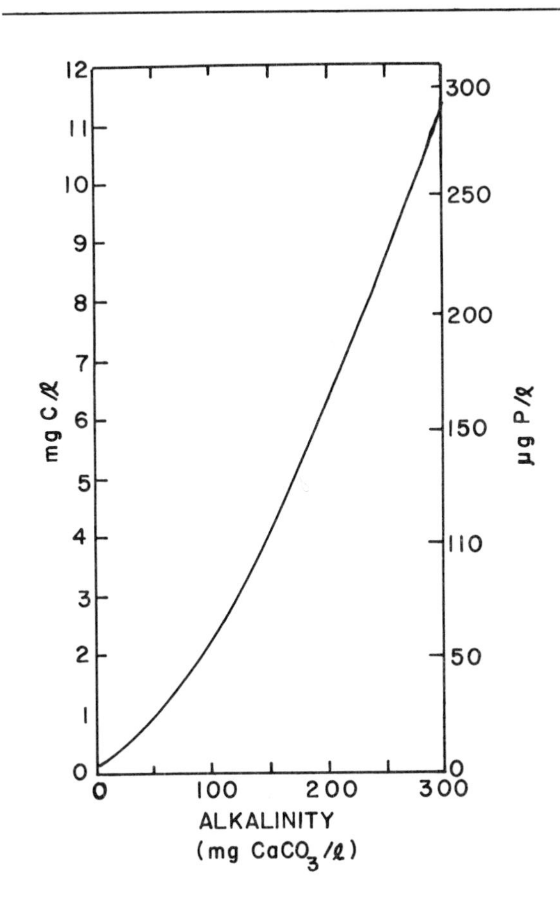

Figure 3.

Amount of inorganic carbon which must be removed from various carbonate-bicarbonate alkalinities to reduce carbon dioxide from 100 to 10 percent saturation with an atmospheric carbon dioxide level of 340 ppm at 20°C given with the amount of phosphorus required for algal fixation of that carbon.

Blue-green algae tend to become more buoyant with photosynthetic stress-induced increase in vacuole formation (Dinsdale and Walsby 1972; Grant and Walsby 1977). King (1970, 1972) suggested that the probability of blue-green algal dominance increased with decreased carbon dioxide concentration of the water, and both King (1970, 1972) and Shapiro (1973) suggested that blue-green algae are better able to function at low carbon dioxide levels than are green algae. King (1972) used this consideration to suggest that the amount of phosphorus required to initiate blue-green algal dominance in a lake increases with the additional carbon dioxide available with increased alkalinity of the water.

From available data, it appears that decreasing carbon dioxide concentrations in productive waters resulting from uptake of carbon dioxide by aquatic plants at rates exceeding recarbonation from atmospheric and respiratory sources play an important role in dictating the type of plant the water will support. The maintenance of carbon dioxide concentrations sufficient to minimize dominance by blue-green algae appears to be determined by light availability, phosphorus, nitrogen, and alkalinity content of the water, and detention time of the water in the pond (King 1976).

The amount of carbon dioxide available for support of desirable plants prior to development of dominance by blue-green algae also is a function of the alkalinity. The addition of only small amounts of phosphorus is required to fix the small amount of carbon necessary to reduce the carbon dioxide to the level favoring blue-green algae in extremely soft waters. In hard waters with high alkalinities, sufficiently more nutrients can be tolerated and much more desirable algae will be produced prior to establishment of the blue-green algae. Thus, it appears that the amount of aquatic photosynthate of a type of widespread value to aquacultural production as a function of nutrient addition is, in turn, a function of the alkalinity of the pond water which is determined by the drainage basin to the pond.

This relationship of pond productivity to the alkalinity of the water is not a recent observation. After their investigation of a great variety of lakes, Birge and Juday (1911) suggested that free carbon dioxide and available bicarbonate were the factors most likely limiting algal production in natural waters. Moyle (1945) related both type and abundance of aquatic plants to the alkalinity of the water, and Ball (1948), Turner (1960), and Hayes and Anthony (1964) found increased fish production with increased alkalinity. King (1970, 1972) discussed the role of alkalinity in the determination of both qualitative and quantitative changes in algal activity. Wright (1960), Wright and Mills (1967), Bartsch and Allum (1957), Welch (1968), and many other investigators have suggested that algal production was limited by the availability of a photosynthetic carbon source associated with the alkalinity in a wide variety of waters.

Moyle (1946) suggested that natural waters with increased alkalinity also would tend to contain increased concentrations of other ions required for photosynthesis. Schindler et al. (1972) present evidence that appreciable amounts of carbon dioxide can be gained from the atmosphere, and that aquatic plant photosynthesis would not be totally limited by an absence of carbon dioxide. However, decreased carbon dioxide levels do result in both decreased specific net carbon fixation rates (King 1980b)and increased probability of blue-green algal dominance (King 1970, 1972; Shapiro 1973).

In aquacultural ponds, the nutrient content is elevated by adding both organic and inorganic fertilizers. In such systems, increased reliance is placed on atmospheric carbon dioxide, and the result often is the development of dominance by blue-green algae, particularly if the alkalinity of the water is low. Under such conditions, the atmosphere continues to provide carbon dioxide, but the resulting photo-synthate is represented largely by blue-green algae.

One of the major vexations in the management of water quality of lakes, ponds, reservoirs, wastewater treatment ponds, and particularly of aquacultural systems is the establishment of blue-green algal dominance of the phytoplankton. In contrast to green algae and diatoms, which are valuable initiators of the aquatic food chain, the blue-green algae are not readily used by the more desirable members of most aquatic systems. Rather, they often serve primarily as an energy source for bacteria and when used by the bacteria, often yield a deleterious impact on the dissolved oxygen resources of the water body. In addition, the gas vacuoles of the blue-green algae increase their buoyancy, causing them to accumulate near the water surface, thereby markedly decreasing light penetration and thus total photosynthetic activity of the pond.

NUTRIENTS

The biogeochemical cycle of phosphorous indicates the conservative nature of this element in the biosphere. Phosphorus has no permanent atmospheric sink but rather tends to be associated with both metal precipitates and exchange sites in both terrestrial and aquatic situations. The natural availability of phosphorus in freshwaters is determined by the same sorts of geologic and climatic factors which determine the general dissolved mineral content. But the propensity of phosphorus to bind, in one way or another, with organic and inorganic particulates allows conservation of this nutrient within freshwater aquatic systems. Cycle and recycle of phosphorus through aquatic ecosystems, particularly through shallow freshwater systems supporting rooted aquatic plants, allows maximum use of the phosphorus available in the production of aquatic photosynthate.

The conservative nature of phosphorus together with the extremely small concentrations required for production of freshwater aquatic plant biomass cause prime focus to be placed on phosphorus in the control of cultural eutrophication. These peculiarities of phosphorus, which are of detriment to the control of cultural eutrophication, are of real value in aquaculture in that only small concentrations of this scarce and often expensive nutrient are required to maintain optimal aquatic plant productivity.

As we noted earlier, the amount of phosphorus which can be added to a freshwater prior to establishment of dominance by blue-green algae is a function of the alkalinity of the water. The role of alkalinity in maintaining pH of the water also is of import in that the solubility of many metal phosphates is related to the water pH.

Once added to a pond system, phosphorus tends to be conserved while other nutrients required to sustain pond productivity can be lost. Nitrogen is of particular import in this regard, in that it is of obvious importance to protein formation and can be lost to the atmosphere both as ammonia, particularly with elevated water pH, and as nitrogen gas following denitrification.

The most probable nutrient sources available for aquaculture in third world countries are animal wastes. Morrison (1961) gives data which yield nitrogen-to-phosphorus weight ratios for the wastes from various animals, ranging from 6.42 for sheep for 2.11 for poultry. For optimal growth, aquatic plants require nitrogen-to-phosphorus ratios in the neighborhood of 7:8, indicating that animal wastes contain enough phosphorus for more aquatic photosynthesis than would be allowed by the nitrogen they contain. Clearly, any reduction in nitrogen availability will result in decreased protein content of the plants.

The addition of animal wastes to a pond stimulates both algae and macrophytes and yields increased productivity. Most commonly, such increased photosynthetic activity results in withdrawal of carbon dioxide from the water at rates faster than it can be replaced from respiration or atmospheric recarbonation, thereby yielding increased pH, with the rate of change being a function of pond alkalinity.

Nutrient-induced increases in aquatic productivity lead to faster rates of biogeochemical nutrient cycle and increase the frequency with which each nitrogen atom will appear as ammonia. The increased rate of nitrogen recycle and the generally elevated pH in ponds where productivity is enhanced by nutrient addition leads to accelerated loss of nitrogen to the air as ammonia gas. Increased photosynthetic productivity also leads to increased probability of establishment of sufficiently reducing conditions at the pond bottom to promote nitrogen loss through denitrification.

The longer the detention time of the water in the pond, the greater is the probability that nitrogen will be lost to the atmosphere. Such loss can lead to significant reduction in the inorganic nitrogen available for plant photosynthesis. In a series of ponds charged with a good quality domestic wastewater, the nitrogen content of the water decreased as an exponential function of detention time as shown in Equation 11 (King 1979).

$$N_t = N_o e^{-0.03t} \qquad (11)$$

where: N_t = total nitrogen concentration (mgN/l) at time t.

N_o = initial total nitrogen concentration (mgN/l).

t = detention time in days.

This loss of nitrogen leads to establishment of dominance by nitrogen-fixing blue-green algae in massive bloom proportions (King 1980a). Bacterial use of these nitrogen-fixing blue-green algae leads to sufficient dissolved-oxygen depletion to cause both summer and winter fish kills (King 1980a).

A major difficulty in maximizing yield and utilization of nutrients is in maintaining sufficient nitrogen and carbon dioxide concentrations to allow total expression of the phosphorus available in a form of aquatic photosynthate of value to the crop of interest being harvested from the pond. A linear series of ponds allows maximum use of the phosphorus added, but nitrogen loss and carbon dioxide uptake from the alkalinity place carbon and nitrogen limits on the aquatic plants. In both cases, this yields blue-green algal dominance in downstream ponds, with the detention time required being a function of the alkalinity, light, and phosphorus content of the water; these interact to control both rate and extent of algal growth (Young and King 1980). The absence of sufficient available nitrogen and the presence of other required growth factors yield nitrogen fixation and protein formation by heterocyst-forming blue-green algae. The ultimate challenge is to channel the protein supplied by such nitrogen fixation into products of value in the human diet. In this regard, some tropical fish may offer an opportunity of converting proteins from nitrogen-fixing blue-green algae to products of value in the human diet.

Although the cell walls of most plants are indigestible by alimentary canal secretions (Prejs and Blaszczyk 1977), blue-green algae can be digested by fish since their cell walls differ in composition from higher plants (Rogers and Perkins 1968). The widely held view that blue-green algae are trophically unimportant does not appear to apply to some tropical fish (Sorokin 1968). *Chanos chanos*, *Etrophs suratensis*, *Haplochromis nigripinnis*, *Mugil cephalus*, and *Tilapia nilotica* are among the herbivorous fishes which consume forms of blue-green algae in the intestine after the cell walls have been lysed by stomach acid (Moriarty 1973) or enzymatic hydrolysis (Hickling 1970; Pillay 1953; Moriarty 1976). *Mugil* spp. also have a gizzard which may increase the efficiency of enzymatic lysis of the blue-green algal cell wall (Pillay 1953; Moriarty 1976).

Table 1

FISH CULTURED OR PROPOSED FOR CULTURE WHICH CONSUME FORMS OF BLUE-GREEN ALGAE

Species	Habitat[a]	Food habits
Chanos chanos - milkfish	M	benthic diatoms and multicellular plants, foraminiferans, molluscs, dead copepods
	B	fry and fingerlings: Bacillariophyceae, Myxophyceae, degraded Chlorophyceae
		adults: diatoms, algae, fish eggs, etc.
	F - B ponds	fry and fingerlings: periphyton
		juveniles (50-100 mm): blue-green algae, diatoms, protozoans, and microcrustaceans mixed with detritus and minerals
		adults: benthic organisms and, reluctantly, filamentous algae
Etroplus suratensis - pearlspot	B - F	omnivorous: mainly blue-green and green algae, detritus, occasionally zooplankton, insects, worms, macrophytes, filamentous algae
Haplochromis nigripinnis	F	bacteria, blue-green algae, zooplankton
Mugil cephalus - grey mullet	F - M	plankton (blue-green, greens and occasional zooplankton), diatoms, aquatic plant soft parts, detritus
Tilapia nilotica - Nile perch	B - F	omnivorous: blue-green algae, detritus, higher plants

[a] M = marine, B = brackish, and F = freshwater environments.

The ability to digest blue-green algae does not indicate that total nutrient requirements can be met by these foodstuffs alone. Fishes which consume blue-green algae often consume bacteria and protozoans associated with the algae (Table 1) which may assist in meeting their nutritional requirements (Odum 1970). Juvenile Tilapia nilotica and Haplochromis nigripinnis can assimilate up to 70-80 percent of the ingested carbon from Anabaena sp. and Microcystis sp. (Moriarty and Moriarty 1973a). Due to diurnal patterns of stomach acid production (Moriarty 1973), daily feeding cycles (Moriarty and Moriarty 1973b) and assimilation patterns (Moriarty and Moriarty 1973a) increase from shortly before dawn to dusk. T. nilotica and H. nigripinnis assimilate an average of about 43 percent and 66 percent of the total ingested blue-green algal carbon per day, respectively. Apparently, juveniles and adults of these species can readily utilize blue-green algae to meet most of their nutritional needs (Moriarty and Moriarty 1973a).

Because of the similarity between the cell wall composition of blue-green algae and bacteria (Rogers and Perkins 1968), enzymatic or acid lyses of blue-green algae by species capable of using bacteria should be possible (Moriarty 1976). Other African cichlids including Tilapia esculenta (Greenwood 1953), T. lencosticta (Moriarty 1973), Sarotherodon mossambica (Fish 1960; Bowen 1976), and Haplochromis spp. (Moriarty 1973) have the ability to retain blue-green algae in the stomach until pH values reach 1.6 or lower and should be able to digest bacteria and blue-green algae. This hypothesis should be thoroughly tested to determine the growth potential of bactivorous and certain herbivorous fishes restricted to a diet composed chiefly of blue-green algae.

Animal Interaction and Feedback

The type of fish a pond contains can exert significant control on both the type of plants the pond supports and the nutrient cycling within the pond. If sufficient zooplankton are present in the pond to control the mass of green algae and diatoms, carbon extraction from the alkalinity may not proceed to the point where blue-green algae dominate, thereby allowing sufficient light to reach the pond bottom to promote macrophyte growth. The addition of zooplanktivorous fish to such systems can reduce the zooplankton control of the algae to the point where large blooms of algae occur (Shapiro et al. 1975; Lynch and Shapiro 1981; Andersson et al. 1978). Such increases in algal mass are often sufficient to force a carbon dioxide level low enough for blue-green algal dominance (King 1970). Development of blooms of algae can reduce light penetration sufficiently to cause the loss of macrophytes from a pond (Spencer 1981). In fact, a pond totally dominated by macrophytes can be altered to total dominance by phytoplankton simply by adding zooplanktivorous fish (Spencer 1981). However, macrophyte and periphyton dominance can be maintained by a sufficient population of pisciverous fish to keep the zooplanktivorous fish in check (Spencer 1981).

The significantly faster growth rates and shorter turnover time of algae would suggest that their dominance over macrophytes would lead to much faster nutrient cycling in the pond. Such accelerated activity would increase the dynamic character of the pond and lead to increased fluctuation of most parameters including dissolved oxygen concentration, light penetration, and pH and nutrient content of the pond water.

The addition of bottom-feeding fish can accelerate the recycling of phosphorus in ponds (Lamarra 1975), but depending on the type of benthic sediments, may also increase turbidity of the water to the point where plant photosynthesis is light-limited.

Obviously, the addition of toxic materials by human use of the water before or after it enters the pond will exert an effect on the system directly dependent on which organisms are affected. For example, Shapiro (1980) cites the work of several investigators (Hurlbert et al. 1972: Cook and Conners 1963; Shane 1948) which indicate development of heavy algal blooms associated with elimination of zooplankton by the addition of a variety of toxic materials.

OTHER BUFFERS

While in most freshwaters the carbonate-bicarbonate alkalinity serves simultaneously as the only significant pH buffer and reserve carbon source for aquatic photosynthesis, algal activity in nutrient enriched waters can stress this buffer, allowing pH values of 11 or higher (King 1970). In addition to the importance of the alkalinity in controlling both pH and potential productivity, other buffers exercise control in some freshwater systems. Chief among these are organic and mineral acids which not only exert an influence of their own but also interact with the carbonate-bicarbonate system to exercise control of the biotic activity of such waters.

The carboxyl-laden organic acids common to many waters (Shapiro 1957; Ghassemi and Christman 1968) produced from both terrestrial and aquatic vegetation (Novak et al. 1975) represent a significant ion exchange mechanism, which interacts to destroy carbonate-bicarbonate alkalinity (Yager 1971) in the manner shown in Equation 12.

$$\text{R-COOH} + M^+ + HCO_3^- \qquad \text{R-COO}(M^+) + CO_2 + \text{HOH} \tag{12}$$

In those basins where the accrual of such organic acids from both terrestrial and aquatic production exceed the accrual rate of cations from the weathering of the rocks and soil of the basin, carbonate-bicarbonate alkalinity will be destroyed and pH will be controlled by the organic acids. Since the pK's of these acids lie between about 4 and 5, the pH of the water will be below the level tolerated by methane-forming bacteria, and any accumulated organic matter will tend to be preserved in a manner leading to formation of peat. While they may not reach this bog state, highly colored waters in basins which supply few cations are not particularly productive. The low pH is buffered by the organic acids which tend to chelate metals and metal-phosphate complexes, while the very low carbonate-bicarbonate alkalinity offers little reserve carbon dioxide for aquatic photosynthesis in these often light-limited waters.

Mineral acids associated with the oxidation of iron pyrite exposed by strip mining or contained in "cats clay" of marine origin or areas with exposed sulfide-rich shales pose a severe problem to aquaculture. Oxidation of exposed sulfides leads to acid formation in the manner shown in Equation 13 (Singer 1970).

$$FeS_{2(s)} + 3.75\ O_2 + 3.5\ \text{HOH} \qquad Fe(OH)_{3(s)} + 2\ SO_4^= + 4\ H^+ \tag{13}$$

However, the resulting sulfuric acid reacts with various soil minerals in the manner shown for the overall reaction for Kaolinite in Equation 14 (King et al. 1974).

$$3\ FeS_{2(s)} + 2\ Al_2SiO_5\ (OH)_{4(s)} + 11.25\ O_2$$

$$+ 10.5\ HOH \quad Fe(OH)_{3(s)} + 6\ SO_4^{=} + Al^{+++}$$

$$+ 4\ H_4SiO_4^{o} + 2\ HOH \qquad (14)$$

This process yields some pH rise over that for sulfuric acid but there is little change in titratable acidity, since both 2 Al^{+++} and 6 H^+ require six equivalents of base for neutralization. Waters in basins which allow these reactions have a very low pH, are often extremely acid, and are buffered by high concentrations of aluminum and/or iron.

Such waters can be recovered to an alkaline condition by adding organic material to a sufficient depth to allow establishment of sufficiently reducing microzones to favor bacterial sulfate reduction (King et al. 1974). Under the low pH conditions, the resulting sulfide is lost to the air. This loss of hydrogen sulfide represents a biologically mediated titration of the acidity and, as pH increases, aluminum is precipitated as a hydroxide. Once the aluminum is precipitated, continued hydrogen sulfide loss results in increased formation of bicarbonate as shown in Equations 15 and 16.

$$CO_2 + HOH \quad HCO_3^- + H^+ \qquad (15)$$

$$H_2S_{(g)} \quad HS^- + H^+ \qquad (16)$$

Since the pK_1 for Equation 15 is 6.4 and the pK_1 for Equation 16 is 7.0, continued formation and release of hydrogen sulfide after the aluminum is precipitated is accompanied by continued bicarbonate increase. The result is a biologically mediated recovery of an extremely acid water to one which has a positive carbonate-bicarbonate alkalinity (King et al. 1974).

The problem with such a solution to mineral acids in aquacultural ponds lies in the need to promote anaerobic conditions at the pond bottom to generate the sulfides from sulfates required for recovery while maintaining an anaerobic seal to prevent further oxidation of pyrites in the benthic clays. Continued acid addition to the pond from the basin also would be a problem. However, organic materials are usually much more abundant than limestone in the areas in which such ponds are located and thus biological recovery may offer more potential than direct chemical neutralization of such acid waters. Where other buffers are sufficient to maintain pH in an alkaline range, the sulfide resulting from sulfate reduction represents an energy source for carbon fixation by chemotrophic bacteria in a form which should be useful to the aquatic food chain (Howarth and Teal 1980).

SUMMARY

The type and amount of aquacultural product is dependent on the type and amount of aquatic photosynthesis in those ponds to which outside foodstuffs are not added. The type and amount of aquatic photosynthesis is determined by interactions between available light, pond morphology, local temperature, and the alkalinity and nutrient content of the water, the animal community present, and detention time of the water in the pond. Since many of these parameters are dictated by local land use and geological and climate conditions peculiar to the drainage basin of individual ponds, it is neither particularly instructive nor worthwhile to attempt generalization of aquacultural potential of a large region without a good bit of base data from specific sites. Rather, emphasis should be aimed at developing an understanding of the exact mechanisms associated with the biological, chemical, and physical factors which interact and feedback to set limits on the ability of freshwater systems to produce products of value. An understanding of these ecological limits would allow better design and operation of aquaculture systems in a manner tailored to fit local environmental constraints on a worldwide basis.

LITERATURE CITED

Ball, R.C. 1948. A summary of experiments in Michigan lakes on the elimination of fish populations with rotenone, 1934-1942. Trans. Am. Fish. Soc. 75:139.

Bardach, J.E., J.H. Ryther, and W.O. McLarney. 1972. Aquaculture, the Farming and Husbandry of Freshwater and Marine Organisms. Wiley-Interscience, New York. 868 pp.

Bartsch, A.F., and M.O. Allum. 1957. Biological factors in treatment of raw sewage in artificial ponds. Limnol. and Oceanogr. 2:77.

Birge, E.A., and C. Juday. 1911. The inland lakes of Wisconsin. The dissolved gases of the water and their biological significance. Wisc. Geol. and Nat. Hist. Surv. Bull. 22:1.

Bowen, S.H. 1976. Mechanism for digestion of detrital bacteria by the cichlid fish, Sarotherodon mossambica (Peters). Nature, Lond. 260:137.

Cook, S.F., and J.D. Conners. 1963. The short-term side effects of the insecticidal treatment of Clear Lake, Lake County, California, in 1962. Annal. Entomol. Soc. Amer. 56:819.

Craig, J. 1978. Carbon dioxide and growth limitation of a submerged aquatic plant. M.S. Thesis, Michigan State University, East Lansing.

Dinsdale, M.T., and A.E Walsby. 1972. The interrelations of cell turgor pressure, gas vacuolation and buoyancy in a blue-green alga. J. Exp. Bot. 23:561.

Fish, G.R. 1960. The comparative activity of some digestive enzymes in the alimentary canal of Tilapia and perch. Hydrobiologica 15:161.

Ghassemi, M., and R.F. Christman. 1968. Properties of the yellow organic acids of natural waters. Limnol. and Oceanogr. 13:583.

Grant, N.G., and A.E. Walsby. 1977. The contribution of photosynthate to turgor pressure rise in the planktonic blue-green alga Anabaena flos-aquae. J. Exp. Bot. 28:409.

Greenwood, P.H. 1953. Feeding mechanisms of the cichlid fish, Tilapia esculenta Graham. Nature, Lond. 172:207.

Hayes, F.R., and E.H. Anthony. 1964. Productive capacity of North American lakes as related to the quantity and trophic level of fish, the lakes dimensions, and the water chemistry. Trans. Am. Fish. Soc. 93:53.

Hicking, C.F. 1970. Estuarine fish farming. Adv. Mar. Biol. 8:119.

Howarth, R.W., and J.M. Teal. 1980. Energy flow in a salt marsh ecosystem: the role of reduced inorganic sulfur compounds. Am. Nat. 116:862.

Hurlbert, S.H., M.S. Mulla, and H.R. Willson. 1972. Effects of an organophosphorus insecticide on the phytoplankton, zooplankton, and insect populations of freshwater ponds. Ecol. Monogr. 42:269.

Kempe, S. 1979. Carbon in the freshwater cycle, pp. 317-342. In B. Bolin, E. Degens, S. Kemp, and P. Ketner (eds), The Global Carbon Cycle. John Wiley and Sons, New York.

King, D.L. 1970. The role of carbon in eutrophication. Jour. Water Poll. Control Fed. 42:2035.

________. 1972. Carbon limitation in sewage lagoons. Special Symposium, Vol. 1 - Nutrients and Eutrophication. Am. Soc. Limnol. and Oceanog. 98 pp.

________. 1976. Changes in water chemistry induced by algae, pp. 73-84. In E.F. Gloyna, J.F. Malina, and E.M. Davis (eds), Ponds as a Wastewater Treatment Alternative. The Center for Research in Water Resources, The University of Texas-Austin.

________. 1979. The role of ponds in land treatment of wastewater, Vol. 2, pp. 191-198. In State of Knowledge in Land Treatment of Wastewater. International Symposium, U.S. Army Cold Regions Research and Engineering Lab., Hanover, NH.

________. 1980a. Some ecological limits to the use of alternative systems for wastewater management. In Aquaculture Systems for Wastewater Treatment: Seminar Proceedings and Engineering Assessment. EPA 430/9-80-006.

________. 1980b. Some cautions in applying results from aquatic microcosms, pp. 164-191. In John P,. Giesy (ed), Microcosms in Ecological Research; Selected papers from a symposium held at Augusta, GA, Nov. 8-10, 1978. Technical Information Center, U.S. Dept. of Energy, Springfield, VA.

________, and J.T. Novak. 1974. The kinetics of inorganic carbon limited algal growth. Jour. Water Poll. Control Fed. 46:1812.

________, J.J. Simmler, C.S. Decker, and C.W. Ogg. 1974. Acid strip mine lake recovery. Jour. Water Poll. Control Fed. 46:2301.

Lamarra, V.A. 1975. Digestive activities of carp as a major contributor to the nutrient loading of lakes. Verh. Internat. Verein. Limnol. 19:2461.

Liehr, S. 1978. Interacting carbon and light limits to macrophyte growth. M.S. Thesis, Michigan State University, East Lansing.

Lim, C., J. Sukhawongs, and F.P. Pascual. 1979. A preliminary study of the protein requirements of Chanos chanos (Forskal) fry in a controlled environment. Aquaculture 17:195.

Lynch, M., and J. Shapiro. 1981. Predation, enrichment, and phytoplankton community structure. Limnol. and Oceanogr. 26(1):86-102.

Mertz, E.T. 1972. The protein and amino acid needs, pp. 106-143. In J.E. Haver (ed), Fish Nutrition. Academic Press, Inc., New York.

Moriarty, D.J.W. 1973. The physiology of digestion of blue-green algae in the cichlid fish, Tilapia nilotica. J. Zool., Lond. 171:25.

______________. 1976. Quantitative studies in bacteria and algae in the food of the mullet, Mugil cephalus c. and the prawn, Metapenaeus bennette (Racek and Dall). J. Exp. Biol. Ecol. 22:131.

Moriarty, C.M., and D.J.W. Moriarty. 1973a. The assimilation of carbon from phytoplankton by two herbivorous fishes: Tilapia nilotica and Haplochromis nigripinnis. J. Zool., Lond. 171:41.

_______________, and D.J.W. Moriarty. 1973b. Quantitative estimation of the daily ingestion of phytoplankton by Tilapia nilotica and Haplochromis nigripinnis in Lake George, Uganda. J. Zool., Lond. 171:15.

Morrison, F.B. 1961. Feeds and Feeding, abridged. The Morrison Publ. Co., Clinton, Iowa. 696 pp.

Moyle, J.B. 1945. Some chemical factors influencing the distribution of aquatic plants in Minnesota. Am. Midland Nat. 34:402.

___________. 1946. Some indices of lake productivity. Trans. Am. Fish. Soc. 76:322.

Novak, J.T., A.S. Goodman, and D.L. King. 1975. Aquatic weed decay and color production. Jour. Am. Waterworks Assn. 67:136.

Odum, W.E. 1970. Utilization of the direct grazing and plant detritus food chains by striped mullet, Mugil cephalus, pp. 227-240. In J.H. Steele (ed), Marine Food Chains. University of California Press, Berkeley.

Pillay, T.V.R. 1953. Food, feeding habits, and alimentary tract of grey mullet. Proc. Natn. Inst. Sci. India 19:777.

Prejs, A., and M. Blaszczyk. 1977. Relationships between food and cellulase activity in freshwater fishes. J. Fish. Biol. 11:447.

Rogers, H.J., and H.R. Perkins. 1968. Cell Walls and Membranes. Spon Ltd., (London). 436 pp.

Schindler, D.W., G.J. Brunskill, S. Emerson, W.S. Broecker, and T.H. Peng. 1972. Atmospheric carbon dioxide: its role in maintaining phytoplankton standing crops. Science 177:1192.

Shane, M.S. 1948. Effects of DDT spray on reservoir biological balance. Jour. Am. Waterworks Assn. 40:333.

Shapiro, J. 1957. Chemical and biological studies on yellow organic acids of lake waters. Limnol. and Oceanogr. 2:161.

_________. 1973. Blue-green algae: why they become dominant. Science 179:382.

_________. 1980. The importance of trophic level interactions to the abundance and species composition of algae in lakes, pp. 105-116. In J. Barica and L.R. Mur (eds), Developments in Hydrobiology; v. 2. Hypereutrophic Ecosystems. Dr. W. Junk bv Publishers, The Hague.

_________, J. Lamarra, and M. Lynch. 1975. Biomanipulation: an ecosystem approach to lake restoration. In P. Brezonik and J. Fox (eds), Water Quality Management Through Biological Control. Rep. No. ENV-07-75-1, University of Florida, Gainesville.

Singer, P.C. 1970. Oxygenation of ferrous iron. FWQA. Rept. No. 14010.

Sorokin, Y.I. 1968. The use of ^{14}C in the study of aquatic animals. Mitt. int. Verein. Theor. Angew. Limnol. 16:1.

Spencer, C.N. 1981. Regulation of algae and macrophytes in eutrophic ponds by interacting biological, chemical, and physical factors. M.S. Thesis, Michigan State University, East Lansing. 78 pp.

Turner, W.R. 1960. Standing crop of fishes in Kentucky farm ponds. Trans. Am. Fish. Soc. 89:333.

Welch, H.E. 1968. Use of modified diurnal curves for the measurement of metabolism in standing water. Limnol. and Oceanogr. 13:679.

Wright, J.C. 1960. The limnology of Canyon Ferry Reservoir: III. Some observations on the density dependence of photosynthesis and its cause. Limnol. and Oceanogr. 5:356.

Wright, J.C., and I.K. Mills. 1967. Productivity studies on the Madison River, Yellowstone National Park. Limnol. and Oceanogr. 12:568.

Yager, J.G. 1971. The role of organic acids in lake dystrophication. M.S. Thesis, University of Missouri-Columbia.

Young, T.C., and D.L. King. 1980. Interacting limits to algal growth: light, phosphorus, and carbon dioxide availability. Water Research 14:409.

FISH/PLANKTON INTERACTIONS

by

Eugene Leslie Torrans

INTRODUCTION

The conversion of solar energy to chemical energy (carbon compounds) by phytoplankton is the essential first link in most aquatic food chains. Phytoplankton is collected and utilized by the primary consumers (zooplankton) which serve as a major food source for a wide variety of organisms, including fish. The biomass of zooplankton in intensively managed static-water culture ponds normally ranges from 2 to 20 g/M^3, and can be as high as 2 kg/M^3 in localized areas for short time periods (Table 1). The biomass of phytoplankton is usually several times higher than the zooplankton.

This high plankton biomass is an undesirable by-product of the intensive culture of channel catfish in the U.S., and necessitates both aeration and dilution to maintain good water quality. However, in most other areas of the world, plankton provide a tremendous food source for a variety of planktivorous fishes. A thorough understanding of fish/plankton interactions can lead to management practices that could greatly increase fish production. The intention of this paper is to summarize the current knowledge and practices involving fish/plankton interactions, and to indicate areas that could bear further examination.

UTILIZATION OF PLANKTON BY FISH LARVAE AND FRY

The lack of fish fry or fingerlings is often a major constraint to developing a successful "grass roots" aquaculture program in many developing countries. Well-trained extension agents could alleviate this problem by establishing private (village-level) sources of fish such as the tilapias (=Sarotherodon spp.). This method is far more cost-effective than centralized government hatcheries. However, the spawning of many popular species of cultured fish requires special facilities and a degree of sophistication that may not be available at the village level. A government facility or skilled private producer is usually necessary to provide the fry of these species.

Hatching success and subsequent fry survival (especially of many cyprinid

Table 1

ZOOPLANKTON STANDING CROP IN FISH PONDS

Nutrient input	Dominant organism by mass	Zooplankton standing crop, g wet weight/m^3
Nitrogen-phosphorous fertilizer (Lyubimova 1974)	Cladocera	0.4 - 13.1
Nitrogen-phosphorous fertilizer (Zaikina 1973)	Cladocera	6.2 - 13.02
Mineral fertilizer and feed (Krazhan et al. 1978)	Cladocera/rotifers	2.8 - 17.0
Phosphorous/nitrogen chemical (Szlauer 1979)	Cladocera	329
Organic fertilizer and feed (Yesipova 1976)	Zooplankton	10 - 2000

species) are extremely variable, ranging from 0 to nearly 100 percent. Poor hatching success, or failure of the fish to spawn at all, may be due to a variety of factors, including:

(1) crowding and/or poor nutrition of brood stock

(2) immature brood stock

(3) hormone injections and forced ovulation of "green fish" or fish that have begun to reabsorb the eggs

(4) physical stress on brood fish

(5) poor water quality in spawning ponds

(6) failure to maintain proper physical/chemical water quality for development of fertilized eggs and fry.

However, even with a viable spawn and good hatching success, the gradual loss of fry can result in 100 percent mortality. This gradual loss is known as the "dwindles" and is largely related to the abundance and type of food available to the fry, as well as the presence of both planktonic and insect predators of the fry.

Larval Food Requirements

Most fish go through a post-hatching developmental period similar to that described for the grass carp (*Ctenopharyngodon idella*) (Fishery Ministry of the U.S.S.R. 1970). At hatching, the pre-larvae are motile, pigmentless, and obtain nutrients entirely from the yolk sac (endogenous feeding). The pre-larval stage lasts from two to five days, depending upon water temperature, and during this period pigmentation develops as do the various other structures. Upon completion of the pre-larval stage, the highly motile larvae are capable of ingesting food:

Larval Stage I - Mixed (endogenous-exogenous) feeding of the larvae

Larval Stage II - Feeding of the larvae exclusively on external food (exogenous).

Successful feeding during this early stage of development is essential for larval survival (Spectorova et al. 1974). Although cyprinids can survive extended food deprivation (silver carp larvae can withstand ten days of starvation), larvae that are starved for the first few days may never learn to feed, even if they are subsequently supplied with abundant food (Lirski et al. 1979). Thus, it is imperative that larvae successfully feed at this early stage so as to mature and develop a normal feeding behavior.

Since larvae need food items that are visible, suspended in the water, small enough to be ingested, nutritious, and palatable, it is no surprise that larvae of most fish feed initially on zooplankton. Although smaller phytoplankton and larger zooplankton may be important first food for some species (Arnemo et al. 1980; Hiatt 1944; Lasker 1975; Parameswaran et al. 1974; Roy and Gupta 1973), rotifers, copepod nauplii and copepodites, protozoa and small cladocera in the 50-200 µ range are the predominant first foods of most species (Table 2).

Food Selectivity

Larvae and fry of most species are visual particulate feeders, catching and consuming individual prey items. A typical feeding sequence would involve 1) searching the visual field to locate prey items, 2) fixation on prey item and approach, 3) attack on prey by biting or snapping aided with suction from the buccal pump to draw the prey item in, and 4) acceptance or rejection of the item based on size, texture and taste. While chemo-senses are used to evaluate captured prey items, they are apparently not used to locate prey by most species of fish larvae or fry (McMahon and Tash 1979).

While searching behavior may be random, there is often a definite selection of particular items from a mixed prey population. This is usually expressed as Ivlev's Electivity Index,

$$E = \frac{ri - pi}{ri + pi},$$

which compared the relative abundance of food items in the intestine (ri) to those in the water (pi) (Ivlev 1961). The values of E range from -1.0 (total avoidance) to +1.0 (total selection).

Food preference and feeding success of fish fry are determined by a combination of several biological and physical factors:

Prey Size Prey size is the major factor controlling prey selectivity by fish larvae and fry. With very small prey items, visual detection is difficult and return/effort of feeding is low. Conversely, there is a maximum size at which prey items can be physically captured and ingested. Fry will usually eat a size range of plankton at a low absolute prey abundance. However, as the prey density increases, the smaller items are sequentially dropped from the diet

Table 2

FIRST FOOD OF VARIOUS CULTURED AND NON-CULTURED FISH SPECIES

Fish species	Fish age/size	Food item
Aristichthys nobilis (bighead carp)	Stage II	Phytoplankton
	Stage III	Rotifers & copepod nauplii
	Stage IV	Copepodites & small cladocera (Lazareva et al. 1977)
Tilapia nilotica	Fry	Rotifers, copepods, detritus aufwuchs, hydracarines (Moriarty et al. 1973)
Hypophthalmichthys molitrix (silver carp)	-----	Rotifers (Lirski et al. 1979)
	5 days	Phytoplankton & zooplankton (Omarov and Lazareva 1974)
Ctenopharyngodon idella (grass carp)	4-5 days/6-8 mm	Rotifers (50-150 μ) (Tamas and Horwath 1976)
Cyprinus carpio (Common carp)	4-5 days/6-8 mm	Rotifers (50-150 μ) (Tamas and Horwath 1976)
	1-3 days	Copepod nauplii (Dabrowski et al. 1978)
Catla catla	Larva & young fry	Phytoplankton (Hora and Pilay 1962)
	1-15 days	Cladocera, rotifers Copepods (Chakrabarty et al. 1973)
Labeo rohito	1-15 days	Nauplii & copepodites (Chakrabarty et al. 1973)
Cirrhina mrigala	1-15 days	Nauplii & copepodites (Chakrabarty et al. 1973)
Puntius pulchellus	18-23 mm	Zooplankton (David and Rahman 1975)
Micropterus salmoides (largemouth bass)	15 mm	copepods & cladocera (Rogers1968)
Sea bream	3 days	Nauplii, copepodites & copepods < 100 u (Stephen 1976)
Green back gray mullet	< 12 mm	Zooplankton (Chan and Chua 1979)
Black sea turbot	-----	Rotifers (Spectorova et al. 1974)
Mullet (12 species)	10 mm	Zooplankton (Blaber and Whitefield 1977)
Chanos chanos (milkfish)	Fry	Phytoplankton (Hora and Pilay 1962)

(Werner and Hall 1974). Since the reaction distance of fish is generally a linear function of prey size (Confer and Blades 1975; Confer et al. 1978), and the return/effort is higher when feeding on larger items, fry will usually select the larger items from a mixed prey population. As fish grow, they will usually select progressively larger prey items (Arthur 1976; Detwyler and Houde 1970; Gorbunova and Lipskaya 1975; Stephen 1976; Tarmas and Horwath 1976). Rotifers and copepod nauplii and copepodites are in the optimum size range for most stage I-II larvae.

Prey Density Given prey items of acceptable size, larval survival is a function of prey density. The critical level for very limited survival of most fish species is between 30 and 200 prey items/liter (Table 3). Survival usually increases with increasing prey density up to about 1000 items/liter. This appears to be the minimum prey level for good survival of most species. The optimum prey density decreases with increasing prey size within the acceptable size range for the species (Dabrowski 1976).

Apparently conflicting results are often seen with respect to larval growth and survival as a function of prey size. The survival of silver carp is better with rotifers as prey, while growth is faster with copepods as prey (Lirski et al. 1979; Tarmas 1975). This is explained by the fact that since silver carp initially begin feeding on rotifers, a high rotifer density is necessary during the first 11 days to obtain high survival. At the end of this developmental period, the larvae can consume larger copepods and, in fact, require them for good growth.

Activity of Prey Prey movement will increase the chances of visual detection by fish larvae. However, zooplankters capable of very strong swimming may avoid capture by the particulate (suction) feeding larvae (Janssen 1976). The probability of being captured by a simulated fish-suction intake ranges from P=0.76 to P=0.96 for cladocera, and only P=0.07 to P=0.28 for copepods (Drenner et al. 1978). Since copepods are negatively rheotactic and stronger swimmers, their successful avoidance of fish predators may result in an apparent selection for cladocera by fish larvae.

Larval Movements Movement of fish through water increases the frequency of prey encounters. For example, whitefish larvae search a total of 188 liters of

Table 3

EFFECT OF PREY DENSITY ON SURVIVAL OF LARVAL FISH

Species	Prey item	Prey density	Results
Bay anchovy	Copepod nauplii	50/L	6% survival (Houde 1978a)
	Copepod nauplii and copepodites	107/L	10% survival (Houde 1978b)
	Copepod nauplii	1000/L	43% survival (Houde 1978a)
Lined sole	Copepod nauplii	50/L	2% survival (Houde 1978a)
	Copepod nauplii and copepodites	130/L	10% survival (Houde 1978b)
	Copepod nauplii	1000/L	48% survival (Houde 1978a)
Sea bream	Copepod nauplii and copepodites	34/L	10% survival (Houde 1978b)
Atlantic cod	Rotifers, nauplii and copepodites	200/L	Died in 12-13 days (Buckley 1979)
	Rotifers, nauplii and copepodites	1000/L	Protein increased 9.3%/day (Buckley 1979)
Black sea turbot	Rotifers	100-1000/L	No change in feeding within this range (Spectorova et al. 1974)
Whitefish	<u>Cyclops</u> <u>sp</u>	200-260/L	Optimum (Dabrowski 1976)

water in a 10-hour daylight period, consuming 60 to 216 organisms/day (Dabrowski 1976).

Light/Turbidity Larvae of most fish species are diurnal visual particulate feeders (Arthur 1976; Burbidge 1974; Gorbunova and Lipskaya 1975). These larvae will commence feeding in the morning, with both the percent of feeding fish and the number of organisms in the larvae's stomach increasing during the day (Last 1978). The peak feeding time for silver carp is 1300 hours (Omarov and Lazareva 1974). High turbidity will reduce the feeding rate, but the prey size selected by fish appears to be independent of turbidity.

Transparency of Prey Because fish larvae are visual feeders, the increased transparency of prey reduces capture success (Greze 1963; Zaret and Kerfoot 1975). However, this may not be as important in practical situations as prey size is in determining the selectivity and capture efficiency of fish larvae.

Utilization of Plankton by Larvae and Fry

Zooplankton are high in protein and essential amino acids (Sadykhov et al. 1975), and are easily digested. Fry normally consume 40-80 percent of their body weight daily (Gorbunova and Lipskaya 1975; Stephen 1976), assimilating from 89-98 percent of the ingested food (Miliyenko 1979; Ogino and Watanabe 1978). Larvae and fry will usually feed throughout the day, digesting a food batch in three to five hours (Spectorova et al. 1974). The rate of digestion varies; e.g., cladocera are digested much more rapidly than copepods. The relative rates of digestion may result in erroneous estimates of food selectivity due to the more rapid disappearance of certain species from the gut (Gannon 1976).

The outstanding qualities of live plankton (movement, size, protein content, palatability, digestibility) are impossible to match with an artificial ration. Although a wide variety of feeds have been tried (Chakrabarty et al. 1973; Dabrowski et al. 1978; Ivanchenko and Ivanchenko 1969; Khan and Mukopadhyay 1973), all have resulted in either increased mortality (due to poor acceptability by fry) or decreased growth rate (due to nutritionally incomplete diets) when compared to live plankton.

With the present state of the art, it is nearly impossible to rear larvae and fry entirely on an artificial ration. However, artificial (non-plankton) food may be useful during two developmental stages:

1) When eggs are hatched in artificial containers, it is often advantageous to feed the larvae before releasing them into the pond. If they are to be fed for only a day or two, hard-boiled egg yolk, crumbled and dissolved in the water, has been found to be satisfactory (Ivanchenko 1969).

2) Once the fry have begun to grow on the natural plankton in the pond, they may be gradually trained to take a supplemental feed if the particle size is acceptable.

Management of Fry Ponds

The zooplankton biomass in a well-fertilized culture pond may be as high as 2kg/M^3 (Yesipova et al. 1976) yet the survival and growth of the fry may be very poor. This is the result of two interrelated factors:

1) The size structure of zooplankton in an established population may be biased in favor of the larger cladocerans and copepods. If sufficient small plankters (in particular rotifers and nauplii) are not present, larvae and fry may starve in an abundance of larger zooplankton. Those that do survive grow faster once they are able to consume these larger organisms.

2) A high density of large predaceous copepods can reduce larvae and fry numbers directly through predation.

Both these problems can apparently be eliminated by treating a pond with Dylox (=foschlcr, flibol, neguvon, dipterex, trichlorophon) or Baytex several days prior to stocking. When properly applied at the correct dose, this organophosphate kills crustaceans without harming rotifers (Tarmas 1975). A week or two later, the ponds can be inoculated with cladocera to promote the growth of advanced fry (Tarmas and Horwath 1976). Applications of quicklime (CaOH) at 500-1500 kg/ha may have a similar inhibiting effect upon the crustacean zooplankton (Krazhan et al. 1978).

Areas for Future Research

Although many basic interactions between fish larvae and/or fry and plankton are well understood, details are lacking for most of the fish species cultured worldwide. To correct this situation, specific information should be obtained relating to the preferred first foods, optimum prey density, and changes in food habits during the early development of these species. The techniques of fry pond management can then be "fine tuned" to greatly increase fingerling production. The development of a simple, nutritionally complete, palatable fry food would be a boon to production of fry requiring artificial hatching, particularly the cyprinids.

UTILIZATION OF PLANKTON BY ADULT FISH

The predominant food of nearly all larval fish is zooplankton, in particular rotifers and smaller crustaceans. However, the food habits of different fish species rapidly diverge during the early life history (Table 4). While some species retain planktivorous habits (Durbin and Durbin 1975; Holanov and Tash 1978; Janssen 1976; Lazareva et al. 1977; Omarov and Lazareva 1974) others turn to larger animals (Rogers 1968), vegetation (David and Rahman 1975; Hora and Pilay 1962) or varying degrees of omnivory (Hora and Pilay 1962; Khan and Siddiqui 1973; McBay 1962). Early divergence of food habits, together with differences in habitat selection plays a vital role in reducing competition for food, both within and among species.

Feeding Structures

The development of various types of feeding and digestive structures, particularly the gill rakers, epibranchial organ and lengthening of the intestine are indicative of planktivorous food habits in the adult.

Gill Rakers The morphology of the gill rakers is an obvious indication of a fish's feeding habits. Most carnivores, herbivores, and omnivores have few, relatively short gill rakers. Although they particulate feed on plankton as juveniles, as they increase in length, spacing between the rakers increases to the point that they can no longer effectively trap plankton. Large individual plankters may still be utilized, but for the most part, such fish turn to larger sized food items.

Planktivorous fish have numerous long, closely spaced gill rakers that act as a sieve to filter water and trap plankton. To compensate for the increased spacing that occurs with growth, the gill rakers of some species develop either secondary projections or microspines, which keep the pore size relatively small (Iwata 1976; Starostka and Applegate 1970; Torrans et al. 1980), or they develop a series of micro-gill rakers at the base of the gill filaments (Whitehead 1959). Silver carp have the most specialized gill raker system of any fish presently being cultured. When the fish are only 26 mm long, the gill rakers fuse together to form a net (Iwata 1976). The gill rakers are completely fused together in adult fish and are comprised of two layers: an internal straining portion of fine filaments and an external

Table 4

AGE-RELATED CHANGES IN THE FOOD HABITS OF FIVE SPECIES OF CULTURED FISH

Species	Size	Food habits
Micropterus salmoides (Rogers 1968)	5-10 mm	Copepods and Cladocera
	15 mm+	Midge larva and pupae
Labeo rohita (Khan and Siddiqui 1973)	fingerling (< 100 mm)	Zooplankton, some phytoplankton
	adults (>300 mm)	Phytoplankton, macro-vegetation organic matter
Aristichthys nobilis (Lazareva et al. 1977)	Stage II	Phytoplankton
	Stage III	Zooplankton
	Stage IV	Larger plankton
Hypophthalmichthys molitrix (Omarov and Lazareva 1974)	Stage II	Phytoplankton
	Stage III-V	Zooplankton
	1 month	Phytoplankton
Putius pulchellus (David and Rahman 1975)	18-22 mm	Zooplankton
	30-50 mm	Decaying plant, leaves, filaments, algae, diatoms
	305-705 mm	Aquatic vegetation

thicker support (Wiliamovski 1972). The gill raker net has pores less than 20 µ in diameter, allowing adult silver carp to capture smaller particles than fry can.

A large number of species have a more limited filtering capacity. These facultative planktivores are able to subsist on plankton when there is a high density of relatively large plankters. When plankton is limited, they readily consume other items (Starostka and Applegate 1970; Torrans and Clemens 1981).

Epibranchial Organ The epibranchial, or palatal, organ is a down growth from the roof of the pharynx. Covered with both taste buds and mucus cells (Torrans et al. 1980), it partially occludes the buccal cavity and is useful in the selection, capture and manipulation of plankton trapped by the gill rakers. Many planktivores have some type of palatal organ (Kuwatani and Kafuka 1978; Torrans et al. 1980), but the highest degree of development appears in the silver carp (Boulenger 1901; Wiliamovski 1972).

Lengthening Of The Intestine Most planktivores have a high intestine length/body ratio, particularly those that feed at least in part on phytoplankton (Wiliamovski 1972). This is viewed as an adaptation for the digestion of plant material. While in most species the length of the intestine increases gradually with age, the intestine of the milkfish (Chanos chanos) undergoes heteronomous growth (Hiatt 1944).

Feeding Behavior

Planktivorous fish utilize a variety of feeding mechanisms to capture plankton, depending on prey size and density. The northern anchovy (Engraulis mordax) captures Artemia adults (3.7 mm long) by biting (particulate feeding) and nauplii (0.65 mm long) by filter feeding. Feeding activity is half biting and half filtering when Artemia adults comprise 2 percent of the total plankton biomass, and is entirely biting when adult Artemia exceed 7 percent of the prey biomass (O'Connell 1972).

Threadfin shad (Dorosoma petenense) likewise utilize both particulate and filter feeding. Small particles (<0.39 mm) are filtered while larger items are taken individually. Particulate feeding decreases with decreasing light intensity, and is not possible with less than bright moonlight. However, shad can filter feed 24 hours a day, using chemosenses to locate prey (McMahon and Tash 1979).

Alewives (Alosa pseudoharengus) employ three modes of feeding: particulate, "gulping", and filtering (Janssen 1976). The mode employed depends upon prey size, prey density and fish size. Larger prey elicit size-selective particulate feeding. As prey size decreases and prey density increases, alewives employ "gulping" (slightly size-selective) and filter feeding (not size-selective). Filtering is more common with large individuals of this species at high prey densities, while particulate feeding is typical of the smaller individuals.

Food Selection

The specific feeding method used by a fish (and the food it ingests) is a combination of that species' functional morphology, the light intensity, and the density and composition of the plankton. The food actually taken by a fish may not, in fact, reflect the preferred food of that species, but rather the most energy-efficient method of nutrient intake given the actual availability of food in the environment (Table 5).

Amidst an abundance of plankton, the food items selected by a planktivorous species can be predicted on the basis of the morphology of its feeding structure. Silver carp graze primarily on phytoplankton in the 8-100 µ size range (Cremer and Smitherman 1980; Spataru 1978; Wiliamovski 1972), while bighead carp select zooplankton and larger phytoplankton in the 17-3000 µ range (Cremer and Smitherman 1980; Moskul 1977). Most planktivorous fish show a positive selection fôr larger prey items within their acceptable size range (Anderson et al. 1978; Brooks 1968; Confer and Blades 1975; Confer et al. 1978; Cramer and Marzolf 1970; Dodson 1970; Durbin and Durbin 1975; Hutchinson 1971; O'Brian 1979; Ware 1973; Wells 1970; Werner and Hall 1974). This is advantageous since feeding on larger organisms is usually more energy efficient. However, in cases of very low plankton biomass, even supposedly stenophagous planktivores will take a variety of foods, including detritus, plant materials, supplemental food and small fish (Cremer and Smitherman 1980; Kohler and Ney 1980; Lazareva et al. 1977; McBay 1962; Omarov and Lazareva 1974; Spataru and Zorn 1978). The degree of omnivory varies considerably among species, and may play a large role in the success of some species (Brooks 1968; Kudrinskaya 1978).

Filter-feeding planktivores normally have a daily feeding period with a peak in late afternoon (Moriarty and Moriarty 1973a; Moskul 1977; Omarov and Lazareva

Table 5
FOOD HABITS OF ADULT PLANKTIVOROUS FISH

Species	Food Habits
Aristichthys nobilis	Zooplankton, suspended organic substances (Moskul 1977) Zooplankton, up to 82-98% detritus (Lazareva et al. 1977) 67% phytoplankton in cages, 69% detritus in ponds (Cremer and Smitherman 1980)
Hypophthalmichthys molitrix	Phytoplankton <20 µ (Spataru 1978; Wiliamovski 1972) 84% phytoplankton, 15% detritus (Cremer and Smitherman 1980) Phytoplankton, zooplankton (Moskul 1977) 10-60% phytoplankton, 10-30% zooplankton (Kajak et al. 1977) 90% detritus (Omarov and Lazareva 1974)
Chanos chanos	"Lablab" in Philippine ponds (Lijauco et al. n.d.) Diatoms in Hawaiian ponds (Hiatt 1944)
Tilapia nilotica	Phytoplankton (Moriarty et al. 1973; Moriarty and Moriarty 1973a; Moriarty and Moriarty 1973b) Blue-green algae (Moriarty 1973) Zooplankton; detritus at times of low zooplankton biomass (Spataru and Zorn 1978) Filamentous algae (McBay 1962)

1974). The total amount of food consumed daily by a planktivore is affected by a combination of factors: temperature, size of fish and plankton, density, and species composition of the plankton. The daily food consumption of planktivores is normally 10-20 percent of the body weight (Moriarty and Moriarty 1973b; Yashouv 1971). However, the food intake of the silver carp may be as low as 1-3 percent when the plankton concentration is low (Sparatu 1978), and they may cease feeding altogether when there are dense blooms of blue-green algae (Kajak et al. 1977).

Assimilation Of Plankton

The digestibility of plankton is extremely variable, ranging from 43-89 percent, depending upon the fish species and the plankton composition (Gannon 1976; Moriarty et al. 1973; Moriarty and Moriarty 1973a; Ogino and Watanabe 1978). The food conversion efficiency (weight of food eaten to weight of fish produced) of planktivores normally ranges from 5:1 to 18:1 on a wet weight basis (Moskul 1977; Ogino and Watanabe 1978), but the digestibility of various plankters varies considerably among the planktivores (Roy and Gupta 1973).

Silver carp are unable to digest blue-green algae, while Tilapia nilotica can assimilate from 70-80 percent (Kajak et al. 1977; Moriarty 1973; Moriarty and Moriarty 1973a; Spataru and Zorn 1978). The high assimilation of blue-green algae by T. nilotica is made possible by the secretion of a stomach acid, which lyses the algal cells. The acid secretion follows the feeding cycle, increasing during the day. Assimilation of blue-green algae is greatest at the end of a feeding period, when acid concentration is greatest and the food retention time is longest (Moriarty 1973; Moriarty and Moriarty 1973a). Other fish species that consume and digest blue-green algae, such as the milkfish and the mullets, may likewise use acid secretions to lyse algal cells (Moriarty 1973).

Although planktivores' intestines are long relative to body length (Cremer and Smitherman 1980; Hiatt 1944), a plankton food batch may pass through an adult fish in as little as five to eight hours (Moskul 1977). This rapid digestion of plankton complicates feeding studies of planktivores.

Grazing by planktivorous fish can cause a tremendous decrease in the density and composition of the plankton community (Oviatt et al. 1972; Schroeder 1975). Selective predation on zooplankton may result in up to a 75 percent decrease in zooplankton biomass (Torrans and Clemens 1981) and an increase in the phytoplankton (Anderson et al. 1978).

In a fish-free system, the larger crustacean zooplankters are usually dominant over the smaller direct predation. A planktivorous fish that selects larger organisms can result in increases and dominance of the smaller zooplankters (Burbidge 1974; Cramer and Marzolf 1970; Dodson 1970; Hutchinson 1971; O'Brian 1979; Wells 1970). Thus, size-selective predation results in a major change in plankton species and size distributions.

Research Areas

Although the dynamics of the predator-prey system of fish and plankton has been extensively studied by aquatic ecologists in natural systems, most studies in warmwater culture ponds have been limited to particle size or species selection as reflected in stomach contents. Detailed investigations of the filtering efficiency (percentage retention of plankton/volume filtered) of commercially cultured fish are lacking, as in information on the digestibility and food conversion efficiency of various plankton/detritus diets. We must understand the comparative capture, conversion and growth efficiencies of common planktivorous fish in order to maximize the flow of energy from plankton to edible fish flesh. Energy flow models that are developed should not be based entirely on academic ideas, but must be tempered by practical consideration, i.e., large-scale availability of fingerlings, suitability of the chosen species to local culture techniques, and marketability of the final product.

APPLIED ASPECTS

In most areas of the world, primary agricultural products are in such high demand for direct human consumption that their use as a supplemental fish food is precluded. In addition, large areas of the world, primarily Africa, have no cultural tradition of confined livestock feeding, using instead the open range. Even small animals, such as chickens and goats, are allowed to forage for their food. Thus it is difficult, even where limited quantities of agricultural or domestic by-products are available, to train rural farmers to feed their fish.

However, since fish forage for natural food, it is somewhat easier to convince rural farmers to fertilize their ponds to increase the natural food for the fish. Fertilization is the most cost-effective means of increasing fish production since it requires only a periodical, rather than daily, effort. The most efficient means of converting this natural food to fish flesh is through primary or secondary consumers (planktivorous fish). There is a strong correlation between fish production and primary productivity (Boyd 1979; Kavlac 1976; Liang et al. 1981; McConnell 1965; McConnell et al. 1977; Shpet and Kharitonova 1975), which explains the worldwide popularity of fish that are primary and secondary consumers.

The application of fertilizer to ponds stocked with fish that feed low on the food chain (milkfish, common carp, silver carp, bighead carp or _Tilapia nilotica_) may increase the production from two to ten times over unfertilized ponds (Hopkins and Cruiz 1980; Kavlac 1976; LeBrasseur and Kennedy 1972; Moav et al. 1977; Schoonbee et al. 1979). Production from 1500 to over 6000 kg/ha/yr is possible using only fertilizer (primarily organic manure) as the energy input. Fish yields can be increased in three ways without feeding.

Add Organic Fertilizer to Increase Both Primary Productivity and the Carrying Capacity of the Pond

Fish production increases linearly with organic input up to the point that water quality becomes limiting for the species being cultured (Hopkins and Cruiz 1980). _Tilapia nilotica_ and common carp are both very tolerant of poor water quality (particularly low dissolved oxygen) and respond well over a wide range of organic loading. Fertilization is usually the easiest means of increasing production, but the type, quantity, frequency and method of application of the organic manure must be considered to maximize returns.

Reduce the Length of the Harvest Cycle

Fish stocked in a pond (assuming there are enough, and/or that there is reproduction) will eventually grow to reach the carrying capacity of the pond. The carrying capacity is largely determined by the nutrient input, the water quality, and the fish species being cultured. Fish that are acceptable to consumers at a smaller size can be stocked at very high rates and will reach the carrying capacity sooner, allowing a farmer to get several complete harvests in a year. This is presently being done in the Philippines, where _Tilapia nilotica_ of 50-100 grams have good consumer acceptance (personal observation). Farmers there stock 20,000 fry/hectare in fertilized ponds and harvest 1000-2000 kg/ha in three months. The annual production thus ranges from 4-8000 kg/ha, remarkable for monoculture in non-fed ponds. In addition, since the fish are harvested before reproduction gets out of hand, all the fish are of marketable size, thus eliminating the major problem with tilapia culture.

Polyculture

Polyculture is the rearing together of more than one fish species (sometimes as many as six or seven) in a pond. In theory, several species (each with slightly different food habits) will more fully utilize the natural food in a pond, resulting in increased production. Often supplemental feed is given to one or more of the species (usually the grass carp or common carp) to provide the energy subsidy that drives the system. This culture system is well established in Asia. In Taiwan, silver carp, gray mullet, bighead carp, grass carp, common carp, black carp and sea perch are stocked together. In other areas, various other cyprinids (*Labeo* spp., *Catla catla*), *Tilapia* spp., and predatory species, such as *Clarias* spp. or *Ophiocephalus* spp. are sometimes used. In Taiwan, the stocking rates presently used are the result of many years of experience, and are based on the amount of food available to each trophic level of fish (Tang 1970). With large energy subsidies, this system can result in fish production of from 6000-8000 kg/ha/year.

While polyculture is well established and very successful in parts of Asia, it cannot (and perhaps should not) be applied to other areas in the world that are struggling to start an aquaculture program. In many areas of the world, both the private and government infrastructures are so poorly developed that it is difficult for farmers to get even fingerlings of *Tilapia nilotica*, a very prolific fish. The establishment of a dependable supply of one or two other species would be a monumental undertaking, and may not be advisable. Where the need for aquaculture is the greatest, very small fish usually have high consumer acceptance. In this areas, tilapia can be stocked heavily and harvested in a few months (as opposed to a 12-month harvest cycle for most cyprinids), resulting in an annual production as high as the Chinese polyculture system. This may not be glamorous, but is probably the most efficient means of producing a large volume of high quality fish for a protein-starved people.

Once the private sector has successfully grasped the basic principles of fish production (fry production, stocking, fertilization, feeding, harvesting and sale), a two-species approach may be useful. A planktivorous species is used (*Tilapia nilotica* or the silver or bighead carp) to graze on plankton (Cremer and Smitherman 1980; Hopkins and Cruiz 1980; Moav et al. 1977; Moskul 1977; Tal and Ziv 1978), and an omnivore/detrivore is added (usually the common carp) which feeds upon and recycles organic detritus from the pond bottom. In this situation, the growth of both species will probably be better than when either is raised alone (Schoonbee et al. 1979).

Research Areas

The local geographical, cultural and political constraints should be fully understood before beginning an aquaculture research program if the program is to have any effect on the private sector. It may do more harm than good if a highly productive but elaborate scheme is developed at a government research center that cannot be applied by the private farmers due to local constraints. Too often, techniques or programs are proposed in developing countries by expatriate advisors because they are effective in the U.S., and not because they are needed in that country.

SUMMARY

The constraints to fish production vary considerably among developing countries. In most areas of the world, particularly in Africa, the most serious problem is not a lack of technology but the transfer of the existing technology to the rural farmers. The lack of trained personnel, poor communications and the limited resources of fisheries departments in developing countries accounts for most of the disparity between the known and applied technology. In other areas, however, private farmers are applying the current state of the art and are limited by basic research and development.

The stated goal of the Title XII Aquaculture CRSP is basic research and development on state-of-the-art constraints to fish production in developing countries. This is necessary for any industry to achieve maximum growth, but it must be complemented with a grassroots extension program if it is to have any real impact on the private sector. In addition, basic R & D should employ appropriate technology so results can be easily transferred to the private sector.

The following is a nonprioritized list of research topics concerning fish/plankton interactions. The relative necessity for basic research on any of these topics will be determined by local cultural, geographical and technical considerations.

1. Preferred first food of fish larvae and fry.

2. Prey density necessary for optimum larval and fry survival.

3. Predation on fish larvae and fry by zooplankton.

4. Management of plankton populations in fry ponds (fertilization and chemical treatments).

5. Development of a complete ration for larvae and fry.

6. Changes in food selection with fish age/size.

7. Feeding mechanisms (histological, morphological and behavioral adaptations) of adult planktivores.

8. Species and particle size selection of adult planktivores.

9. Feeding efficiency (water volume filtered and capture efficiency) of planktivores.

10. Assimilation of plankton by fish (digestibility, assimilation and conversion of various plankton diets as affected by the species and size of fish).

11. Effect of fish predation on plankton populations.

12. Energy-flow diagrams for pond systems with various nutrient inputs and fish stocks--least-cost fish production based on locally available inputs.

LITERATURE CITED

Anderson, G., H. Berggren, G. Cronberg, C. Gelin. 1978. Effects of planktivorous and benthivorous fish on organisms and water chemistry in eutrophic lakes. Hydrobiologia 59(1):9-15.

Arnemo, R., C. Puke, N.G. Steffner. 1980. Feeding during the first weeks of young salmon in a pond. Arch. Hydrobiol. 89(1/2):265-273.

Arthur, D.K. 1976. Food and feeding of larvae of three fishes occurring in the California current, Sardinpos sagax, Engraulis mordax, and Trachurus symmetricus. Fishery Bulletin 74(3):517-530.

Blaber, S.J.M., and A.K. Whitefield. 1977. The feeding ecology of juvenile mullet (Mugilidae) in south-east African estuaries. Biol. J. Linn. Soc. 9:277-284.

Boulenger, G.A. 1901. On the presence of a superbranchial organ in the cyprinoid fish Hypophthalmichthys. Ann. Mgmt. Natl. Hist. 7(8):186-188.

Boyd, C.E. 1979. Water Quality in Warm Water Fish Ponds. Craftsman Printers, Inc. Opeleka, Alabama. 359 pp.

Brooks, J.L. 1968. The effects of prey size selection by lake planktivores. Syst. Zool. 17:272-291.

Buckley, L.J. 1979. Relationships between RNA-DNA ratio, prey density, and growth rate in Atlantic cod (Gadus morhua) larvae. J. Fish. Res. Board. Can. 36:1497-1502.

Burbidge, R.G. 1974. Distribution, growth, selective feeding, and energy transformations of young-of-the-year blueback herring, Alosa aestivalis (Mitchill) in the James River, Virginia. Trans. Amer. Fish. Soc. 103:297-311.

Chakrabarty, R.D., P.R. Sen, D.K. Chatterjee, and G.V. Kowtal. 1973. Observations on the relative usefulness of different feed for carp spawn and fry. J. Inland Fish. Soc. India 5:185-188.

Chan, E.H., and T.E. Chua. 1979. The food and feeding habits of greenback grey mullet, Liza subvirdis (Valenciennes), from different habitats and at various stages of growth. J. Fish. Biol. 15:165-171.

Confer, J.L., and P.L. Blades. 1975. Omnivorous zooplankton and planktivorous fish. Limnol. Oceanogr. 20(4):571-579.

__________, G.L. Howick. M.H. Corzette, S.L. Kramer, S. Fitzgibbon, and R. Landesburg. 1978. Visual predation by planktivores. Oikos 31:27-37.

Cramer, J.D., and G.R. Marzolf. 1970. Selective predation on zooplankton by gizzard shad. Trans. Am. Fish. Soc. 99:320-332.

Cremer, M.C., and R.O. Smitherman. 1980. Food habits and growth of silver and bighead carp in cages and ponds. Aquaculture 20:57-64.

Dabrowski, K.R. 1976. How to calculate the optimal density of food for fish larvae. Env. Biol. Fish. 1(1):87-89.

Dabrowski, K., H. Dabrowski, and C. Grudiewski. 1978. A study of the feeding of common carp larvae with artificial food. Aquaculture 13:257-264.

David, A., and M.F. Rahman. 1975. Studies on some aspects of feeding and breeding of Puntius pulchellus (Day) and its utility in culturable waters. J. Inland Fish. Soc. India 7:226-238.

Detwyler, R., and E.D. Houde. 1970. Food selection by laboratory-reared larvae of the scaled sardine Harengula pensacolae (Pisces, Clupeidae) and the bay anchovy Anchoa mitchilli (Piscel, Engraulidae). Mar. Biol. 7:214-222.

Dodson, S.L. 1970. Complementary feeding niches sustained by size selective predation. Limnol. Oceanogr. 15:131-137.

Drenner, R.W., J.R. Strickler, and W.J. O'Brian. 1978. Capture probability: The role of zooplankter escape in the selective feeding of planktivorous fish. J. Fish. Res. Board Can. 35:1370-1373.

Durbin, A.G., and E.G. Durbin. 1975. Grazing rates of the Atlantic Menhaden Brevoortia tyrannus as a function of particle size and concentration. Mar. Biol. 33:265-277.

Fishery Ministry of the U.S.S.R., All-Union Scientific Research Institute of Pond Fishery. 1970. Manual on the biotechnology of the propagation and rearing of phytophagous fishes. Translated from Russian and distributed by Div. of Fish. Res. Bur. Sport Fish. and Wildl., Washington, D.C. 49 pp.

Gannon, J.E. 1976. The effects of differential digestion rates of zooplankton by Alewife, Alosa pseudopharengus, on determinations of selective feedings. Trans. Am. Fish. Soc. 105:89-95.

Gorbunova, N.N., and N.Y. Lipskaya. 1975. Feeding of larvae of blue marlin, Makaira nugricans (Pisces, Istiophoridae). J. Ichthyol. 15:95-101.

Greze, V.N. 1963. The determination of transparency among planktonic organisms and its protective significance. Dorklady Acad. Sci. USSR Biol. Sci. Sect. 151:956-968.

Hiatt, R.W. 1944. Food chains and the food cycle in Hawaiian fish ponds - Part I. The food and feeding habits of mullet (Mugil cephalus), milkfish (Chanos chanos), and the tenpounder (Clops machnata). Trans. Amer. Fish. Soc. 74:250-261.

Holanov, S.H., and J.C. Tash. 1978. Particulate and filter feeding in threadfin shad, Corosoma petenense, at different light intensities. J. Fish. Biol. 13:619-625.

Hopkins, K., and E.M. Cruiz. 1980. High yields but still questions: Three years of animal fish farming. ICLARM Newsletter 3(4):12-13.

Hora, S.L. and T.V.R. Pillay. 1962. Handbook of fish culture in the Indo-Pacific region. F.A.O. Fisheries Biology Technical Paper No. 14. 204 pp.

Houde, E.D. 1978a. Food concentration and stocking density effects on survival and growth of laboratory-reared larvae of bay anchovy Anchoa mitchilli and lined sole Archirus lineatus. Mar. Biol. 43:333-341.

__________. 1978b. Critical food concentrations for larvae of three species of subtropical marine fishes. Bull. Mar. Sci. 28(3):395-411.

Hutchinson, B.P. 1971. The effect of fish predation on the zooplankton of ten Adirondack lakes, with particular reference to the alewife, Alosa pseudoharengus. Trans. Amer. Fish. Soc. 100(2):325-335.

Ivanchenko, L.A., and O.F. Ivanchenko. 1969. Transition to active feeding by larval and juvenile white sea herring (Clupea harengus pallasi natio maris-albi Berg) in artificial conditions. Dorklady Acad. Sci. USSR 184:207-209.

Ivlev, V.S. 1961. Experimental Ecology of Feeding of Fish. Yale University Press.

Iwata, K. 1976. Morphological and physiological studies on the phytoplankton feeders in Cyprinid fishes--I. Developmental changes of feeding organs and ingestion rates in Kawachibuna (Carassius auratus curvieri), silver carp (Hypophthalmichthys molitrix) and Nigorobuna (C. auratus grandoculis). Jap. J. Limnol. 37(4):135-147.

Janssen, J. 1976. Feeding modes and prey size selection in the alewife (Alosa pseudoharengus). J. Fish. Res. Board Can. 33:1972-1975.

__________. 1976. Selectivity of an artificial feeder and suction feeders on calanoid copepods. Amer. Mid. Nat. 95(2):491-493.

Kajak, Z., I. Spodniewska, and R.J. Wisniewski. 1977. Studies on food selectivity of silver carp (Hypophthalmichthys molitrix (Val.)). Ekologia Polska 25(2):227-239.

Kavlac, J. 1976. Effect of fertilization and food on the pond production of tilapia in Zambia. Comm. Inland Africa 4:642-658.

Khan, H.A., and S.K. Mukopadhyay. 1973. Effect of yeast and cobalt chloride in increasing survival rate of hatchlings of Heteropneustes fossils (Bloch). Indian J. Anim. Sci. 43(6):540-542.

Khan, R.A., and A.W. Siddiqui. 1973. Food selection by Labeo rohita (Ham.) and its feeding relationship with other major carps. Hydrobiologica 43(3):429-442.

Kohler, C.C., and J.J. Ney. 1980. Piscivority in a landlocked alewife (Alosa pseudoharengus) population. Can. J. Fish. Aquat. Sci. 37:1314-1317.

Krazhan, S.A., N.N. Khritonova, K.I. Bne'ko, S.A. Isayeva, N.M. Mikulina, Z.G. Kolorina. 1977. Quantitative dynamics of zooplankton and zoobenthos in Ukrainian finishing ponds at different fish-stocking densities. Hydrobiol. J. 13(1):31-38.

____________, _____________, and S.A. Isayeva. 1978. The effect of the liming of fish-breeding ponds on some crustacean plankters. Hydrobiol. J. 14(1):31-33.

Kudrinskaya, O.I. 1978. The extent of food availability for larvae of different species depending on the development of food supply. J. Ichthyol. 18:243-250.

Kuwatani, Y., and T. Kafuka. 1978. Morphology and function of epibranchial organ studied and inferred on milkfish. Bull. Freshwater Fish. Res. Lab. 28:221-236.

Lasker, R. 1975. Field criteria for survival of anchovy larvae: The relation between inshore chlorophyll maximum layers and successful first feeding. Fishery Bulletin 73(3):453-462.

Last, J.M. 1978. The food of four species of pleuronectiform larvae in the eastern English channel and southern Northern Sea. Mar. Biol. 45:359-368.

Lazareva, L.P., M.O. Omarov, A.N. Lezina. 1977. Feeding and growth of the bighead, Aristichthys nobilis, in the waters of Dagestan. J. Ichthyol. 17:65-71.

LeBrasseur, R.J., and O.D. Kennedy. 1972. The fertilization on Great Central Lake. II. Zooplankton standing stock. Fishery Bulletin 70(1):25-36.

Liang, Y., J.M. Melack, and J. Wang. 1981. Primary production and fish yields in Chinese ponds and lakes. Trans. Amer. Fish. Soc. 110(3):346-350.

Lijauco, M., J. V. Juario, D. Baliao, E. Grino, and O. Quinito. (No date). Milkfish culture in brackishwater ponds. Aquaculture Dept. SEAFDEC. Tigbauan, Iloilo, Philippines. 17 pp.

Lirski, A., B. Onoszkiewsicz, K. Opuszynski, and M. Wozniewski. 1979. Rearing of cyprinid larvae in new type flow-through cages placed in carp ponds. Pol. Arch. Hydrobiol. 26(4):545-559.

Lyubimova, T.S. 1974. Effect of fertilization on zooplankton growth in carp nursery ponds in the Southern Urals. Hydrobiol. J. 10(1):24-29.

McBay, L.G. 1962. The biology of Tilapia nilotica Linneaus. Proc. Ann. Conf. Southeastern Assoc. Game and Fish. Comm. 15:208-218.

McConnell, W.J. 1965. Relationship of herbivore growth rate to gross photosynthesis in microcosms. Limnol. Oceanogr. 10:539-543.

_____________, S. Lewis, and J.E. Olsen. 1977. Gross photosynthesis as an estimator of potential fish production. Trans. Amer. Fish. Soc. 106:417-423.

McMahon, T.E., and J.C. Tash. 1979. The use of chemosenses by threadfin shad, Dorosoma petenense to detect conspecifics, predators and food. J. Fish. Biol. 14:289-296.

Miliyenko, A.V. 1979. The assimilation of food by juvenile fishes. Hydrobiol. J. 15(6):105-106.

Moav, R., G. Wohlfarth, G.L. Schroeder, G. Hulata, and H. Barash. 1977. Intensive polyculture of fish in freshwater ponds. I. Substitution of expensive feeds by liquid cow manure. Aquaculture 10:25-43.

Moriarty, D.J.W. 1973. The physiology of digestion of blue-green algae in the cichlid fish, Tilapia nilotica. J. Zool. Lond. 171:25-39.

_____________, J.P.E.C. Darlington, I. G. Dunn, C.M. Moriarty, and M.P. Tevlin. 1973. Feeding and grazing in Lake George, Uganda. Proc. R. Soc. Lond. B. 184:299-319.

Moriarty, D.J.W., and C.M. Moriarty. 1973. The assimilation of carbon from phytoplankton by two herbivorous fishes: Tilapia nilotica and Haplochromis migripinnis. J. Zool. 171:15-23.

Moriarty, C.M., and D.J.W. Moriarty. 1973. Quantitative estimation of the daily ingestion of phytoplankton by Tilapia nilotica and Haplochromis migripinnis. J. Zool. 171:41-55.

Moskul, G.A. 1977. Feeding of two-year-old silver carp and bighead in foraging lagoons of the Krasnodar Area. Hydrobiol. J. 13(3):37-41.

O'Brian, W.J. 1979. The predator-prey interaction of planktivorous fish and zooplankton. Amer. Scientist 67:572-581.

O'Connell, C.P. 1972. The interrelation of biting and filtering in the feeding activity of the Northern anchovy (Engraulis mordax). J. Fish. Res. Board Can. 29:285-293.

Ogino, C. and T. Watanabe. 1978. Nutritive value of proteins contained in activated sludge, phytosynthetic bacteria and marine rotifer for fish. J. Tokyo Univ. Fish. 61(2):101-108.

Omarov, M.O., and L.P. Lazareva. 1974. The food of the silver carp in Dagestan ponds and lakes. Hydrobiol. J. 10(4):80-83.

Oviatt, C.A., A.L. Gall, and S.W. Nixon. 1972. Environmental effects of Atlantic menhaden on surrounding waters. Chesapeake Science 13:321-323.

Parameswaran, S., C. Selvaraj, and S. Radhakrishnan. 1974. Observations on the biology of Labeo gonius (Hamilton). Indian J. Fish. 21:54-75.

Rogers, W.A. 1968. Food habits of young largemouth bass (Micropterus salmonides) in hatchery ponds. Proc. Southeastern Assoc. Game and Fish Comm. 21:543-553.

Roy, D.C., and A.B. Gupta. 1973. Studies on the effect of feeding Cirrhina mrigala (Ham) and Labeo rohita (Ham) with Chlamydomonas species. Labdev. J. Sci. Tech. 11-B(3-4):53-58.

Sadykhov, D.A., I.B. Bogatova, and V.I. Filatov. 1975. The amino acids in certain representatives of the fresh-water zooplankton. Hydrobiol. J. 11(6):39-42.

Schoonbee, H.J., V.S. Nakani, and J. Prinsloo. 1979. The use of cattle manure and supplementary feeding in growth studies of the Chinese silver carp in Transkei. South Africa Journal Science 75:489-495.

Schroeder, G.L. 1975. Some effects of stocking fish in waste treatment ponds. Water Research 9:591-593.

Shpet, G.I., and N.N. Kharitonova. 1975. A multipoint rating system for carp fattening ponds. Hydrobiol. J. 11(2):36-43.

Spataru, P. 1978. Gut contents of silver carp - Hypophthalmichthys (Val.), and some trophic relations to other fish species in a polyculture system. Aquaculture 11:137-146.

Spectorova, L.V., T.M. Aronovich, S.I. Doroshev, and V.P. Popova. 1974. Artificial rearing of the black sea turbot larvae (Scophthalmus maeoticus). Aquaculture 4:329-340.

Starostka, V.J., and R.L. Applegate. 1970. Food selectivity of bigmouth buffalo, Ictiobus cyprinellus, in Lake Poinsett, South Dakota. Trans. Amer. Fish. Soc. 99(3):571-576.

Stephen, W.P., Jr. 1976. Feeding of laboratory-reared larvae of the sea bream Archosargus rhombiodalis (Sparidae). Mar. Biol. 38:1-16.

Szlauer, L. 1979. Methods of utilization of waste effluents from mineral fertilizer plants in carp fry cultures. Pol. Arch. Hydrobiol. 26(1/2):231-246.

Tal, S. and I. Ziv. 1978. Culture of exotic species in Israel. Bamidgeh 30:3-11.

Tarmas, G. 1975. Controlled production of rotifer zooplankton in fingerling ponds. Hydrobiol. J. 11(6):53-56.

__________, and L. Horvath. 1976. Growth of cyprinids under optimal zooplankton conditions. Bamidgeh 28:50-56.

Tang, Y.A. 1970. Evaluation of balance between fishes and available fish foods in multispecies fish culture ponds in Taiwan. Trans. Amer. Fish. Soc. 99(4):708-718.

Torrans, E.L., and Clemens, H.P. 1981. Commercial polyculture of bigmouth buffalo and channel catfish in Oklahoma. Proc. Southeastern Assoc. Game and Wildl. Agencies 35:554-561.

Torrans, E.L., H.P. Clemens, and M.R. Whitmore. 1980. Morphology of filter-feeding structures in the bigmouth buffalo Ictiobus cyprinellus. Okla. Soc. Elec. Micros. Newsletter 2(1):11-12.

Ware, D.M. 1973. Risk of epibenthic prey to predation of rainbow trout (Salmo gairdneri). J. Fish. Res. Board Can. 30:787-797.

Wells, L. 1973. Effects of alewife predation of zooplankton populations in Lake Michigan. Limnol. Oceanogr. 15:556-565.

Werner, Earl E., and D.J. Hall. 1974. Optimal foraging and the size selection of prey by the bluegill sunfish (Lepomis macrochirus). Ecology 56:1042-1052.

Whitehead, P.J. 1959. Feeding mechanisms of Tilapia nigra. Nature 184:1509-1510.

Wiliamovski, A. 1972. Structure of the gill apparatus and the subrabranchial organ of Hypophthalmichthys molitrix Val (silver carp). Bamidgeh 24(4):87-98.

Yashouv, A. 1971. Interaction between the common carp (Cyprinus carpio) and the silver carp (Hypophthalmichthys molitrix) in fish ponds. Bamidgeh 23:85-92.

Yesipova, M.A., L.M. Solov'yeva, and I.V. Glazacheva. 1976. Optimum zooplankton biomass in fish ponds, Hydrobiol. J. 12(2):52-53.

Zaikina, A.I. 1973. Development of zooplankton population in sturgeon ponds during the first cycle of fry growth. Hydrobiol. J. 9(5):41-45.

Zaret, T.M., and W.C. Kerfoot. 1975. Fish predation on Bosmina longirostris: Body-size selection versus visibility selection. Ecology 55:32-237.

PART 2

POND CULTURE PRACTICES

POND PRODUCTION SYSTEMS: STOCKING PRACTICES IN POND FISH CULTURE

by

Yung C. Shang

Pond management practices vary not only between species but also between different socio-economic and cultural settings. In this and the following four papers the purpose is to review and synthesize pond management practices in the areas of 1) stocking, 2) fertilization, 3) feeds and feeding, 4) water quality, and 5) disease and predator control of major fresh- and brackish-water species (e.g., carps, milkfish, mullet, tilapia, and walking catfish), to establish credible guidelines for an effective pond management system.

INTRODUCTION

In aquaculture "stocking materials" (or "fish seed") generally refers to the fry, post larvae, and fingerlings of the species to be cultured. Fish seed are usually obtained from two sources: (1) natural waters, e.g., milkfish, grey mullet (to a large extent), and eel; and (2) hatcheries, e.g., Chinese major carps, Indian major carps, rainbow trout, channel and walking catfish, tilapia, grey mullet (to a limited extent), and snakehead (Shang 1981).

The term "stocking rate" has a variety of interpretations. In many countries it is defined as the number of fish seed stocked per unit water surface, with due consideration to the size and/or age group of the stocking material. In others, it is the total weight of fish seed stocked per unit water surface regardless of the age groups of the seed (Rabanal 1968). Various degrees of intergrading these two stocking concepts are encountered in the literature.

FACTORS AFFECTING THE STOCKING RATE OF A POND

One of the most important pond-management practices is the stocking of the appropriate species and quantity of fish. A fish pond can only support a certain quantity of fish because of its limited space and the amount of natural food available. The latter is affected mainly by the soil conditions and water quality of the pond. This limit is usually called the carrying capacity (or maximum standing crop) of the pond. It is defined as the maximum weight of fish stock that can be sustained by a pond (by either the food produced within the pond or made available to the fish) without gaining or losing weight (Hickling 1962). The carrying capacity of a pond can be increased by fertilization and/or supplemental feeding. The purpose of fertilization is to increase the production of plankton or benthic algae as fish food, while supplemental feeding compensates (directly or indirectly) for nutrients which are in short supply in the pond. In Israel, fertilization increased the carrying capacity of carp ponds almost fourfold and fertilization and supplementary feeding increased it ten to fifteen times (Yashouv 1959; Hepher 1978).

Van der Lingen (1959) revealed that the maximum standing crop of tilapia cultured in Rhodesia is about 2.5 times higher in fertilized ponds (and about seven times higher in fertilized ponds with supplemental feeding) than under natural conditions. Tang (1970) estimated that a heavily fertilized pond in Taiwan produces ten times the amount of phytoplankton and zooplankton than an unfertilized reservoir produces. Shang (1976) indicated that the stocking rate (and hence the total yield) in intensively managed milkfish ponds (with fertilization and/or supplemental feeding) in Indonesia and the Philippines is about three to four times higher than that of extensively managed ponds.

Aeration and running-water systems usually increase the amount of dissolved oxygen and, therefore, increase the carrying capacity of a pond. For example, in the running-water ponds of the Philippines, common carp fry are stocked at rates of 280,000 to 850,000/ha as compared to 50,000/ha in still water (Bardach et al. 1972). Kawamoto (1957) and Chiba (1965) have shown that the yield and growth of carps in ponds with running water is 100 to 1000 times higher than in ponds with still water because of the elimination of growth-inhibiting metabolites and the continuous supply of oxygen. Basing his study on common carp culture in several countries under different management systems, Hepher (1978) demonstrated that the stocking rate and the yield/ha increased with fertilization, supplemental feeding, and running water systems.

A further increase in the stocking rate of a pond can be achieved by polyculture (stocking a number of species in the same pond) and by stock manipulation (methods used to manage the fish population in the pond).

STOCKING PRACTICES

A sound stocking technique is to balance fish populations with fish foods available in the pond. To estimate the right stocking rate requires basic information on the quality and quantity of natural foods available in a given period of time as well as the feeding rate of the different age (or size) groups of the species concerned. This sort of information varies and is not always available. The most widely used stocking technique is based on common sense and experience, as shown in the following formula:

Number of fish to be stocked =

$$\frac{\text{Size of fish pond (ha) x expected yield (kg) per ha based on experiences of previous years}}{\text{Expected avg. weight of individual fish when harvested} - \text{Avg. weight of individual fry or fingerling when stocked}}$$

+ expected mortality (no.)

+ additional number of stocking due to fertilization and feeding

The estimated stocking rate with fertilization and feeding is respectively based on the additional amount of natural food available in the pond and the food conversion ratio (units of feed to produce one unit of fish).

Based on this empirical stocking method, several fish-stocking practices of varying complexity have been used for various species in different regions. These stocking practices may be divided into two management systems: monoculture and polyculture.

Monoculture

Monoculture, which is practiced in many countries, is the stocking of a single species in a pond. Within a monoculture system, there are several stocking practices that affect the fish production of a pond.

Mono-size Stocking This is simply to stock one species of the same size in a pond and to harvest all the fish at marketable size. This practice has some disadvantages. If stocking density is too great, the fish would be overcrowded when they reach adult size. Thus growth and survival rates would be reduced. On the other hand, if stocking density is low, the water space and the natural food in the pond will not be efficiently utilized during the earlier part of the rearing period. Understocking is a common practice. For example, many extensive milkfish operators in Indonesia and the Philippines use low stocking densities and, as a result, achieve only a low level of production (Shang 1976). A recent production function analysis of the milkfish industry in the Philippines based on cross-section data revealed that a 0.6 percent increase in milkfish production can be expected with every one percent increase in the fry stocking rate (Chong et al. 1982).

Fry rearing in nursery ponds usually uses mono-size stocking practice. The stocking rates of fry for various species in different countries are summarized in Table 1.

Multi-stage Stocking An alternative stocking practice that avoids the disadvantages mentioned above is multi-stage stocking, where fish of uniform size are stocked in progressively larger ponds as more space is needed. This method takes advantage of the maximum growth potential of the fish; their density can be adjusted as they are transferred to larger ponds. The smaller ponds are then prepared for the rearing of succeeding batches of younger fish. This enables the farmers to undertake a continuous cycle of stocking and harvesting. This stocking method has been practiced by some milkfish farms in the Philippines with encouraging results (Tang 1972; Rabanal 1968). However, it can only be practiced by relatively large-scale operations with several pond units. An example of multi-stage stocking is presented in Table 2.

Multi-size Stocking This practice also avoids the stocking density problems of mono-size stocking. Fish ponds frequently produce a variety of fish food, and the feeding habits of the young and adult fish are often quite different. Consequently, the stocking rate (and, hence, the total yield of a pond) can be increased by stocking different age groups of a fish species to more efficiently utilize the available forage. This practice requires periodic harvesting of the marketable-size fish. The pond may then be restocked with smaller fish to replace the ones removed. A typical example of this stocking practice (in the milkfish culture in Taiwan) is shown in Table 3.

The initial stocking of the ponds in April involves fingerlings of sizes varying from 5 to 100 g in weight at about 5,000 fish/ha. Fish harvest begins at the end of May and at least eight harvests are made until the middle of November. Stocking of newly captured fry is made from April to August. The new fry stocked in July (and after) do not reach marketable size by November and are put into wintering ponds for next years stocking. Under this management system,

Table 1

STOCKING RATE OF FRY BY COUNTRY AND SPECIES

Country and species stocked	Stocking rate, no. of fry/ha	Age or size	Source
JAPAN			Bardach et al. 1972
Common carp	3,000-15,000	Fry	
Grass & Silver carp	1,400-1,800/m^2	10-20 mm	
INDONESIA			
Common carp	60,000	3 wks. old	Bardach et al. 1972
Milkfish	300,000	Fry	Korringa 1976
PHILIPPINES			
Common carp	50,000	8-10 mm	Bardach et al. 1972
Milkfish	30-50/m^2	Fry	Korringa 1976
TAIWAN			
Chinese carp	300,000/0.1 ha	Fry	Chen 1976
	70,000-80,000/0.1 ha	2-3 cm	
Milkfish	70,000-150,000	Fry	Korringa 1976
ISRAEL			Bardach et al. 1972
Mullet	10,000-20,000	Fry	
USA			Bardach et al. 1972
Common carp	250,000 (experimental pond)	3-4 wks.	
Catfish	3,300-4,400	Fingerlings	
	1,650-2,200	Fingerlings	
	6,000-110,000 (running water)	Fingerlings	
USSR			Bardach et al. 1972
Common carp	10-80/m^2	Fingerlings	
INDIA			Bardach et al. 1972
Common carp	1.25×10^6 - 2.5×10^6	2 days old	
CHINA			Food and Agriculture Organization 1979
Grass carp	20,000/0.066 ha	3 cm	
	4,000-5,000/0.066 ha	4.8 cm	
Bighead	15,000/0.066 ha	3 cm	
	4,000-6,000/0.066 ha	6 cm	
Mud carp	27,000/0.066 ha	3 cm	
	9,000/0.066 ha	5.8 cm	
Silver carp	20,000/0.066 ha	3 cm	
	800-1,000/0.066 ha	6-9.5 cm	

a total of 15,000 fish can be stocked per ha with an annual yield of more than 2,000 kg of marketable fish. The high yield of milkfish ponds in Taiwan, when compared to those in Indonesia and the Philippines, may be attributed to the use of multiple-size stocking. This practice has been adapted for use in other countries for both fresh- and brackish-water culture.

Mono-sex Stocking The major problem of culturing a species such as tilapia, which reproduces in the rearing pond, is that the pond will soon become overpopulated with small unmarketable fish. One way to solve this problem is to stock a single sex of the species. In this case, no reproduction is possible.

Double Cropping This is the practice of stocking two species in the same pond but in different seasons to take advantage of their different thermal requirements. The stocking of channel catfish during the summer and rainbow trout during the winter in the United States increases the total production and profit of a pond (Brown 1979).

Table 2

A TYPICAL CASE OF MULTI-STAGE STOCKING OF MILKFISH PONDS OF DIFFERENT SIZE[a]

Stages of growing	Ponds required			Initial stocking rates			No. of days cultured	Expected population at the end growing stage	
	Size, ha	No.	Total area, ha	No/ha	No/kg	Kg/ha		No/kg	Kg/ha
Stage I	1.0	1	1.0	18,000	400	45	42	50	360
Stage II	3.0	1	3.0	6,000	50	120	46	13	480
Stage III	3.0	3	9.0	2,000	13	160	56	3	650

[a] After Tang 1972

The stocking rates (based on the available information) in monoculture systems by species and by country are summarized in Table 4.

Polyculture

A fish pond, especially a freshwater pond, usually produces a variety of food organisms in different layers of the water. Therefore, stocking species (or different size classes of a given species) that have complementary feeding habits or that feed in different zones will efficiently utilize space and available food in the pond and increase total fish production.

Polyculture originated in China thousands of years ago and has been improved with the passing of time. It is the practice of polyculture that enables Chinese fish culturists to achieve high productivity per unit area. To assure efficient use of a given pond ecosystem, Chinese fish culturists usually polyculture the following fish: (1) the grass carp (Ctenopharyngodon idellus) which roams in all strata of the water and feeds mainly on higher aquatic plants; (2) the silver carp (Hypothalmichthys molitrix), a midwater dweller that prefers phytoplankton as food; (3) the bighead carp (Aristichthys nobilis), also a midwater dweller, which consumes zooplankton; (4) the black carp (Mylopharyngodon piceus), a bottom-dwelling carnivore that feeds on mollusks; and (5) the mud carp (Cirrhinus motilorella), a bottom-dwelling omnivore that feeds on benthic animals and detritus. Other fish

Table 3

A TYPICAL PRACTICE OF MULTIPLE-SIZE STOCKING OF MILKFISH IN SAME PONDS IN TAIWAN[a]

Month of stocking	Approximate average weight of fingerlings stocked, g	Approximate number of fingerlings stocked
April	5-100 g	5,000
May	0.05	2,500
June	0.06	2,500
July	0.06	2,500
Aug-Sept	0.06	2,500
TOTAL		15,000

[a] From Chen 1976

Table 4

STOCKING RATES IN MONOCULTURE SYSTEMS BY COUNTRY AND SPECIES

Country and species stocked	Stocking rate, No. of fish/ha	Size of fish	Source
PUERTO RICO			Fram and Pagan-Font 1978
T. nilotica	10,000	15.9 g	
T. horhorum	10,000	53.9 g	
U.S.A.			Bardach et al. 1972
T. nilotica	20,000 (experimental pond)	-	
Java tilapia	50,000 (experimental pond)	-	
Grass carp	1,700	30-40 cm	
Catfish	1,540-3,300	200-450 g	
NIGERIA			Bardach et al. 1972
Common carp	25,000-30,000	30-50 mm	
PHILIPPINES			
Milkfish	1,500-6,500	-	Shang 1976
T. nilotica	10,000-20,000	-	Bardach et al. 1972
Common carp	50,000 (running water) 5,000 (stagnant water)	60-180 mm 20-50 g	
U.S.S.R.			Bardach et al. 1972
Grass carp	500-800 160-240 40,000-50,000	Yearling 3-4 years up to 5 g	
INDONESIA			Shang 1976
Milkfish	4,000-6,000	-	
TAIWAN			
T. nilotica	20,000	20-30 mm	Chen 1976
Catfish	30-60/m^2	25-35 mm	Chen 1976
Mullet	4,000-10,000	-	Chen 1976
Milkfish	15,000	-	Shang 1976
THAILAND			Ling 1977
Catfish	80-100/m^2 60-80/m^2 40-60/m^2	16 g 20 g 25 g	
JAPAN			Suzuki 1979
Common carp	1/m^2 0.5-1/m^2	40-100 g 80-100 g	

often reared with the Chinese carps are common carp (Cyprinus carpio), grey mullet, tilapia, and bream (Parabramis pekinensis). Grass carp is the major species usually stocked when plants are abundant in the pond. In plankton-rich ponds, silver carp and bighead carp are the major species usually stocked. In deeper ponds, the productivity of the bottom water is substantially reduced and the stocking of bottom dwellers is usually low. In cold regions where temperatures rule out the mud carp, other bottom-feeding omnivores like common carp and bream are usually stocked. Various combinations of species used in polyculture in China are given in Table 5.

Table 5
VARIOUS COMBINATIONS OF SPECIES USED IN POLYCULTURE IN CHINA[a]

Species	Composition by no. of fish stocked, %
1. Silver carp as main species	
Silver carp	65.0
Bighead carp	10.0
Grass carp	12.0
Common carp	5.0
Wuchan fish (bream)	8.0
2. Black carp as major species	
Black carp	42.0
Grass carp	24.2
Silver carp	12.4
Bighead carp	7.4
Wuchan fish	7.4
Common carp	3.4
Golden carp	3.2
3. Grass carp as major species	
Grass carp	55.0
Silver carp	16.0
Bighead carp	10.0
Others	19.0

[a]From Tapiador et al. 1977

In other Asian countries, either the above-mentioned species or native fish with similar food habits are stocked. Grass, bighead, silver, and common carps are usually polycultured in Malaysia, Singapore, and Thailand. Mullet and the above-mentioned Chinese carps are polycultured in Hong Kong and the Philippines (Table 6). In these countries, species that are more adapted to colder waters (such as black carp) are not usually raised. The distinctive feature of polyculture stocking in Hong Kong and the Philippines is the use of mullet, which are abundant in those regions.

In Indonesia, native species like tambakan (Helostoma temmincki), tawes (Puntius javanicus), nilem, (Osteochilus hasselti), and gourami (Osphronemus goramy), are usually polycultured with common carp and tilapia (Table 7). The tambakan is a plankton feeder, while the tawes consumes coarse pond vegetation, playing the role that the grass carp performs in other countries; the gourami is a herbivore that feeds on softer vegetable materials along the pond margins; the nilem eat plankton, periphyton, and softer or decayed vegetation.

In India and Pakistan, the species most commonly used for polyculture are catla (Catla catla), rohu (Labeo rohita), mrigal (Cirrhina mrigala), and calbasu (Labeo calbasu). These are shown in Table 8. Like Chinese carps, these species have suitable complementary feeding habits and behavior. The catla feeds on the surface strata; the mrigal is a bottom feeder; the rohu is a water column feeder; and the calbasu is a bottom feeder that eats mollusks. In many cases, these native species are polycultured with silver, grass, and common carps (Central Inland Fisheries Research Institute 1978).

The essential feature of the polyculture system in Asia is the stocking in small ponds of as many as eight fish species with different feeding habits. In well-managed Asian ponds, an average annual production of 3,000 kg/ha can be achieved.

In European countries, polyculture is based largely on natural foods. It is further characterized by big fish ponds and is dominated by common carp. Common carp and tench (Tinca tinca) are polycultured in Yugoslavia; common carp and roach (Rutilus rutilus) in France; common carp with cyprinus carpio x crucian carp hybrids, goldfish, bream, and sterlet (Acipenser rughenus) in the Soviet Union (Bardach et al. 1972).

In Africa, little information on polyculture is available. Balarin and Hatton (1979) indicated that in the past T. rednalli and S. macrochir were commonly polycultured; in more recent times, however, common carp, Helerotis niloticus, Clarias lazera, and tilapia have become important species for polyculture.

The Israeli polyculture system features the intensive management of moderate-sized ponds with fertilization and supplemental feeding. Common carp, tilapia, and/or grey mullet are usually cultured together (Table 9). In many cases, these species are polycultured with chinese carps.

In general, the selection and combination of species for polyculture depends mainly on the compatability of the species, the availability of natural and supplementary foods, the suitability of environmental conditions, the availability of fish seed, and the demand for a price of fish.

There are two polyculture stocking practices that result in high yield, namely multi-age (or size) stocking and multi-stage stocking:

Multi-age polyculture is a practice where different fishes of different sizes are reared in the same pond from fingerling to marketable size. The practice, like the multi-size stocking in a monoculture system, requires periodic harvesting of the market-size fish and subsequent restocking with small fish. An example is provided in Table 10.

Multi-stage polyculture involves moving different species of fish through a series of ponds as they grow from fingerling to marketable size, with fish sorted in the ponds according to size.

Table 6
STOCKING COMBINATIONS IN AN INDONESIAN POND[a]

Species	Composition under two stocking practices[b] (%) I	II
Based on Tilapia as Major Species		
Tilapia	35	40
Common carp	30	10
Tambakan	--	20
Tawes	--	15
Nilem	20	15
Gourami	15	--
Based on Tambakan and Tawes As Major Species		
Tambakan	50	37.5
Tawes	10	37.5
Common Carp	20	12.5
Nilem	20	12.5

[a]From Rabanal 1968
[b]Stocking practice I (in stagnant ponds) and II (in ponds with water inflow)

The stocking rate decreases as fish become larger. An example is given in Table 11.

Carnivorous species are sometimes used as predators in a polyculture system, especially when cultured species reproduce in rearing ponds. For example, in tilapia culture, Clarias sp., Lates niloticus, Bagrus docmac, Micropterus salmoides, Hemichromis fasciatus, freshwater Jack Dempsey, and Cichlasoma managuense were used as predators (Balarin and Hatton 1979; Dunseth and Bayne 1978; Suffern 1980). In channel catfish culture at Auburn University, largemouth bass (Micropterus salmoides) were used as predators (Swingle 1968). In European carp culture, pike (Esox lucius) or pike-perch (Lucioperca sandra) have been used to eliminate both small carp that resulted from unexpected spawning and wild fish that entered the pond. The optimal ratio of predator to prey is determined by comparing the size of prey with the size and voraciousness of the predator (Balarin and Hatton 1979). The less voracious the predator, the higher its stocking rate needs to be. In tilapia culture, for example, two to three percent of the total stock usually consists of predators like Hemichromis fasciatus and Lates niloticus (Huet 1972; Pruginin 1975); when Clarias species are used as predators, though, 5-10 percent is the effective ratio (Meecham 1975).

Polyculture requires a delicate stocking balance to minimize both interspecies and intraspecies competition. Interspecies competition in the carp/mullet/tilapia combination in Israel resulted in a 28 percent reduction in productivity of each species relative to their productivity in monoculture; however, there was a 30 percent increase in overall fish yields (Balarin and Hatton 1979). Another study (Spataru and Hepher 1977) observed that common carp stocked in high density with tilapia were preying on the tilapia fry. On the other hand, a high stocking density of tilapia will inhibit carp growth (Yashouv 1969). Observations in Yugoslavia revealed that carp-tench competition in densely stocked ponds depressed carp production (Yashouv 1968). These experiences in temperate-water culture may apply to tropical regions. Thus, although polyculture usually results in higher overall fish production per unit of pond, a careful choice of stocking sizes and densities is necessary.

FUTURE RESEARCH NEEDS

One of the major problems encountered in intensive aquaculture is the inadequate supply of suitable seed stock in many areas. This problem is particularly severe in those species where artificial

Table 7

STOCKING RATE IN POLYCULTURE SYSTEMS IN SELECTED ASIAN COUNTRIES, BY SPECIES

	Common Carp	Bighead Carp	Grass Carp	Silver Carp	Grey Mullet	Java Carp	Gold-fish	Milk-fish	Others	Source
Hong Kong	1500/km^2 (2.5-3.0 cm)	975/km^2 (7.5-10 cm)	1200/km^2 (7.5-10 cm)	100/km^2 (7.5-10 cm)	17500/km^2 (2.5-3.5 cm)		875/km^2 (2.0-2.5 cm)			Indo-Pacific Fisheries Council 1977
Philippines	500/ha (400-500 g)	300ha/ (100-150 g)	600/ha (100-200 g)	2000/ha (200-300 g)	3000/ha (400-500 g)			1500/ha (300-	100/ha 400 g)	Bardach 1972
Malaysia	1250/ha (5 cm)	125/ha (12-15 cm)	500/ha (12-15 cm)			2500/ha (4 cm)			625/ha	Indo-Pacific Fisheries Council 1977
	1250/ha ha (5 cm)	500/ha ha (12-15 cm)	125/ha ha (12-15 cm)	250/ha ha (10-12 cm)						Indo-Pacific Fisheries Council 1977
Singapore										
In nursery	-- (30 g)	250/ha (40 g)	1250/ha (30 g)	250/ha (40 g)						Hora and Pillay 1962
In fattening pond	225/ha	150/ha (2000 g)	450/ha (1500 g)	175/ha (1800 g)						

Table 8
STOCKING COMBINATION PER HECTARE USED IN INDIA AND PAKISTAN[a]

Species	Stocking rate (8-13 cm) under two stocking practices	
	I	II
Catla	1,875	1,875
Rohu	3,750	3,750
Mrigal	625	625
Calbasu	--	625

[a]From Rabanal 1968

Table 9
STOCKING COMBINATION PER HECTARE USED IN ISRAEL

Tilapia	Mullet	Common Carp	Source
1050	600 (30-70 g)	1200	Bardach et al. 1972
5000 (2 g)	--	3000 (37 g)	Yashouv 1958
1500 (50 g)	1000 (50-100g)	2500 (200 g)	Korringa 1976

breeding has not been successful. When the supply of seed depends on natural sources, its availability fluctuates. In many cases, there is a shortage of seed when it is needed for stock manipulation in order to increase fish production. The solution to this problem is obvious: promote the domestication process. Research in this area is needed. Meanwhile, short-range efforts should be made to improve the efficiency of fry handling and to increase the survival rate during the pond rearing period.

Current stocking rates require basic information on the quality and quantity of fish available in the pond, plus the food preferences and feeding rates of the

Table 10
EXAMPLE OF MULTI-STAGE POLYCULTURE IN TAIWAN[a]

Species stocked	Stocking rate No./ha	Initial average size cm	Month of stocking
Central Taiwan			
Common Carp	1000	2.5	March-April
Grass Carp	50	7-12	March-April
Silver Carp	800	7-12	March-April
Mullet	2000	5	March-April
Mud Carp	1000	5	March-April
Bighead	100	7-12	March-April
Southern Taiwan			
Tilapia mossambica	2000	3	February
Silver Carp	1000	10-13	February-April
Common Carp	2000	3-4	March
Mullet	2000	5	March
Mud Carp	1500	7-10	March
Bighead	400	10-13	March
Grass Carp	200	12-15	March
Walking Catfish	500	5	May
Snakehead	500	10	June

[a] From Chen 1976

Table 11

EXAMPLE OF MULTI-STAGE POLYCULTURE IN CHINA[a]

Stage	Size of fish, g	Stocking density/ha				Rearing period, days
		Bighead	Grass Carp	Silver Carp	Mud Carp	
1	14-65	6,750	1,500			40
	14-80			3,750		
	5-20				112,500	
2	65-225	2,100	6,750			150
	80-500			600		
					37,500	
3	225-500	900	3,000			150
	500-1,000			300		
	60-270				12,750	
4	500-1,200	375	1,050			

[a] From Tapiador et al. 1977

different age groups of the species concerned. This sort of information varies from place to place and is not usually available. Research on these aspects is necessary.

The stocking information used in this section was based on various studies conducted some time ago. The systematic collection of more up-to-date stocking data for various species in different countries under different culture systems is needed.

LITERATURE CITED

Balarin, J.D., and J.P. Hatton. 1979. Tilapia: A Guide to Their Biology and Culture in Africa. University of Stirling, Stirling, Scotland. 174 pp.

Bardach, J.E., J.H. Ryther, and W.O. McLarney. 1972. Aquaculture, the Farming and Husbandry of Freshwater and Marine Organisms. Wiley-Interscience, New York. 868 pp.

Brown, E.E. 1979. Fish production costs using alternative systems and the economic advantages of double-cropping, pp. 235-239. In T.V.R. Pillay and W.A. Dill (eds), Advances in Aquaculture. Fishing News Books, Ltd., Surrey, England.

Central Inland Fisheries Research Institute. 1978. First 138 Case Studies of Composite Fish Culture in India, Bulletin No. 23. Central Inland Fisheries Research Institute, West Bengal, India. 155 pp.

Chen, T.P. 1976. Aquaculture Practices in Taiwan. Fishing News Books, Ltd., Surrey, England. 176 pp.

Chiba, K. 1965. Studies on the carp culture in running water ponds. Fish production and its environmental conditions in a certain fish farm in Gunma Prefecture. Bull. Freshwat. Fish. Res. Lab. 15(1):13-33.

Chong, K.C., I.R. Smith, and M.S. Lizarondo. 1982. Economics of the Philippine Milkfish (Chanos chanos Forskal) Resource System. Resource Systems Theory and Methodology No. 3. The United Nations University.

Dunseth, D.R., and D.R. Bayne. 1978. Recruitment control and production of Tilapia aurea (Steindachner) with the predator, Cichlasoma managuense (Gunther). Aquaculture 14:383-390.

Food and Agriculture Organization. 1979. Aquaculture Development in China. FAO Aquaculture development and Coordination Programme. 65 pp.

Fram, M.J., and F.A. Pagan-Font. 1978. Monoculture yield trials of an all-male hybrid tilapia (female Tilapia nilotica x male T. hornorum) in small farm ponds in Puerto Rico, pp. 53-64. In R.O. Smitherman, W.L. Shelton and J.H. Grover (eds), Symposium on Culture of Exotic Fishes. Fish Culture Section, American Fisheries Society.

Hepher, B. 1978. Ecological Aspects of Warm-water Fishpond Management, pp. 447-468. In S.D. Gerking (ed), Ecology of Freshwater Fish Production. John Wiley & Sons, New York.

Hickling, C.F. 1962. Fish Culture. Faber and Faber, London, England. 295 pp.

Hora, S.L., and T.V.R. Pillay. 1962. Handbook on Fish Culture in the Indo-Pacific Region. Food and Agriculture Organization of the United Nations, Rome, Italy. 204 pp.

Huet, M. 1972. Textbook of Fish Culture: Breeding and Cultivation of Fish. Fishing News Books, Ltd., Surrey, England. 436 pp.

Indo-Pacific Fisheries Council (IPFC). 1977. Proceedings of the 17th Session (Section III, Symposium on the Development and Utilization of Inland Fishery Resources). FAO Regional Office for Asia and the Far East, Bangkok, Thailand. 500 pp.

Kawamoto, N.Y. 1957. Production during intensive carp culture in Japan. Prog. Fish-Cult. 19(1):26-31.

Korringa, P. 1976. Farming Marine Fishes and Shrimps. A Multidisciplinary Treatise. Elsevier Scientific Publishing Company, New York. 208 pp.

Ling, S. 1977. Aquaculture in Southeast Asia, A Historical Overview. University of Washington Press, Seattle, Washington. 108 pp.

Meecham, K. 1975. Aquaculture in Malawi. FAO/CIFA Symposium on Aquaculture in Africa, Accra, Ghana. CIFA/75/SCI. 6 pp.

Newton, S.H., J.C. Dean, and A.J. Handcock. 1978. Low intensity polyculture with Chinese carps, pp. 137-141. In R.O. Smitherman, W.L. Shelton and J.H. Grover (eds), Symposium on Culture of Exotic Fishes. Fish Culture Section. American Fisheries Society.

Pruginin, Y. 1975. Species Combination and Stock Densities in Aquaculture in Africa. FAO/CIFA Symposium on Aquaculture in Africa, Accra, Ghana. CIFA/75/SR6. 5 pp.

Rabanal, H.R. 1968. Stock manipulation and other biological methods of increasing production of fish through pond fish culture in Asia and the Far East. FAO Fisheries Report 44(4):274-288.

Shang, Y.C. 1976. Economics of Various Management Techniques for Pond Culture of Finfish. South China Sea Fisheries Development and Coordinating Programme. 32 pp.

__________. 1981. Aquaculture Economics: Basic Concepts and Methods of Analysis. Westview Press, Inc. Boulder, Colorado. 153 pp.

Sparatu, P., and B. Hepher. 1977. Common carp predating on Tilapia fry in a high density polyculture fishpond system. Bamidgeh 29(1):25-28.

Suffern, J.S. 1980. The potential of Tilapia in United States aquaculture. Aquaculture Magazine Sept.-Oct.:14-18.

Suzuki, R. 1979. The culture of common carp in Japan, pp. 161-166. In T.V.R. Pillay and W.A. Dill (eds), Advances in Aquaculture. Fishing News Books, Ltd., Surrey, England.

Swingle, H.S. 1968. Biological means of increasing productivity in ponds. FAO Fisheries Report 44(4):243-257.

Tang, Y.A. 1970. Evaluation of balance between fishes and available fish foods in multispecies fish culture ponds in Taiwan. Trans. Am. Fish. Soc. 99:708-718.

__________. 1972. Stock manipulation of coastal fish farms, pp. 438-453. In T.V.R. Pillay (ed), Coastal Aquaculture in the Indo-Pacific Region. Fishing News Books, Ltd., Surrey, England.

Tapiador, D.D., H.F. Henderson, M.H. Delmendo, and H. Tsutsui. 1977. Freshwater Fisheries and Aquaculture in China. FAO Fisheries Technical Paper No. 168. 84 pp.

Van der Lingen, M.I. 1959. Some Preliminary Remarks on Stocking Rate and Production of Tilapia Species at the Fisheries Research Centre, pp. 54-62. In Proc. 1st. Fish Day in S. Rhodesia, Aug. 1957, Govt. Printer, Salisbury.

Yashouv, A. 1958. On the possibility of mixed cultivation of various tilapia with carp. Bamidgeh 10:2.

__________. 1959. Studies on the producivity of fishponds. I. Carrying capacity. Proc. Gen. Fish. Coun. Medit. 5:409-419.

Yashouv, A. 1968. Mixed fish culture--an ecological approach to increase pond productivity. FAO Fisheries Report 44(4):258-273.

__________. 1969. The fish pond as an experimental model for study of interactions within and among fish populations. Verh. int. Ver. Limnol. 17:582-593.

POND PRODUCTION SYSTEMS:
FERTILIZATION PRACTICES IN WARMWATER FISH PONDS

by

Randolph Yamada

INTRODUCTION

The importance of fertilizers for enhancing pond production in modern fish culture is indisputable. However, in most tropical countries suitable fertilization practices (actual mixtures, quantities, and application schedules) have not yet been determined. The estimation of required nutrients for a pond fertilization program depends on the pond's morphology, hydrology, environs, bottom materials, and water quality; on the type of fish cultured; and on the type of fertilizer employed. In addition, fertilizer recommendations for one location may be unsuitable for another location. Nevertheless, some benefits of fertilization are shown in Table 1.

All fertilizers may be classified as either organic or inorganic. Inorganic fertilizers are nutrients in simple inorganic compound form whose primary components contain at least one of the following: nitrogen, phosphorus, and/or potassium (N-P-K). Possible secondary nutrients include calcium, magnesium, and sulfur. Trace elements include copper, zinc, boron, manganese, iron, and molybdenum (Boyd and Lichtkoppler 1979). Commercial inorganic fertilizers used in pond culture are the same as those for agricultural crops. Table 2 shows their typical composition.

Organic fertilizers are animal manures and plant wastes containing 40-50 percent carbon by dry weight (Woynarovich 1975). Unlike chemical fertilizers, these materials usually have a low N-P-K-content (Table 3) and must therefore be used in large quantities.

INORGANIC FERTILIZER

At present, there is no universally accepted method of fertilization. However, in a citation from Lahnovitch (1968), Seymour (1980) lists three schools of inorganic usage: (1) "American" - mixed nitrogen, phosphorus, and potassium fertilizer, (2) "Soviet" - nitrogen and phosphorus fertilizer, and (3) "German" - non-nitrogenous fertilizers.

Table 1

FISH PRODUCTION INCREASE WITH FERTILIZATION VERSUS NO FERTILIZATION

Species	Increase, %	Fertilizer	Culture	Country	Source
Tilapia mossambica	440	Superphosphate	Freshwater	South Africa	van der Lingen 1967
Tilapia sp.	370	Phosphate	Freshwater	Malaysia	Hickling 1962
Tilapia sp., Ctenopharyngodon idella, Puntius javanicus	386-530	Phosphate	Freshwater	Malaysia	Hickling 1962
Cyprinus carpio	172-832	Superphosphate/ ammonium sulfate	Freshwater	Israel	Hepher 1962
Mugil cephalus	167	Phosphate	Brackish-water	United Arab Republic	El Zarka and Fahmy 1968
Mugil cephalus, Cyprinus carpio, Chinese carp, tench	96	Superphosphate	Freshwater	Taiwan	Lin 1968
Indian and Chinese carp polyculture	133-219	Cowdung and N-P-K	Freshwater	India	Sinha 1979

Table 2

COMPOSITION OF SOME COMMON INORGANIC FERTILIZER MATERIALS[a]

Material	Composition, %		
	N	P_2O_5	K_2O
Ammonium nitrate	33-35	-	-
Ammonium sulfate	20-21	-	-
Calcium metaphosphate	-	62-64	-
Calcium nitrate	15.5	-	-
Ammonium phosphate	11-16	20-48	-
Muriate of potash	-	-	50-62
Potassium nitrate	13	-	44
Potassium sulfate	-	-	50
Sodium nitrate	16	-	-
Superphosphate (ordinary)	-	18-20	-
Superphosphate (double or triple)	-	32-54	-

[a] From Boyd 1979

Phosphorus

It has been conclusively demonstrated by German investigators that phosphorus is the most important nutrient supplement usually present in minimal supply (Wahby 1974). Very small quantities are found in natural waters, with a concentration of only 1 ppm or 1 mg/l being considered optimal for planktonic growth (Hora and Pillay 1962). A complex network of factors is responsible for the dynamics of the phosphorus cycle and its fixation: dilution; assimilation by macrophytes, planktonic and benthic algae, bacteria, and fungi; and adsorption to bottom deposits.

Ponds with a history of manuring have a lower rate of fixation than ponds that have not been manured (Hickling 1962). Hepher (1958) assumed this occurs primarily because the adsorption of phosphates by mud decreases as it becomes saturated by phosphate-containing fertilizers. Except in certain instances, such as those reported from Yugoslavia by Fijan (1967), where water inflow contains sufficient concentrations of phosphorus, its addition will normally be beneficial.

Table 3

AVERAGE COMPOSITION OF SEVERAL MANURES[a]

	Composition, %			
	Water	N	P	K
Dairy cows	79	0.5	0.1	0.5
Fattening cattle	78	0.7	0.2	0.5
Sheep	64	1.1	0.3	1.1
Sheep[c]	-	0.7	0.3	0.3
Pig	74	0.5	0.2	0.4
Pig	-	0.6	0.2	0.4
Hen	76	1.1	0.4	0.4
Poultry	-	1.6	0.7	0.7

[a] From Schroeder 1980

Nitrogen

The usefulness of nitrogen fertilizers has been questioned and they have been ruled ineffective and/or uneconomical. Although nitrogen is required by all living organisms (proteins contain about 17 percent nitrogen content (Woynarovich 1975)), some investigators argue that it is unnecessary because either (1) phosphorus increases blue-green algae for nitrogen fixation, thus stimulating phytoplankton productivity (Seymour 1980) or (2) denitrifying bacteria breaks down the added nitrogen to nullify its effect (Wahby 1974). Furthermore, although nitrogen fertilizers have a stimulating growth effect on fish food organisms (as in the case of chironomid larvae), it may be demonstrated that equally good results can be obtained by the substitution of phosphorus and potash fertilizers, provided carbohydrate is present (Wahby 1974). Thus in Europe, nitrogen is usually not used.

In contrast, Israel and the United States have found inorganic nitrogen compounds to be quite effective. Woynarovich (1975) attributes this contradiction to differences in stocking density. For example, temperate culture in Europe stocks 1 fish per 20 m^3 and nitrogen needs are generally met by the activities of nitrogen-fixing bacteria and blue-green algae. In the more productive subtropical conditions in Israel, the stocking rate is typically 1 fish per 2 to 5 m^3, thus placing a critical demand on the pond's nitrogen supply. Boyd (1976) in the United States has demonstrated that Tilapia aurea production increased when nitrogen, phosphorus, and potassium fertilizers were used rather than a phosphorus potassium control mixture (Table 4).

Hora and Pillay (1962) suggest 4:4:1 (ppm) as the best N-P-K ratio for optimal plankton production. They further recommend that since ammonia is an index of pollution (unpolluted water contains less than 0.5 ppm total dissolved ammonia), waters should have less than 2 ppm for healthy growth. However, concentrations of 4 ppm at pH 7.3 to 7.5 have not affected the growth of Tilapia, common carp, bighead carp, grass carp, silver carp, and black carp.

Potassium

Seymour (1980) considers potassium treatment to be "groundless", yet Hora and Pillay (1962) consider it an important factor in the stimulation of aquatic flora. The general consensus is that potassium is not a limiting factor for most fish production ponds (Hickling 1962).

Table 4

SUMMARY OF POND FERTILIZATION EXPERIMENTS USING N, P, K COMPARED WITH THOSE USING P AND K OR P ONLY FOR SEVERAL SPECIES OF FISH[a]

Fish	Nitrogen applied, kg/ha	Nitrogen source	Increase in production, %
Cross-bred Tilapia	28	Calcium nitrate	10.1[b]
Hybrid male Tilapia	28	Urea	13.1[b]
Cyprinus carpio	90	Ammonium nitrate	11.7[b]
Cyprinus carpio	90	Ammonium nitrate	13.3[b]
Carassius auratus	90	Ammonium nitrate	13.9[b]
Ictalurus punctatus	90	Ammonium nitrate	0.4[b]
Micropterus salmoides and Lepomis macrochirus	90	Ammonium nitrate	33.6[b]
Tilapia aurea	19.9	Ammonium nitrate	31.3[c]
Tilapia aurea	67.5	Ammonium nitrate	30.0[c]

[a] From Boyd 1976

[b] Increase over P fertilizer only

[c] Increase over P and fertilizer only

ORGANIC FERTILIZERS

In most tropical countries, organic manures are far more commonly used than inorganic fertilizers. Organic manures include the following:

Green Manure

Fertilization through green manure (plant wastes) is sometimes all that is possible in depressed regions of the tropics. Animal manure, compost, and guano are generally in short supply and are reserved for fields and gardens. In ponds, rotting plant tissues often serve as a substrate for a host of aquatic invertebrates that may be a direct source of fish food.

Sewage Water

Water-borne sewage disposal has become a serious pollution problem in many parts of the world. As a partial remedy, these wastes may be recycled into fish flesh. For example, the fertility of brackish-water fishponds in Manila Bay in the Philippines may be attributed to sewage inflow from the shoreline's inhabitants. Nash and Brown (1980) and Edwards (1980) have reviewed the theory, feasibility, and practices of recycling animal, human, and agro-industrial wastes for fishpond fertilization.

Liquid Manure

Liquid manures come from the anaerobic fermentation of animal manure in biogas plants or from mixing fresh manure with water. Liquid manure is a powerful stimulant to phytoplankton growth producing high fish yields. The recommended application is through small, frequent dosages added in the deeper parts of the pond so that little is wasted on shoreline macrophytes (Hickling 1962). Liquid manure from biogas digesters may be easily handled through pipes and canals; animal manures mixed with water require transport, handling, and bulk storage facilities, and their availability is often limited.

Animal Manure

Animal manures are the most common manures used in fish pond work. They may be applied as compost, liquid manure, or in a fresh, untreated form. In China, all three forms are utilized. The Chinese composting procedure consists of mixing cow and chicken manure with plant materials and soft mud in a composting pit and allowing it to ferment for ten days (Tapiador et al. 1977).

PATHWAY OF ORGANIC FERTILIZERS

The pathways of organic material entering the pond food web have been outlined by Tang (1970): (1) the material enters as a source of nutritive substances (e.g., carbon, phosphorus) for photosynthesis in chlorophyll-bearing plants, (2) serves as an organic substrate for microorganisms which, in turn, support a zooplankton population, or (3) it may be directly consumed by fish, crustaceans, or insects.

Tang's (1970) polyculture experiment indicated that only half of the total fish growth (30 kg/ha/d) could be attributed to the consumption of natural food organisms like plankton or insects. The other half came from the direct consumption of organic materials like night soil. In another study, Noriega-Curtis (1979) observed that fish yields considerably exceeded predicted yields based upon primary production models. He concluded that manuring not only enhances auto-trophic production but also promotes fish growth through an "alternative" pathway. Other investigators concluded that this discrepancy in fish yields could be accounted for by the consumption of heterotrophic bacteria and protozoa flourishing on the organic substrate (Schroeder 1978, 1980). If autotrophic production were solely dependent on the amount of solar energy incident to a pond, there would be predictable upper limits to fish growth. However, since heterotrophic production does not suffer from this limitation, the results of Tang (1970) and Noriega-Curtis (1979) could be explained by the presence of additional forage provided by bacteria and protozoa in the dietary regimen of pelagic and bottom feeding fishes.

POND SOILS

There are instances when the productivity of an unfertilized pond is equal to or greater than that of a fertilized one. The reason is often found in the chemical and physical properties of the soil. Pond soils, as outlined by Potter (1976):

1. Store and release plant nutrients.
2. When dried, allow bacteria to mineralize organic bottom deposits.
3. Are the substrate and source of nutrients for bottom dwelling animals and plants.
4. Regulate "water quality" (e.g., dissolved oxygen, ammonia, hydrogen sulfide).

5. Are the substrate of pond and diking material.

6. Can act as a reservoir for harmful pesticides and algicides used in the pond.

Soils in Brackish-water

Where the growth of benthic vegetation is considered, soils are more significant for productivity in brackish-water culture than in a freshwater pond. This is why milkfish culturalists in southeast Asia choose rich alluvial soil sites for their brackish-water ponds. The different soil types and textures are classified in Table 5.

The most suitable algal pastures for Taiwan milkfish culture appear to be silty loams, sandy loams, loam, and silt (Tang and Chen 1967). These soils keep seepage to a minimum and form strong dikes (Chen 1970); sandy and loamy-sandy are poor soils because of their light texture and their low clay and fine silt content, which form soil colloids, the most active portion of the bottom soil. The highest fertility in Indonesia is derived from juvenile volcanic soil rich in nutrients; lowest in fertility are senile lateritic soil, rock, or sand (Djajadiredja and Poernomo 1970). The clay loam soils of the deltaic regions in Bengal are considered very fertile (Hora and Pillay 1962). In general, it appears that soil textures ranging from fine sandy loam to clay loam are the most suitable for brackish-water milkfish culture.

Tang and Chen (1967) state that growth of benthic algae in brackish-water ponds is related to the nitrogen content, which is directly proportional to the amount of organic matter present. Ghosh (1975) demonstrated that there is an increased rate of nitrogen mineralization from manures with increasing salinity. In considering that the optimal ration of N:P for phytoplankton growth is 4:1, Djajadiredja and Poernomo (1970) found that a lack of nitrogen in certain Indonesia brackish-water pond soils was a more limiting factor than phosphate deficiency. In a Taiwan pond, available phosphorus for algal pastures appeared plentiful (42 ppm); however, the addition of organic or inorganic phosphorus, alone or in combination with nitrogen, still improved algal growth (Tang and Chen 1967). The potassium content of sandy soils is often poor; clay soils are richer. However, this is relatively unimportant in brackish-water culture, because saline waters are generally adequate in the essential elements (e.g., potassium in the form of potassium chloride) necessary for algal growth.

Soil pH can control water pH and the availability of algal nutrients. In

Table 5

SOIL TYPES AND TEXTURES[a]

Common names	Texture	Basic soil textural class names
Sandy soil	Coarse	Sandy Loamy sands
	Moderately coarse	Sandy-loam Fine sandy loam Very fine sandy loam
		Loam
Loamy soils	Medium	Silt loam Silt Clay loam
	Moderately fine	Sandy clay loam Silty clay loam Sandy clay
Clayey soils	Fine	Silty clay Clay

[a] From Potter 1976

tropical countries, the soil may often be poor in calcium but rich in ferric, alumina, or manganese compounds that render the waters acid due to the slight colloidal solution produced (Hora and Pillay 1962; Potter 1976). Other reasons for low pH may be the oxidation of sulfur compounds in decaying plant materials (e.g., mangrove muds) or peaty bottoms formed by plant debris that has not decomposed. The best ways to control soil acidity are (1) regular pond draining to maintain productive capacity through alternative periods of mud formation and mineralization (Hora and Pillay 1962), and (2) liming (Potter 1976).

Soils in Freshwater

In stagnant freshwater ponds, nutrients supplied from the pond bottom are usually more significant than those supplied from inflowing water (Hickling 1962). However, fertile soil is generally allocated for agricultural purposes instead of freshwater culture due to its better income.

Liming

Although not strictly a fertilizer, lime can increase phytoplankton production, which in turn leads to increased fish production (see Water Quality Section). The many purposes of liming may be summarized as follows (Hora and Pillay 1962; Woynarovich 1975; Boyd 1979):

1. The toxic and caustic action of liming kills bacteria, fish parasites, and their intermediate hosts.

2. In acid waters where many fish species grow poorly or do not survive, liming neutralizes undesirable iron compounds and buffers the pH to an acceptable alkaline level.

3. Liming reduces the potential of oxygen depletion by making carbon dioxide available for photosynthesis.

4. Liming may precipitate excess dissolved organic material that normally contributes to turbidity (e.g., humic stains), which interferes with light penetration for photosynthesis.

5. Liming improves soil condition and promotes the bacterial breakdown of waste material.

6. Liming assists in the release of nutrients from the soil. For example, bottom mud pH is increased, and this then increases the availability of phosphorus.

Boyd (1979) claims that a total alkalinity of 20 ppm is sufficient for the consistent production of plankton; below this level unlimed ponds experience variable responses to fertilization. There is little or no benefit to excessive liming; in fact, it may cause the precipitation of phosphorus as insoluble calcium phosphate (Hepher 1958) and thus decrease expected fish yields. Liming to neutrality gives a linear relationship between the fish crop and the total amount of phosphate added (Djajadiredja and Poernomo 1970). Therefore, liming needs should always be based on the total alkalinity as well as the nature of the pond soil (Table 6).

The types of lime available are limestone (calcium carbonate), hydrated or slaked lime (calcium hydroxide), and quicklime (calcium oxide). Limestone in the form of finely ground particles (<0.25 mm) has a high neutralizing value and is the standard by which other liming materials are measured. It is also the first choice for fish ponds. If slaked lime or quicklime is used in large quantities, the pH may be so high as to damage the fine tissues coating the gills in fish and thus cause the fish to die.

Lime can be applied on the pond bottom, added to the water at the inlet, or broadcast on the water surface. Hora and Pillay (1962) have recommended that to control the growth of parasites and to improve the pond bottom, the pond bed and not the water column should be treated. For example, slaked lime or quicklime should be applied after a pond has been drying for two weeks. When the control of gill rot disease or the precipitation of organic substances is desired, the water itself should be treated. In tropical countries, a proper fertilization program must be implemented a few weeks after liming to achieve increased pond productivity.

Acid Sulfate Soils

Acid sulfate soils are formed from marine and estuarine sediments that undergo extreme acidification due to the oxidation of sulfides, mainly pyrite (FeS_2). The oxidation is typically caused by drying and aeration from the tidal fluctuations of saline water or from pond draining. These soils are usually located in low coastal areas (e.g., mangroves) where topographic and hydrologic conditions are ideal for

Table 6

LIMING RECOMMENDATIONS FOR AQUACULTURE PONDS[a]

	Lime requirements in hundreds of kg of calcium oxide/ha				
pH of soil	Heavy clay or loam	Loamy sand	Sand	pH of mud	Calcium carbonate required in hundreds of kg/ha
More acid than 4	40	20	12.5	Less than 4	60 - 120
4 - 4.5	30	15	12.5	4.0 - 4.5	48 - 96
4.5 - 5	25	12.5	10.0	4.5 - 5.0	36 - 72
5 - 5.5	15	10	5	5.0 - 5.5	30 - 48
5.5 - 6	10	5	2.5	5.5 - 6.0	16 - 30
6 - 6.5	5	5	0	6.0 - 6.5	14 - 16

[a] From Hora and Pillay 1962

brackish-water milkfish culture. Unfortunately, these areas are also poor in productivity and are generally unsuitable for fish culture because of the following reasons (Singh 1980):

1. Low pH (<4.0) due to the presence of sulfuric acid and iron and aluminum sulfates.

2. The toxic effect, especially to many plants, of excess iron and aluminum.

3. Low nutrient status (e.g., binding of phosphorus to excess iron and aluminum) and micronutrient deficiencies leading to low food web productivity.

4. Poor physical soil conditions.

In tropical Asia, less than 2 million of the 15 million hectares of acid sulfate soils are under cultivation.

When pH is increased through reclamation treatments, concentrations of aluminum and iron are reduced, while the availability of phosphorus is increased. Therefore, indicated pond management recommendations according to Philippine research (Singh 1980) are as follows:

1. Tilling the pond bottom to speed up the oxidation process of pyrite, then leaching to remove acidity. The periodic drying and flushing with seawater of both the pond bottom and dikes is advised; a pH shift of 3.70 to 6.25 was recorded after this treatment.

2. Liming and proper fertilization must be followed up for the healthy growth of benthic algae. Repeated application of nitrogen and phosphorus in small quantities appears more effective than one large dosage.

Table 7 summarizes some of the results of treating acid sulfate soils in brackish-water ponds.

Apparently the Philippines is the only country currently conducting research on the reclamation of acid sulfate soils for fish culture. Although these soils can clearly be improved for fish cultivation, optimal treatments have not yet been developed and the costs involved are unknown.

APPLICATION OF FERTILIZERS

The purpose of fertilization in freshwater ponds is different in concept from that in brackish-water milkfish ponds. In freshwater culture the aim is to increase the production of planktonic algae, whereas in brackish-water culture it is to increase benthic algal growth for browsing milkfish.

Secchi Disc

An important tool to consider at this point is the Secchi disc. Secchi transparency provides crude guidelines for the proper rate and amount of fertilizer treatment. Melack (1976) demonstrated that fish production is, as expected, related to primary productivity. Almazan and Boyd (1978) found that the coefficient of determination of

Table 7

TOTAL FISH PRODUCTION IN PONDS GIVEN VARIOUS LEVELS OF LIME AND CHICKEN MANURE INPUTS FOLLOWING TILLING AND REPEATED LEACHINGS[a]

Treatment	Production before treatment application[b]	Production after treatment application[c]
1. No lime; 2t/ha chicken manure	336.5	419.3
2. 4t/ha lime; 2t/ha chicken manure	478.5	555.0
3. 8t/ha lime; 2t/ha chicken manure	402.7	565.6
4. 4t/ha lime; 8t/ha chicken manure	689.0	929.8
5. 8t/ha lime; 8t/ha chicken manure	412.2	805.0

[a] From Singh 1980

[b] Before tilling and leaching

[c] After tilling and leaching

Tilapia yields was related to primary productivity as measured by Secchi disc transparency ($r^2 = 0.71$), chlorophyll *a* concentrations ($r^2 = 0.89$), actual phytoplankton counts ($r^2 = 0.78$), and through the light-dark bottle method ($r^2 = 0.79$). Since the last three methods are not practical for the fish culturalist, the Secchi disc is the most suitable index of plankton abundance when plankton is the primary source of pond turbidity. Stickney (1979) recommends a depth of 30 cm to achieve and maintain proper fertilization. When a Secchi disc is not available, the rule of thumb is to submerge one's arm up to the elbow and note the disappearance of the hand.

Brackish-water Inorganic Fertilization

Presently large-scale brackish-water milkfish culture exists only in Indonesia, the Philippines, and Taiwan. In Indonesia, some inorganic fertilizers like urea are toxic to fish; thus, ammonia-carrying fertilizers are preferred. Since "kelekap" or benthic algae growth usually lasts for only 1.5 to 2 months after chemical fertilization, Djajadiredja and Poernomo (1970) have formulated these application techniques:

1. Prepare the soil by draining and tilling to make it soft and colloidal; this stimulates "kelekap" and prevents the occurrence of the snail pest Cerithidae, which is a food competitor. Biodegradable tea seed cake (200 kg/ha) or tobacco waste (100-200 kg/ha) is also effective in eradicating this snail. Cerithidae shell formation can deplete calcium supplies and has been observed in densities of 700 per square meter (Bardach et al. 1972).

2. After tilling and raking, Bardach et al. (1972) state lime treatments are applied to wet, foul-smelling spots to prevent anaerobic decay and production of hydrogen sulfide. The lime also kills potential predators.

3. Apply fertilizers before stocking with water level at 1-3 cm and salinity at 15-30 ppt. Several investigators favor water depths ranging from 1 to 50 cm. Regardless, the important guidelines here are adequate light penetration to the bottom and the prevention of phytoplankton blooms.

4. Wait at least one week after fertilization before stocking fish lest toxic effects occur. The best time to stock is when "kelekap" is at its maximum growth, which is about 2-3 weeks after fertilization.

These procedures were designed to maintain an adequate supply of algal pasture for 11-12 weeks, the time it takes to rear fry to the advanced fingerling stage of 8-12 cm. In the Visaya Islands, Philippines, fertilizer grades and applications recommended for "lab lab" production (benthic algal growth) are shown in Table 8. The use of inorganic fertilizers in brackish-water milkfish culture has been largely determined empirically.

Although chemical fertilizers are strongly promoted in Taiwan (Bardach et al. 1972), they are rarely used as replacements for organic materials in milkfish culture (Chen 1970). Nevertheless, the recommended conditions for chemical fertilization during the rearing season when pond water levels are full are (1) clear water, (2) salinity greater than 15 ppt, and (3) a cloudless day for maximum light penetration.

Brackish-water Organic Fertilization

Manuring of nursery ponds in Indonesia is not recommended because of algal blooms that cause heavy fry mortality. In contrast, Taiwan nursery ponds are manured because blooms are considered less of a problem in the more temperate climate. Taiwan manuring may consist of rice bran placed in 22 to 30 kg bags, perhaps enriched with human waste, straw or oil cakes, and applied at 400-1000 kg/ha. The bags are then soaked in water, cut open, and the fertilizer spread around (Bardach et al. 1972).

The production ponds of Indonesia are fertilized with green manures stacked in heaps with a topping of mud to prevent drifting away at 2000 kg/ha, while in the Philippines green manure or copra slime is applied at 450-900 kg/ha (Bardach et al. 1972). In Taiwan, commercial growers continue to rely on experience and fertilize with available materials (rice bran; legume seed; soybean or peanut cake; human, pig, or chicken manures; etc.).

Organic material provides a substrate upon which benthic growth may thrive; its benefits therefore are especially realized in new ponds. However, BFAR (1976) of the Philippines concludes that inorganic fertilizers are generally more profitable than organic ones, except for new ponds and others that have low levels of rapidly decaying organic matter. The reason for this is the high cost of assembling, processing, storing, transporting, and applying organic fertilizers. This contrast with the aforementioned Taiwanese practice of rarely using chemical fertilizers is just another example of the different strategies necessitated by local pond conditions and socio-economic environments.

Freshwater Inorganic Fertilization

Hora and Pillay (1962) have recommended chemical fertilizers are best applied by raking them into a drained pond. Alternately, in ponds full of water, fertilizer may be spread from boats or by hand broadcasting. In new ponds, fertilizer should be applied 2 to 3 times per week, then later at monthly intervals. Secchi disc transparency of 45 cm is considered proper management of the phytoplankton population. Care must be taken against overfertilization, which can result in phytoplankton blooms with associated toxicity, oxygen depletion, and shading problems.

Table 8

"LAB-LAB" FERTILIZATION PROGRAM[a]

Fertilizer grade	Rate, kg/ha	Time/method of application
PRE-STOCKING:		
18-46-0	100-200	Broadcast, then admit water immediately into the pond.
18-46-0	50-100	Apply every 10-15 days up to 1 week before stocking.
POST-STOCKING:		
18-46-0	15-25	Apply 1 week after stocking. Repeat every 10-15 days interval up to harvest.

[a] From Ballesteros and Mendoza 1976

In Alabama, Boyd (1979) reports that large fertilizer applications over long intervals are considered wasteful because of phosphorus adsorption to muds and denitrification of nitrogen. Thus, the traditional practice is to broadcast fertilizers over shallow areas every two to four weeks. A more efficient way of preventing phosphorus from being adsorbed is to place fertilizers onto underwater platforms 30 cm below the water surface where currents can distribute the nutrients as they dissolve. Two to four platforms per hectare of pond area are sufficient.

Freshwater Organic Fertilization

The old practice of organic manuring involves either scattering or heaping materials onto the bottom of drained ponds (Hora and Pillay 1962). This has disadvantages when the ponds are filled; scattered manure on the pond bottom is not amenable to aerobic digestion because of the low dissolved oxygen usually present in the muds. In addition, heaps are inefficient because aerobic digestion can only occur on the outer surface, even though widespread deoxygenation is prevented. Conditions become anaerobic and may potentially produce three toxic agents: hydrogen sulfide, ammonia, and methane gas (Schroeder 1980).

Instead, Woynarovich (1975) states that manures should be finely distributed in the water column, where abundant populations of reducing bacteria exist to decompose organic material into simpler compounds for immediate utilization by phytoplankton. The manure should be distributed over as much of the pond as possible; dispersal from a boat or from a hanging basket filled with manure and buoyed by an old car innertube, pulled to and fro from the shore.

Schroeder (1978) has found that as much as 40 percent of the total solids in fresh cow manure can remain suspended in the water column with 50-60 percent of this organic matter composed of inorganic minerals. Within one to two hours, 90 percent of the coarse organic particles will settle to the bottom, resulting in anaerobic digestion when a layer of more than a few mm. accumulates. Therefore, from Israeli experience, the maximum amount of manure that a pond can safely digest without undesirable anaerobic effects is about 70 to 140 kg/ha/d.

Wohlfarth and Schroeder (1979) have reviewed worldwide the frequency of manuring and found that it varies from daily (through weekly, fortnightly, six-weekly, three-monthly, and six-monthly) to annually. This variation is, of course, not unexpected in view of the varying manure types, pond conditions, climates, and such social factors as labor costs, convenience, and need of disposal. Table 9 summarizes the advantages and disadvantages of organic versus inorganic fertilizers.

INTEGRATED AGRICULTURE-AQUACULTURE FARMING

Integrated agriculture-aquaculture farming is best exemplified by the workable patterns that have been practiced for centuries by the Chinese. Human protein needs are supplied by livestock, fowl, or fish, which are fed aquatic plants, crop wastes, and kitchen leftovers. The animal manures, in turn, serve as fertilizer for the vegetable crops and fish ponds. The water from manured ponds is also used for irrigation. Unfortunately, most of these integrated systems are subsistence-level operations based on empirical experience, and there is little detailed information regarding technology, economics, and yields. However, ecologically balanced animal-plantfish farming can produce yields comparable to intensive fish culture if supplementary feeds are used; the nominal cost of manures increases the potential profitability (Delmendo 1980).

The quantity and quality of manures is determined by the animal's total live weight and the type of feed it consumes. The quality (potential biological activity) is reflected by the biochemical oxygen demand (BOD), where a higher BOD implies the rapid digestion and conversion of the organic matter by microorganisms in the receiving waters. For example, poultry manure generally has a higher BOD than cow manure (because of poultry feed's better quality); this would require a great deal more care in application in order to prevent oxygen depletion. Schroeder (1980) has listed the general values of manures in increasing order as cow and sheep manure, followed by a grouping of pig, chicken, and duck manure.

In terms of quantity, Delmendo (1980) has estimated that the total annual tonnage of manure production per animal in China is 6.0 tons for cows, 3.0 tons for pigs, 0.8 tons for goats or sheep, and 0.025 tons for poultry. The total annual amount of organic fertilizer (derived mainly from pigs) is about 1689 million tons, which is equivalent to 8,320,000 tons of nitrogen, 5,092,000 tons of phosphorus, and 9,671,000 tons of potassium. Based upon these values and the information from Table 3, the number of animals required to supply organic nutrients to a pond can be crudely estimated.

Table 9

ADVANTAGES AND DISADVANTAGES OF ORGANIC VERSUS INORGANIC FERTILIZATION IN FISH PONDS

Item	Organic	Inorganic
1. Quality of fertilizer based on N-P-K content	N-P-K value low	N-P-K value high
2. Quality of N-P-K within and between various sources	Highly variable	Always consistent
3. Serving as feed for direct consumption by fish and other fish food organisms	Yes	No
4. Presence of growth factors for promotion of algae growth	Present	Absent
5. Source of organic carbon for autotrophic and heterotrophic production	Yes	No
6. Substrate for benthic algae in brackishwater milkfish culture	Yes	No
7. Effects on physical structure of soil	Improvement	No effect
8. Fish kills from inappropriate application	1. Oxygen depletion from plankton bloom, 2. Toxic H_2S, NH_3, CH_4, 3. Pollution	Oxygen depletion from plankton bloom
9. Cost per unit N-P-K nutrient	Most expensive	Least expensive
10. Cost in assembling, processing, storing, and applying	High in terms of money, labor, facilities, and general unpleasantness	Low in terms of money, labor, facilities and general unpleasantness
11. Integrated agriculture-aquaculture farming	Low input cost using recycled wastes resulting in economically viable operation; costs minimized in (9) and (10)	Not applicable

Pig-fish farming is quite popular, with pig sties often located above or adjacent to fish ponds so that wastes may easily be washed down into the waters for fertilization. Although this practice is employed in many areas of the world, the number of pigs/unit area of pond has been standardized only in China (30-45 pigs/ha of pond) (Tapiador et al. 1977; Delmendo 1980). Additionally, there are systems involving poultry-fish farming (Woynarovich 1980), livestock-fish farming (Wohlfarth and Schroeder 1979), and combination livestock-fowl-fish farming (Delmendo 1980). Table 10 provides some examples of annual production data.

Besides animal-fish farming, there is plant-fish farming, predominantly, rice-fish systems (Coche 1967; Huet and Tan 1980). Two methods are generally practiced: (1) combined fish and rice culture, and (2) the rotational cropping of rice and fish (Cruz 1980). Terrestrial crops (beans, sweet potatoes) may be grown on paddy dikes while aquatic plants (e.g. *Ipomoea*, *Colocasia* sp.) are grown in the water. Since pesticides have become a serious problem (Koesoemadinata 1980), Cruz (1980) has recommended adopting rotational cropping of rice and fish to reduce possible accumulation in fish tissues. Estores et al.

Table 10

ANNUAL FISH PRODUCTION DATA OF WASTE UTILIZATION PROJECTS[a]

Annual production, kg/ha	Fish under culture	Manure source	Country
4,900	Carps	Fluid cowshed manure	Israel
3,500	Polyculture	Ducks	Southeast Asia
4,140	Carps, catfish, largemouth bass, buffalo fish	Swine manure	USA
2,729[b]	Silver carp, bighead carp	Sewage lagoons	USA
3,000	Polyculture	Domestic septic tank system	Java
3,700	Polyculture	Domestic wastewater storage reservoirs	Israel
1,000	Carp	Fish ponds receiving sewage waters	Germany
2,000	Tilapia	Pig manure	Rhodesia
4,000	Tilapia	1,000 ducks	Rhodesia
3,000	Tilapia	Compost and farmyard manure	Madagascar
1,300	Carp	Town sewage effluents	Poland

[a] From Nash and Brown 1980

[b] Area equivalents

(1980) report that the pesticide carbofuran is nontoxic to fish and leaves no residue.

FUTURE RESEARCH NEEDS

Optimizing fish production in a dynamic pond ecosystem by the use of fertilizers is an extremely difficult task. Since a pond's interacting biological, chemical, and physical factors are not always understood or known, the estimation of fertilizer requirements and application rates have often been determined empirically rather than scientifically.

Fertilizer problems meriting attention are currently more basic than applied. Standardized methods are needed to monitor fertilizer pathways (e.g. radioactive tracers) to quantify the production and its utilization by the microbiological and physical community. This should help shed light on the rate of cycling and necessity of potassium and nitrogen nutrients, plus provide comparisons of the short and long-term efficiency of organic versus inorganic fertilizers.

The prospect of future energy costs behooves the further use and development of organic fertilizers in integrated agriculture-aquaculture farming systems. However, it is important to realize that because of local socio-economic-environmental conditions, recommendations in one country may be entirely unsuitable for another country.

LITERATURE CITED

Almazan, G., and C.E. Boyd. 1978. Plankton production and tilapia yield in ponds. Aquaculture 15:75-77.

Ballesteros, O.Q., and S.P. Mendoza, Jr. 1976. Brackish-water fishpond management, pp. 19-27. In Pond Construction and Management. Western Visayas Federation of Fish Producers, Inc., Philippines.

Bardach, J.E., J.H. Ryther, and W.O. McLarney. 1972. Aquaculture: The Farming and Husbandry of Freshwater and Marine Organisms. John Wiley and Sons, New York. 868 pp.

BFAR. 1976. Fertilizers and fertilization of brackish-water fishponds, pp. 97-102. In Pond Construction and Management. Western Visayas Federation of Fish Producers, Inc., Philippines.

Boyd, C.E. 1976. Nitrogen fertilizer effects on production of Tilapia in ponds fertilized with phosphorus and potassium. Aquaculture 7:385-390.

Boyd, C.E. 1979. Water Quality in Warmwater Fish Ponds. Auburn University Agricultural Experiment Station, Alabama. 359 pp.

__________, and F. Lichtkoppler. 1979. Water Quality Management in Pond Fish Culture. Auburn University Experiment Station, Alabama. 30 pp.

Chen, T.P. 1970. Fertilization and feeding in coastal fish farms in Taiwan, pp. 410-416. In T.V.R. Pillay (ed), Coastal Aquaculture in the Indo-Pacific Region. FAO Fishing News Books Ltd. London, England.

Coche, A.G. 1967. Fish culture in rice fields: a worldwide synthesis. Hydrobiologia 30:11-44.

Cruz, de la C.R. 1980. Integrated farming with fish as the major enterprise. Asian Aquaculture 3(3):3-7.

Delmendo, M.N. 1980. A review of integrated livestock fowl-fish farming systems, pp. 59-71. In S.V. Pullin and Z.H. Shehadeh (eds), Integrated Agriculture-Aquaculture Farming Systems. International Center for Living Aquatic Resources Management (ICLARM) Proceedings 4, Manila, Philippines.

Djajadiredja, R., and A. Poernomo. 1970. Requirements for successful fertilization to increase milkfish production, pp. 398-409. In T.V.R. Pillay (ed), Coastal Aquaculture in the Indo-Pacific Region. FAO Fishing News Books Ltd., London, England.

Edwards, P. 1980. A review of recycling organic wastes into fish, with emphasis on the tropics. Aquaculture 21:261-279.

El Zarka, S.E., and F.K. Fahmy. 1968. Experiment in the culture of the grey mullet Mugil cephalus in brackish-water ponds in U.A.R. FAO Fish. Rep. 44(5):255-266.

Estores, R.A., F.M. Laigo, and C.I. Adordioinisio. 1980. Carbofuran in rice-fish culture, pp. 53-57. In S.V. Pullin and Z.H. Shehadeh (eds), Integrated Agriculture-Aquaculture Farming Systems. International Center for Living Aquatic Resources (ICLARM) Proceedings 4, Manila, Philippines.

Fijan, N. 1967. Problems in carp pond fertilization. FAO Fish. Rep. 44(3):114-123.

Ghosh, S.R. 1975. A study on the relative efficiency of organic manures and the effect of salinity on its mineralization in brackish-water fish farm soil. Aquaculture 5:359-366.

Hepher, B. 1958. On the dynamics of phosphorus added to fishponds in Israel. Limnol. Oceanogr. 3:84-100.

__________. 1962. Ten years of research in fish pond fertilization in Israel. I. The effect of fertilization on fish yields. Bamidgeh 14:29-38.

Hickling, C.F. 1962. Fish Culture. Faber and Faber, London. 295 pp.

Hora, S.L., and T.V.R. Pillay. 1962. Handbook on Fish Culture in the Indo-Pacific Region. FAO Fisheries Biology Technical Paper No. 14. 204 pp.

Huet, K.H., and E.S.P. Tan. 1980. Review of rice-fish culture in Southeast Asia, pp. 1-14. In S.V. Pullin and Z.H. Shehadeh (eds), Integrated Agriculture-Aquaculture Farming Systems. International Center for Living Aquatic Resources Management (ICLARM) Proceedings 4, Manila, Philippines.

Koesoemadinata, S. 1980. Pesticides as a major constraint to integrated agriculture-aquaculture farming systems, pp. 45-51. In S.V. Pullin and Z.H. Shehadeh (eds), Integrated Agriculture-Aquaculture Systems. International Center for Living Aquatic Resources Management (ICLARM) Proceedings 4, Manila, Philippines.

Lahnovitch, V.P. 1968. Theoretical Bases for Fertilizing Ponds. REP. FAO/UNDP (TA) (2547), 183 pp.

Lin, S.Y. 1968. Milkfish farming in Taiwan fish culture report. Taiwan Fisheries Research Institute 3:43-48.

Melack, J.H. 1976. Primary productivity and fish yields in tropical lakes. Trans. Am. Fish. Soc. 105:575-580.

Nash, C.E., and C.M. Brown. 1980. A theoretical comparison of waste treatment processing ponds and fish production ponds receiving animal wastes, pp. 87-97. In S.V. Pullin and Z.H. Shehadeh (eds), Integrated Agriculture-Aquaculture Farming Systems. International Center for Living Aquatic Resources Management (ICLARM) Proceedings 4, Manila, Philippines.

Noriega-Curtis, P. 1979. Primary productivity and related fish yield in intensely manured fish ponds. Aquaculture 17:335-344.

Potter, T. 1976. Seminar on fish pond soil quality, pp. 175-186. In Pond Construction and Management, Western Visayas Federation of Fish Producers, Inc., Philippines.

Schroeder, G.L. 1978. Autotrophic and heterotrophic production of microorganisms in intensely manured fish ponds, and related fish yields. Aquaculture 14:303-325.

______________. 1980. Fish farming in manure-loaded ponds, pp. 7386. In S.V. Pullin and Z.H. Shehadeh (eds), Integrated Agriculture-Aquaculture Farming Systems. International Center for Living Aquatic Resources Management (ICLARM) Proceedings 4, Manila, Philippines.

Seymour, E.A. 1980. The effects and control of algal blooms in fish ponds. Aquaculture 19:55-74.

Singh, V.P. 1980. Management of fishponds with acid sulfate soils. Asian Aquaculture 3(4):4-6; 3(5):4-6; 3(6):4-6.

Sinha, V.R.P. 1979. Contribution of supplementary feed in increasing fish production through composite fish culture in India, pp. 565-574. In J.E. Halver and K. Tiews (eds), Finfish Nutrition and Fishfeed Technology. H. Heenemann Gmbh and Co., Berlin. Vol. I.

Stickney, R.R. 1979. Principles of Warmwater Aquaculture. John Wiley and Sons, New York. 375 pp.

Tang, Y.A. 1970. Evaluation of balance between fishes and available fish foods in multispecies fish culture ponds in Taiwan. Trans. Am. Fish. Soc. 99:708-718.

__________, and S.H. Chen. 1967. A survey of the algal pasture soils of milkfish ponds in Taiwan. FAO Fish Rep. 44(3):198-209.

Tapiador, D.D., H.F. Henderson, M.N. Delmendo, and H. Tsutsui. 1977. Freshwater Fisheries and Aquaculture in China. FAO Fisheries Technical Paper No. 168, FIR/T168. 84 pp.

van der Lingen, M.I. 1967. Fertilization in warmwater pond fish culture in Africa. FAO Fish. Rep. 44(3):43-53.

Wahby, S.D. 1974. Fertilizing fish ponds. I. Chemistry of the waters. Aquaculture 3:245-259.

Wohlfarth, G.W., and G.L. Schroeder. 1979. Use of manure in fish farming - a review. Agricultural Wastes 1(3):297-299.

Woynarovich, E. 1975. Elementary Guide to Fish Culture in Nepal. FAO. 131 pp.

_____________. 1980. Raising ducks on fish ponds, pp. 129-134. In S.V. Pullin and Z.H. Shehadeh (eds), Integrated Agriculture-Aquaculture Farming Systems. International Center for Living Aquatic Resources Management (ICLARM) Proceedings 4, Manila, Philippines.

POND PRODUCTION SYSTEMS:
FEEDS AND FEEDING PRACTICES IN WARMWATER FISH PONDS

by

Randolph Yamada

INTRODUCTION

Finfish nutrition research, which is based on studies of the intake, digestion, and metabolic utilization of foods or feed, did not start producing usable results until after World War II. Fish husbandry studies could have been patterned after agricultural work, but progress was slowed because of the adaptations necessitated by a fish's poikilothermic nature and its unique aquatic environment (Utne 1979). Although many deficiencies still exist in the knowledge of feeds and feeding practices, the recent reviews on fish nutrition by Cowey and Sargent (1972), Halver (1972), the National Academy of Sciences (1973, 1977), Braekkan (1977), and Halver and Tiews (1979) are an impressive testimony to the endeavors in this area.

FINFISH NUTRITIONAL REQUIREMENTS

A complete fish diet must provide a suitable energy source and be in proper balance with respect to proteins, carbohydrates, lipids, and the growth factors vitamins and minerals. Precise nutritional requirements are difficult to ascertain because they change with variations in the environment, fish size/age, and reproductive condition. Until recently, a major problem for comparative studies has been the lack of standardization (Harris 1980; Utne 1979). Present knowledge of fish nutrition, primarily derived from studies of the rainbow trout (Salmo gairdneri) and channel catfish (Ictalurus punctatus), concerns requirements of the ten essential amino acids, gross protein levels, water and fat soluble vitamins, and some essential polyunsaturated fatty acids of the omega-3 and omega-6 series. Although very little is known about the nutritional requirements of almost all cultured fish species, it appears there is little variation in nutritional needs within the warmwater and coldwater fishes. The major difference probably concerns the essential fatty acids or lipids (Lovell 1979a). Thus rainbow trout and channel catfish may be used as models from which to refine recommended allowances for other species.

Proteins

Proteins are a significant dietary component because of their cost and their constraints on growth. The gross protein requirements in fish are higher than in warm-blooded animals (Lovell 1979a). Table 1 summarizes the requirement of certain fish species; surprisingly high are the herbivores, Ctenopharyngodon idella and Brycon sp.

Lovell (1979a) states that protein levels of 30-36 percent will probably be adequate for most warmwater fish diets. The optimum level is influenced by several factors:

1. Fish size. Young fish have higher protein requirements than older fish.

2. Physiological function. Less protein is needed for a maintenance diet than for rapid growth.

3. Protein quality. More low-quality protein is needed for maximum growth than high-quality protein.

4. Non-protein energy in the diet. If its diet is deficient in energy, a fish will use part of its protein to meet its energy needs.

5. Feeding rate. Fish fed to less than satiation (e.g., in intensive culture) will benefit more from diets containing a high percentage of protein than fish fed at or near satiation rate.

6. Natural foods. If natural food contributes significantly to daily intake, then protein level in prepared diets may be lower.

7. Economics.

The protein requirements of euryhaline cold-water rainbow trout and coho salmon do not differ between freshwater and 20 ppt; full-strength sea water has not been examined (Zeitoun et al. 1973, 1974). The requirements for euryhaline warmwater species like Tilapia sp. have never been tested in different salinities.

For optimal utilization of dietary protein, the amino acid profile of the feed should closely resemble the ten essential amino acid requirements of the fish. As shown in Table 2, real differences exist between species. Thus it is quite difficult to formulate practical diets for fish whose amino acid requirements are unknown.

Table 1

GROSS PROTEIN LEVELS FOR CERTAIN FISH SPECIES[a]

Species	Crude protein level in diet for optimal growth, g/kg
Rainbow trout (Salmo gairdneri)	400-460
Carp (Cyprinus carpio)	380
Chinook salmon (Oncorhynchus tschawytscha)	400
Eel (Anguilla japonica)	445
Plaice (Pleuronectes platessa)	500
Gilthead bream (Chrysophrys aurata)	400
Grass carp (Ctenopharyngodon idella)	410-430
Brycon sp.	356
Red sea bream (Chrysophrys major)	550
Yellowtail (Seriola quinqueradiata)	550

[a] From Cowey 1979

When a diet is deficient in one or more amino acids, it may be possible to supplement it in the appropriate amounts. Andrews and Page (1974) substituted soybean meal fortified with methionine, cystine, or lysine (the most limiting amino acids) for menhaden meal in a channel catfish diet. The growth and feed efficiency were reduced. A similar result was obtained in young common carp when amino acid mixtures were used to replace the protein components of casein and gelatin (Aoe et al. 1970). In contrast, the chinook salmon is able to grow well with supplemental amino acids (Halver 1957). It is too early yet to attempt to explain why certain species can use free amino acids and others cannot.

The design of practical diets with regard to proteins is a compromise. Very high protein levels will result in the deamination of amino acids and the burning off of carbon residues for energy. Growth may subsequently be depressed by excess excretion of ammonia causing stress or gill damage in a confined, heavily stocked pond. Too little protein will result in the accumulation of fat, producing undesirable changes in the carcass composition.

Carbohydrates

Carbohydrates comprise a broad group of substances that include sugars, starches, gums, and celluloses. The omnivorous channel catfish and common carp, as well as the herbivorous grass carp, are able to digest carbohydrates of plant origin. By contrast, carnivorous fishes do not have the capacity to handle significant quantities of complex carbohydrates in their diet. They can, however, efficiently utilize the simple carbohydrates (glucose, sucrose, lactose) as a primary energy source.

The ability to assimilate starches depends on the enzymatic activity (production of amylase) of the fish. In herbivores, amylase is widespread throughout the entire digestive tract; in carnivores, it is primarily of pancreatic origin.

Although cellulase of bacterial origin is present in the gut of the common

Table 2

AMINO ACID REQUIREMENTS OF CERTAIN FISH[a]

Amino acid	Amino acid requirement, g/kg dry diet					
	Chinook salmon	Japanese eel	Carp	Channel catfish	Gilthead bream	Rainbow trout
Arginine	24	17	16		10.4	12
Histidine	7	8	8			
Isoleucine	9	15	9			
Leucine	16	20	13			
Lysine	20	20	22	12.3	20	
Methionine	16[b]	12[b]	12[b]		16[c]	
Phenylalanine	21[d]	22[d]	25[d]			
Threonine	9	15	15			
Trypotophan	2	4	3	2.4		
Valine	13	15	14			

[a] From Cowey 1979

[b] In the absence of cystine

[c] Methionine + cystine

[d] In the absence of tyrosine

carp (Schroeder 1978), cellulase and galactosidase are not normally secreted by fish. This lack of galactosidase may partially explain the poor growth response of fish fed soybean meal, which contains significant amounts of galactosidic oligosaccharides, raffinose and stachyose. Since oligosaccharides undergo enzymatic hydrolysis during the germination process to yield galactose and sucrose, Chow and Halver (1980) have suggested soaking soybeans for 48 hours prior to meal processing. This recommendation applies to most legume seeds, since a large portion of their carbohydrates are in the form of oligosaccharides.

Lipids

The optimal lipid requirements of almost all warmwater cultured species have not been determined. The establishment of lipid requirements is important because it costs less than proteins and because its high energy levels can spare protein. Homeothermic animal studies and the analysis of the fatty acid composition of fish oils have shown that fish lipids have a low omega-6 and high omega-3 polyunsaturated fatty acid content as compared to mammalian lipids, which have a high omega-6 and low omega-3. Mammals require omega-6 essential fatty acids (EFA) and fish require omega-3 (Cowey 1979; Halver 1979).

The fatty acid patterns differ between species as well as between freshwater and marine fish (Table 3), with marine fish having a higher omega-3 requirement than freshwater fish. Fatty acid patterns are also different in anadromous fish (masu salmon, *Oncorhynchus masu*) and catadromous fish (sweet smelt, *Plecoglosus altivelis*) migrating into different gradient salinities (Cowey 1979; Halver 1979). This is shown in Table 4. These differences may be a result of differences in the fatty acid content of their diets or of the specific dietary requirement related to physiological adaptation to the environment.

Temperature is another factor that appears to affect fatty acid composition in fish. The omega-3 requirement is greater for fish raised at a lower

Table 3

FATTY ACID PATTERNS IN FRESHWATER AND MARINE FISH[a]

Fatty Acid	Freshwater Fish						Marine Fish						
	Sheeps herd	Tullibee	Tullibee Flesh	Maria	Alewife	Rain-bow Trout	Atlantic Herring	Pacific Herring	Atlantic Cod	Chinook Salmon	Mack-erel	Menhadon	Deepsea Smelt
14:0	2.8	4.5	5.5	3.1	6.7	2.1	5.1	7.6	3.7	2.2	4.9	8.0	1.4
16:0	16.6	13.8	17.7	13.2	14.6	11.9	10.9	18.3	12.6	17.0	28.2	28.9	17.2
16:1	17.7	21.5	7.1	16.2	14.7	8.2	12.0	8.3	9.3	4.1	5.3	7.9	11.0
18:0	3.3	2.9	3.0	2.8	1.5	4.1	1.2	2.2	2.3	3.2	3.9	4.0	3.7
18:1	26.1	25.2	18.1	29.1	18.2	19.8	12.6	16.9	22.7	21.4	19.3	13.4	31.4
18:2ω6	4.3	1.9	4.3	2.2	3.7	4.6	0.7	1.6	1.5	2.0	1.1	1.1	0.2
18:3ω3	3.6	2.6	3.4	1.9	3.6	5.2	0.3	0.6	0.6	1.0	1.3	0.9	
18:4ω3	0.9	1.5	1.8	1.3	2.9	1.5	1.5	2.8	0.6	2.0	3.4	1.9	
20:1	2.4	1.3	1.2	1.2	1.6	3.0	16.1	9.4	7.5	5.4	3.1	0.9	4.8
20:4ω5	2.6	1.7	3.4	2.4	2.4	2.2	0.4	0.4	1.4	0.9	3.9	1.2	2.5
20:4ω3	0.7	0.8		1.1	1.5		0.4		0.6				
20:5ω3	4.7	6.2	5.9	5.5	8.2	5.0	7.4	8.6	12.9	6.7	7.1	10.2	3.6
22:1	0.3	0.3	2.8	0.3	0.4	1.3	19.8	11.6	6.2	9.4	2.8	1.7	2.5
22:5ω6	0.4	0.5		0.9	1.3	0.6	0.4		0.3	0.6		0.7	2.5
22:5ω3	2.0	1.8	3.3	2.4	1.5	2.6	1.1	1.3	1.7	2.3	1.2	1.6	0.3
2222:6ω3	2.0	3.8	13.3	7.8	6.0	19.0	3.9	7.6	12.7	16.1	10.8	12.8	15.0
ε sat	25.5	23.2	27.2	20.6	24.9	18.1	17.8	20.1	19.7	22.4	37.0	40.9	22.3
ε mono	49.1	49.6	33.6	48.7	36.5	32.3	61.5	46.2	47.1	40.3	30.5	23.9	49.7
ε ω5	8.5	5.4	9.9	6.7	9.4	8.0	1.9	2.0	3.7	4.2	5.0	3.0	5.7
ε ω3	14.3	17.0	31.1	20.4	24.2	33.3	14.6	20.9	29.1	28.1	23.8	27.4	19.9
ω6/ω3	0.59	0.32	0.32	0.33	0.39	0.24	0.13	0.10	0.13	0.15	0.21	0.11	0.28
	Mean ω6/ω3		0.37 ± 0.12					0.16 ± 0.06					

[a] From Castell 1979

Table 4

CHANGES IN FATTY ACID COMPOSITION IN MIGRATING FISH[a]

Fatty Acid	Sweet smelt				Masu salmon			
	April Marine		May Freshwater		May Freshwater		June Marine	
	TG	PL	TG	PL	TG	PL	TG	PL
14:0	8.0	2.3	10.0	8.6	5.2	1.9	5.7	2.2
16:0	21.6	22.6	18.7	31.8	19.9	30.1	20.0	27.0
16:1	10.0	3.2	17.0	11.3	11.6	4.5	8.7	2.9
18:0	2.8	4.4	2.9	8.1	4.6	4.0	3.9	5.9
18:1	12.8	9.6	11.5	18.9	23.3	11.2	21.7	13.5
18:2	2.8	0.9	4.3	1.5	3.9	1.3	1.7	0.6
18:3ω3	3.0	0.0	5.1	0.9	3.0	1.2	1.3	0.5
18:4ω3	5.1	1.0	4.3	0.7	1.4	0.4	2.3	0.5
20:1	1.1	0.5	—	—	3.0	0.6	6.7	1.8
20:4ω6	1.4	1.3	1.5	1.3	1.0	2.3	0.6	0.9
20:4ω3	1.9	0.7	1.8	0.7	1.5	1.3	1.2	0.9
20:5ω3	8.2	10.9	6.3	1.4	4.2	8.5	7.0	7.6
22:1	—	—	—	—	1.9	—	4.2	0.5
22:5ω6	—	—	1.1	—	—	—	—	—
22:5ω3	1.4	1.5	1.2	1.1	1.8	2.1	2.4	2.2
22:6ω3	12.1	34.5	5.2	2.1	6.7	26.3	9.0	31.6
ε sat	34.9	31.8	35.1	53.8	31.9	37.5	31.0	36.0
ε mono	27.4	16.1	32.0	35.9	43.0	18.6	43.1	19.2
ε ω 6	4.4	2.2	7.2	3.2	5.7	4.0	2.3	1.5
ε ω 3	31.7	49.4	23.9	6.9	18.6	39.8	23.2	43.3
ω6/ω3	0.14	0.04	0.30	0.46	0.31	0.10	0.10	0.03

[a] From Castell 1979

temperature (e.g., rainbow trout); warmwater common carp, channel catfish, and Tilapia may do better with a mixture of omega-6 and omega-3 fatty acids (Table 5). Generally, fish tend to utilize omega-3 over omega-6. High omega-6 diets undergo alteration of the omega-6:omega-3 ratio in favor of omega-3 fatty acids in the tissue lipids (Cowey 1979; Halver 1979).

Although omega-9 series can be synthesized, it appears that a critical balance of omega-3, -6, and -9 must be maintained under a particular set of environmental conditions for optimal metabolic function. Since EFA in tissues and organs are so highly influenced by diet and the specific EFA needs, the ovaries and eggs of a fish probably best represent the EFA requirements of the species.

Excessive dietary lipid may result in nutritional diseases like fatty liver or cause large fat deposition in the muscle and viscera, thus producing off-flavors, spoiling the quality of the fish, and reducing its dress-out weight percentage.

The addition of omega-3 polyunsaturated fatty acids in fish diets creates storage problems. Lipids are very labile to oxidation; the nutritional level of proteins and vitamins may be reduced and the oxidative products may be lethal. The addition of alphatocopherol acetate or vitamin E provides a sparing antioxidant effect on the lipids. A further reduction in lipid oxidation may be achieved by storing finished feed in airtight containers at low temperatures with minimum exposure to UV radiation.

Growth Factors

Vitamins and Minerals The vitamin requirements, which play a major role in fish physiology, are related to species, size, environmental conditions, and the amount of physiological stress encountered. Most fish have requirements for eleven water-soluble vitamins and at least three of the four fat-soluble vitamins (Halver 1979). These requirements are summarized in Table 6, which gives the vitamin requirements for growth in certain fish species.

Table 5

TEMPERATURE EFFECT UPON FATTY ACID COMPOSITION[a]

Fatty acid	Mosquito fish[b]		Guppies[b]		Guppies[c]		Goldfish intestine		Catfish Liver[d] Beef tallow		Catfish Liver[d] Menhaden oil	
	14-15C	26-27C	14-15C	26-27C	17C	24C	3C	32C	20C	33C	20C	33C
14:0	1.3	1.6	3.9	3.7	1.5	0.9			0.6	1.1	0.8	1.3
16:0	14.7	16.0	19.2	22.5	22.9	36.0	15.6	17.3	16.4	18.9	17.4	18.1
16:1	20.0	19.8	10.1	14.1	15.9	8.9	2.2	0.9	4.2	4.6	2.1	3.0
18:0	5.4	6.5	10.4	7.7	8.2	9.8	12.4	19.5	9.1	6.7	10.6	10.7
18:1	31.8	30.8	26.6	25.7	18.3	15.0	7.7	11.9	45.4	56.1	26.5	40.0
18:2ω6	7.3	7.9	15.0	8.0	Tr	Tr	14.3	21.1	1.7	1.7	2.2	2.0
18:3ω3	Tr	Tr	0.1	1.7	1.4	0.8	--	--	2.4	1.6	1.0	1.9
18:4ω3	0.4	1.0	0.8	1.3	--	--	--	--	--	--	--	--
20:1	5.0	5.1	2.5	3.6	--	--	2.2	1.2	--	--	--	--
20:2	--	--	--	--	--	--	1.6	4.2	--	--	--	--
20:3	--	--	--	--	--	--	3.9	6.4	6.5	2.6	1.2	0.4
20:4ω6	4.0	4.5	1.5	2.7	2.0	2.0	13.7	6.0	3.8	0.9	2.6	1.1
20:5ω3	1.2	1.2	0.5	0.7	4.8	4.6	--	--	1.4	0.9	8.5	5.5
22:1	--	--	--	--	--	--	--	--	--	--	--	--
22:4ω6	0.4	--	0.3	--	1.3	1.0	--	--	0.6	0.7	0.4	0.5
22:5ω6	--	--	--	--	--	--	--	--	+	+	+	+
22:5ω3	2.1	1.4	1.5	0.6	6.1	7.3	3.0	2.9	1.3	1.7	3.6	2.8
22:6ω3	5.9	3.6	5.1	4.0	16.5	11.5	18.2	5.0	2.8	0.6	22.0	10.4
ε Sat.	21.4	24.1	33.5	33.9	32.6	46.7	28.0	36.8	26.1	26.7	28.8	30.1
ε Mono	56.8	55.7	39.2	43.4	34.2	23.9	12.1	14.0	--	--	--	--
ε ω 6	11.3	12.4	16.5	10.7	3.2	3.0	28.0	27.1	7.1	4.5	5.0	4.1
ε ω 3	9.6	7.2	8.0	8.3	28.8	24.2	21.2	8.4	8.3	4.5	35.9	21.2
ω5/ω3	1.18	1.72	2.06	1.30	0.11	0.12	1.32	3.23	0.86	1.00	0.14	0.19

[a] From Castell 1979

[b] fed trout pellets

[c] fed Artemia salina

[d] fed casein based artificial diet with 10% lipid supplement as noted.

Table 6

VITAMIN REQUIREMENTS OF FISH[a,b]

Vitamin, (mg/kg) dry diet	Carp	Channel catfish	Eel	Sea bream	Turbot	Yellow-tail
Thiamin	2-3	1-3	2-5	R[c]	2-4	R
Riboflavin	7-10	R	R	R	R	R
Pyridoxine	5-10	R	R	2-5	R	R
Pantothenate	30-40	25-50	R	R	R	R
Niacin	30-50	R		R	R	R
Folacin		R	R		R	
Cyanocobalamin		R		R		
myo-Inositol	200-300	R		300-500		
Choline	500-600	R		R	R	
Biotin	1-15	R	R		R	
Ascorbate	30-50	30-50		R		R
Vitamin A	1000-2000 IU	R				R
Vitamin E[d]	80-100	R				R
Vitamin K	R	R				R

	Rainbow trout	Brook trout	Brown trout	Atlantic salmon	Chinook salmon	Coho salmon
Thiamin	10-12	10-12	10-12	10-15	10-15	10-15
Riboflavin	20-30	20-30	20-30	5-10	20-25	20-25
Pyridoxine	10-15	10-15	10-15	10-15	15-20	15-20
Pantothenate	40-50	40-50	40-50	R[c]	40-50	40-50
Niacin	120-150	120-150	120-150	R	150-200	150-200
Folacin	6-10	6-10	6-10	5-10	6-10	6-10
Cyanocobalamin	R	R	R	R	0.015-0.02	0.015-0.02
myo-Inositol	200-300	R	R	R	300-400	300-400
Choline	R	R	R	R	600-800	600-800
Biotin	1-1.5	1-1.5	1.5-2		1-1.5	1-1.5
Ascorbate	100-150	R	R	R	1-1.5	1-1.5
Vitamin A	2000-2500 IU	R	R		R	R
Vitamin E[d]	R	R	R		40-50	R
Vitamin K	R	R	R		R	R

[a] From Halver 1979

[b] Fish fed at reference temperature with diets at about protein requirement.

[c] R - required.

[d] Requirement directly affected by amount and type of unsaturated fat fed.

Minerals have a great diversity of uses within the fish body, yet they have been largely neglected in studies of fish nutrition because they are difficult to quantify. Fish have the ability to absorb ions not only from their diet but also by ion exchange across the gills and skin (Lall 1979). The trace elements, which are not yet clearly defined, should be incorporated into artificial diets used in intensive culture conditions. Since the exact trace requirements are not known, Chow and Schell (1980) recommend arbitrary levels that are based upon land animal requirements. Table 7 summarizes information on the mineral requirements of fish.

Table 7

MINERAL REQUIREMENTS OF FISH[a]

Mineral element	Principal metabolic activities	Deficiency symptoms	Requirement, g/kg dry diet
Calcium	Bone and cartilage formation; blood clotting; muscle contraction	Not defined	5
Phosphorus	Bone formation; high energy phosphate esters; other organo-phosphorus compounds	Lordosis, poor growth	7
Magnesium	Enzyme co-factor extensively involved in the metabolism of fats, carbohydrates and proteins	Loss of appetite, poor growth, tetany	0.5
Sodium	Primary monovalent cation of intercellular fluid; involved in nerve action and osmoregulation	Not defined	1-3
Sulphur	Integral part of sulphur amino acids and collagen; involved in detoxification of aromatic compounds	Not defined	3-5
Chlorine	Primary monovalent anion in cellular fluids; component of digestive juice (HCl); acid-base balance	Not defined	1-5
Iron	Essential constituent of haeme in haemoglobin, cytochromes, peroxidases, etc.	Microcytic, homochronic anaemia	0.05-0.10
Copper	Component of haeme in haemocyanin (of cephalopods); co-factor in tyrosinase and ascorbic acid oxidase	Not defined	1-4
Manganese	Co-factor for arginase and certain other metabolic enzymes; involved in bone formation and erythrocyte regeneration	Not defined	0.02-0.05
Cobalt	Metal component of cyanocobalamin (B_{12}). Prevents anaemia; involved in C_1 and C_3 metabolism	Not defined	0.005-0.01
Zinc	Essential for insulin structure and function; co-factor of carbonic anhydrase	Not defined	0.03-0.10
Iodine	Constituent of thyroxine; regulates oxygen use	Thyroid hyperplasia (goiter)	0.10-0.30
Molybdenum	Co-factor of xanthine, oxidase, hydrogenases and reductases	Not defined	(trace)
Chromium	Involved in collagen formation and regulation of the rate of glucose metabolism	Not defined	(trace)
Fluorine	Component of bone appetite	Not defined	(trace)

[a] From Chow and Schell 1980

FEED EFFICIENCY

A discussion of the ability of fish to efficiently convert food into edible flesh requires a brief statement on nutritional bioenergetics.

Bioenergetics

Bioenergetics is concerned with the energy transformations in living organisms. It has recently been reviewed by Webb (1978), Fischer (1979), Braaten (1979), and Smith (1980). Many energy budget models have been proposed for different fish species, with the general energy budget equation expressed as follows (Braaten 1979):

$$C = F + U + \Delta B + R$$

where

$$R = R_s + R_d + R_a$$

- C = energy value of food consumed
- F = energy value of feces
- U = energy value of materials excreted in the urine or through the gills or skin
- ΔB = total change in energy value of materials of body (growth)
- R = total energy of metabolism: this can be subdivided as follows: $R_s + R_d + R_a$
- R_s = energy equivalent to that released in the course of metabolism of unfed and resting fish (standard metabolism)
- R_d = additional energy released in the course of digestion, assimilation and storage of minerals consumed (including SDA).
- R_a = additional energy released in the course of swimming and other activity.

The ability to isolate individual components in the above equation has improved in the past ten years due to better experimental techniques. Braaten (1979) has reviewed the current methodologies.

The metabolic energy requirements of fish are less than those of mammals and birds for the following reasons (Lovell 1979a; 1979b; Smith 1980):

1. Poikilothermic fish do not have to expend energy to maintain a constant body temperature.

2. Fish exert relatively little muscle activity or energy to maintain position in the water compared to land animals.

3. The excretion of nitrogen waste requires less energy in fish than in homeothermic animals.

Thus fish can synthesize more protein per calorie of energy consumed than poultry or livestock. Lovell (1979b) concludes the primary advantage of fish over other domesticated animals is the lower energy cost of protein gain rather than any superior food conversion efficiency.

Approximately 60 percent of energy in fish feeds is utilized for maintenance; the remaining 40 percent is used for growth (Hepher 1975). It is important that the correct quality and quantity of energy sources (proteins, carbohydrates, and lipids) be incorporated into diets. The maintenance ration must have priority over the growth ration in order to ensure normal basal metabolism. Consequently, Hickling (1962) recommends that fish should be subject to as little disturbance as possible to avoid reduced growth rates. Trout culture is the exception; here the need for water currents for oxygenation outweighs the weight-gain factor.

Protein is used very efficiently by fish, having metabolizable energy = 4.5 kcal/g (Smith 1980), but it is the most expensive energy source in manufactured diets and should be kept to the minimum consistent with good growth and food conversion. Some natural foods contain a greater percentage of protein than is required for growth (Hepher 1975). Much of this excess protein is wasted. On the other hand, insufficient protein retards growth. Fish eat to satisfy their metabolic requirements and cease feeding when their caloric needs are met (Lovell 1979a). Consequently, if their diet contains too much energy in relation to protein, they will not meet their daily protein needs for optimum growth, even if they feed to satiation. Furthermore, there will be a tendency for increased fattiness in the fish (Nose 1979).

Carbohydrates are the cheapest source of energy. Their value depends on the type of carbohydrate and the processing to which it has been subjected. The metabolic energy values for fish range from near zero for cellulose to about 3.8 kcal/g for easily digested sugars (Smith 1980). It appears that warmwater fishes are better able to use starches than the coldwater fishes (Lovell 1979a).

Fats are the long-term storage products for energy metabolism. They are generally well digested and utilized by fish, having a metabolizable energy value of about 8.5 kcal/g (Smith 1980).

Food Conversion Ratio

In general, fish can convert food into body tissue more efficiently than can farm animals. As previously mentioned, Lovell (1979a, 1979b) attributes this superiority to the ability of fish to better assimilate diets with higher percentages of protein because of their lower energy requirements.

Feed efficiency has traditionally been given special attention and, according to the literature, is usually expressed in one of two ways (Utne 1979):

Food Conversion Ratio = (Feed Intake/ Weight Gain)

OR

Conversion Efficiency = (Weight Gain/Feed Intake) x 100

For consistency in discussion, the term food conversion ratio (FCR) will be used throughout the text.

The FCR assumes that all food has been consumed and that the same units of measurement are used (Reay 1979). These ratios are often expressed in terms of dry weight of food:wet weight of fish, which is why 1:1 ratios are frequently reported. However, not all the food may be consumed (at least by the targeted fish population), nor is the consumption of natural food from the pond included. Consequently, Hepher (1975) uses the term "apparent food conversion ratio."

In addition to the FCR, Swingle (1968) expressed pond conversion values (S) as:

S = Feed Added in Pond/Net Fish Production from Pond

It is again assumed that most of the feed is consumed by fish. S values for channel catfish have ranged from 1.2 to 2.0 (Boyd 1979).

In conventional fish culture, Hepher (1975) and Schroeder (1980) state that supplemental feed in the form of grains (e.g., sorghum) to produce five to ten tons of common carp/ha/year should have a FCR of 2.5 to 3.5. For fishmeal-enriched pellets containing 25 percent protein, the FCR ranges from 2.0 to 2.5. In polyculture, where natural food is utilized more efficiently, the FCR of supplemental feeds should be lower. For manured polycultured ponds in Israel, common carp, silver carp, and Tilapia hybrids were grown in liquid cow wastes and produced a manure conversion ratio of about 6.6 kg dry manure/kg wet fish weight and about 4.5 kg organic matter in manure/kg wet fish weight (Noriega-Curtis 1979). Interestingly, Schroeder (1973) found that the most important factor affecting FCR in manured ponds was the abundance of natural food in the form of heterotrophs instead of autotrophs.

The nutritional composition and the FCR of different feed materials are listed in Tables 8 and 9 respectively. These tables have often been presented in other publications (e.g., Ling 1967) and provide excellent relative comparisons.

According to Hickling (1962) and Hora and Pillay (1962) the factors affecting FCR are:

Physical environment:

1. Increased water temperature increases food consumption.

2. Decreased dissolved oxygen reduces food consumption (e.g., Tilapia have reduced appetite at D.O. < 1.5 mg/l).

3. Increased water acidity reduces food consumption.

Food:

4. Size of food. A finely divided food has a better conversion rate than a coarsely divided one.

5. Amount of food. An unlimited food supply may pass through the gut faster than a limited one and be only partially digested.

6. Composition of food. Foods with a high water or woody tissue content (e.g., leaves and potatoes) will have a less favorable conversion rate.

7. Method of presentation. Certain foods provide a better conversion rate if they are mixed than if they are presented separately (see Table 9).

Fish:

8. Size of fish. A small fish grows fast because its gut capacity is high relative to its body; this ratio decreases with increasing weight of fish.

9. Sexual maturity. Gonadal development reduces growth.

10. Stocking rate. Excessively high rates (crowding) will slow growth.

11. Fish species. Herbivores utilize some food materials better than carnivores.

Thus FCRs are crude empirical estimates and for local application only, especially when predicting the next fish crop.

FINFISH DIETS

Wild fish seldom show signs of nutritional diseases (Lovell 1979b). However, under intensive pond culture, fish cannot rely solely on the limited natural food produced in ponds and must depend for healthy growth on diets that may be classified as complete or supplemental. These diets are evaluated on four basic criteria: 1) acceptability to

Table 8

FOOD CONVERSION RATIO OF SOME FEED MATERIALS[a]

Foods of animal origin		Foods of plant origin			
Gammarus	3.9-6.6	Lupin seeds	3-5	Mill sweepings	8.0
Chironomids	2.3-4.4	Soyabeans	3-5	Rice flour	8.0
Housefly maggots	7.1	Maize	4-6	ditto	6-12
Fresh sea fish	6-9	Cereals	4-6	Manioc leaves	13.5
Fish flour	1.5-3.0	All cereals	5	ditto	10-20
Freshwater fish	2.9-6.0	Potatoes	20-30	Manioc flakes	17.6
Fresh meat	5-8	Potatoes	15	Household scraps	18.5
Liver, spleen and abbatoir offals	8	Maize	3.5	Colocasia leaves	23.7
		Cottonseed	2.3	Banana leaves	25.0
Prawns and shrimp	4-6	Cottonseed cake	3.0	Manioc flour	49.4
White cheese	10-15	ditto	2.5-5	Napier grass	48.0
Dried silkworm pupae	1.8	Groundnut cake	2.7	Manioc rind	50.7
		ditto	2-4		
		Ground maize	3.5		
		Ground rice	4.5		
		Oil palm cakes	6.0		
		ditto	6-12		

Food mixtures			
Fresh sardine, mackerel scad, dried silkworm pupae	5.5	2/3 groundnut cake, 1/3 manioc leaves	3.5
Liver of horse and pig, sardine, silkworm pupae	4.5	1/2 manioc leaves, 1/2 ground rice	11.0
Silkworm pupae silkworm feces, grass, soyabean cake, pig manure, night soil	4.1	2/5 manioc leaves, 3/5 manioc cosettes	12.8
Cortland Trout diet No. 6	7.1	Cottonseed and manioc flour	5.1
Raw silkworm pupae, pressed barley, Lema and Gammarus	2.55	Leaves and fresh manure roots	26.8
		Manioc leaves and household scraps	25.2

[a] From Hickling 1962

Complete Diets

Dry or moist diets that contain all the essential nutrients in the correct proportions for fish growth are called complete diets. They usually contain two to four times the protein that terrestrial animal feeds contain, and they rely on fish meal as the principal protein source. Complete diets are normally fed to luxury fish like trout in intensive culture conditions where natural foods are limited or nonexistent; however, fish cultured under high-density pond conditions may also need a complete diet.

The first complete diets were made for salmon hatcheries in the early 1960s (Nose 1979). At present, the formulas for most commercially prepared dry diets are generally unavailable to the public, but many institute formulas may be found in the literature (Halver 1972; National Academy of Sciences 1973, 1977). Examples are shown in Table 10a, b.

Supplemental Diets

Since warmwater fishes are usually less valuable than the coldwater species (e.g., salmonids), they are grown more cheaply in ponds where natural food is available (Hepher 1975). However, net fish yields on natural food alone are too low to cover fixed economic costs (Tal and Hepher 1967) and supplemental feeding is normally necessary; it typically comprises about 50 percent of total production expenditures (Collins and Delmondo 1976). If local waste materials like manure are available, costs may be significantly reduced.

Supplemental feeding is based upon three criteria: 1) the amount of natural food in the pond, 2) the nutritional requirements of the fish population in the pond, and 3) fish density. These points may be significantly influenced by other factors (seasonal condition, water temperature, fish size, etc.) that can cause enormous variation in the feeding requirements. Since no satisfactory

Table 9

NUTRITIONAL COMPOSITION OF SOME FEED MATERIALS[a]

Fodder	Average composition, %							Digestible nutrients, %				Mineral comp., %		
	Dry matter	Crude protein	Oil (ether extract)	Carbohydrate (nitrogen free extractives)	Crude fiber	Ash	True protein	Dig. crude protein	Dig. oil	Dig. fiber	Dig. carbohydrate (nitrogen free extractives)	C_aO	P_2O_s	K_2O
Green fodders														
Guinea grass, cut at 3-week intervals	23.0	2.9	0.2	10.3	6.6	3.0	--	2.5	0.1	5.3	8.9	0.06	0.07	--
Roadside grasses (Malaya)	23.0	2.4	0.5	12.5	6.0	1.6	--	1.3	0.2	3.7	7.8	0.28	0.09	--
Lallang grass (Imperata arundinacea), cut at intervals of 4 weeks (Malaya)	36.4	4.3	0.7	17.1	11.7	2.6	--	2.4	0.3	7.1	10.6	0.07	0.16	--
Kudzu (Pueraria phseoloides), leaves and stems (Malaya)	19.1	3.8	0.4	7.9	5.5	1.5	--	2.9	0.2	3.3	6.5	0.19	0.08	--
Pueraria javanlca (Congo)	20.0	4.4	0.4	6.5	7.2	1.5	--	--	--	--	--	--	--	--
Centrosema pubescens, leaves and stems (Malaya)	24.3	5.4	0.6	8.5	7.5	2.3	--	4.1	0.4	4.5	6.9	0.19	0.11	--
Water-Kangkong (Ipomaea reptans), leaves and stems (Malaya)	7.5	2.1	0.2	2.9	0.9	1.4	--	1.8	0.1	0.8	2.8	0.13	0.17	--
Sweet potato (Ipomaea batata)	13.0	1.6	0.4	6.8	2.3	1.6	--	--	--	--	--	0.21	0.10	--
Sweet potato (Ipomaea batata) (Malaya), Vines	13.3	2.5	0.3	6.5	2.5	1.5	--	1.9	0.2	1.5	5.3	0.14	0.09	--
Manioc, tapioca (Manihot utilitissima), leaves and stems (Malaya)	23.1	4.5	1.2	11.8	3.9	1.7	--	2.7	0.8	2.1	7.7	0.28	0.19	--
Manioc, tapioca (Manihot Utilitissima), leaves and stems (Congo). First leaf	27.2	9.4	0.8	6.2	9.3	1.5	--	--	--	--	--	0.26	0.21	--

Continued next page

Table 9 continued

Fodder	Average composition, %							Digestible nutrients, %				Mineral comp., %		
	Dry matter	Crude protein	Oil (ether extract)	Carbo-hydrate (nitro-gen free extrac-tives)	Crude fiber	Ash	True protein	Dig. crude protein	Dig. oil	Dig. fiber	Dig. carbo-hydrate (nitro-gen free extrac-tives)	C_aO	P_2O_s	K_2O
ditto (Congo). First six leaves	27.3	8.8	0.9	6.2	9.8	1.7	--	--	--	--	--	0.32	0.28	--
ditto. Old leaves	27.6	6.8	1.5	7.5	9.9	2.0	--	--	--	--	--	0.36	0.31	--
Coco-yam (Colocasia sp.), leaves (Malaya)	12.1	2.3	0.7	6.1	1.4	1.6	--	1.5	0.3	0.8	4.6	0.21	0.07	--
Queensland Lucerne (Stylosanthes) (Malaya)	24.0	4.0	0.4	9.6	7.6	2.4	--	2.3	0.1	1.6	4.5	0.52	0.31	--
Velvet bean (Mucuna utilis), leaves and stems (Malaya)	16.6	5.8	0.5	6.4	2.4	1.5	--	4.4	0.3	1.4	5.2	0.21	0.16	--
Velvet bean (Mucuna utilis)	18.0	3.4	0.5	7.2	5.8	1.1	--	--	--	--	--	0.21	0.12	--
Maize (Zea mays), leaves and immature cobs (Malaya)	20.0	3.0	0.6	12.2	2.3	1.7	--	1.8	0.4	1.4	7.9	0.08	0.15	--
Maize (Zea mays), leaves (Europe)	19.4	1.7	0.5	10.4	5.6	1.2	1.3	1.0	0.3	3.1	6.7	--	--	--
Pistla stratioias (Malaya)	7.1	1.4	0.3	2.6	0.9	1.9	--	1.2	0.2	0.8	2.5	0.20	0.06	--
Water Hyacinth (Eichhornia crassipes) (Malaya)	5.9	1.0	0.1	2.4	1.2	1.2	--	0.7	0.1	0.6	2.2	0.19	0.05	--
Myriophyllum (Minnesota)	13.6	2.4	0.2	6.8	1.8	2.5	--							
Potamogation (Minnesota)	22.7	3.3	0.4	10.5	4.9	3.7	--							
Ceratophyllam (Minnesota)	14.3	2.4	0.3	6.0	2.0	3.2	--							
Seeds and roots														
Maize (Zea mays), (Malaya)	86.2	8.8	4.3	70.4	1.3	1.4	--	7.0	2.6	0.5	64.8	0.02	1.00	--

Continued next page

Table 9 continued

Fodder	Average composition, %							Digestible nutrients, %				Mineral comp., %		
	Dry matter	Crude protein	Oil (ether extract)	Carbo-hydrate (nitro-gen free extrac-tives)	Crude fiber	Ash	True protein	Dig. crude protein	Dig. oil	Dig. fiber	Dig. carbo-hydrate (nitro-gen free extrac-tives)	C_aO	P_2O_s	K_2O
Maize (Europe)	87.0	9.9	4.4	69.2	2.2	1.3	9.4	7.9	2.7	0.8	63.7	0.02	0.82	0.40
Maize flaked (Europe)	89.0	9.8	4.3	72.5	1.5	0.9	9.4	9.4	2.0	0.5	70.4		0.60	0.25
Oats (Europe)	87.0	10.4	4.8	58.4	10.3	3.1	9.5	8.0	4.0	2.6	44.9	0.14	0.81	0.55
Barley (Europe)	85.0	9.0	1.5	67.4	4.5	2.6	8.5	6.8	1.2	2.5	61.7	0.07	0.84	0.57
Rye (Europe)	87.0	11.6	1.7	69.8	1.9	2.0	10.7	9.6	1.1	1.0	64.2	--	--	--
Wheat (Europe)	87.0	12.2	1.9	69.3	1.9	1.7	11.0	10.3	1.2	0.9	63.8	0.05	0.55	0.60
Rice, hulled (Malaya)	88.7	8.4	2.1	76.7	0.7	0.8	--	7.3	1.1	0.3	74.4	0.01	0.65	--
Broken rice (white), (Malaya)	88.6	7.5	0.5	79.9	0.2	0.5	--	6.5	0.3	0.1	77.5	0.02	0.14	--
Lupin, sweet (yellow), (Europe)	87.0	41.8	5.5	24.8	10.4	4.5	38.7	37.8	4.6	9.5	18.8	0.29	1.24	1.43
Groundnuts (Arachia hypogora)	94.0	26.8	44.9	17.5	2.6	2.2	24.9	24.1	40.3	0.2	14.7	--	--	--
ditto (decorticated), (Malaya)	92.7	30.8	44.3	13.6	1.6	2.4	--	27.7	39.9	0.1	11.4	0.10	0.90	--
Cottonseed (Bombay)	91.0	17.8	19.3	29.7	19.9	4.3	16.5	12.2	16.7	15.1	14.9	--	--	--
ditto (Egyptian)	91.0	19.6	23.8	21.4	21.2	5.0	18.2	13.4	20.6	16.1	10.7	--	--	--
Soybean (Glyeine)	90.0	33.2	17.5	30.5	4.1	4.7	29.9	29.5	15.8	1.7	20.8	--	--	--
ditto (Malaya)	88.5	38.5	16.4	24.1	4.9	4.6	--	34.3	14.4	1.8	16.1	0.37	1.30	--
Acorns, fresh	50.0	3.3	2.4	36.3	6.8	1.2	2.8	2.7	1.9	4.1	32.6	--	--	---
Potatoes	23.8	2.1	0.1	19.7	0.9	1.0	1.6	1.1	--	--	17.7	0.03	0.18	0.60

Continued next page

Table 9 continued

Fodder	Average composition, %							Digestible nutrients, %				Mineral comp., %		
	Dry matter	Crude protein	Oil (ether extract)	Carbo-hydrate (nitro-gen free extrac-tives)	Crude fiber	Ash	True protein	Dig. crude protein	Dig. oil	Dig. fiber	Dig. carbo-hydrate (nitro-gen free extrac-tives)	C_aO	P_2O_s	K_2O
Sweet potatoes (Ipomaea batata), tubers (Malaya)	25.4	1.3	0.1	22.7	0.8	0.5	--	0.7	0.1	0.3	20.4	0.03	0.07	--
Manioc, tapioca (Manihot utilitissima, whole roots, fresh (Malaya)	37.6	0.4	0.2	35.5	0.8	0.7	--	0.1	0.1	0.6	19.9	0.03	0.05	--
Manioc refuse, fresh (Malaya)	20.0	0.4	0.1	17.6	1.6	0.3	--	0.1	0.1	1.3	9.9	0.04	0.05	--
Oil cakes														
Coconut cake, single pressing (Malaya)	83.7	17.3	16.3	42.8	7.5	4.8	--	13.5	16.0	4.8	35.5	0.06	1.30	--
Coconut cake	90.0	21.2	7.3	44.2	11.4	5.9	19.7	16.6	7.1	7.2	36.6	0.16	1.27	2.41
Groundnut cake (decorticated) (Malaya)	92.2	47.9	10.9	25.0	3.6	4.8	--	43.1	9.9	1.3	21.3	0.10	1.22	--
Groundnut cake (Congo)	87.9	44.4	7.2	22.9	9.0	4.6	--	--	--	--	--	--	--	--
Groundnut cake (decorticated)	90.0	4.54	6.0	26.4	6.5	5.7	42.5	40.5	5.4	0.5	22.4	0.20	1.30	1.30
Cotton cake (decorticated)	90.0	41.1	3.0	26.4	7.8	6.7	39.6	35.3	7.5	2.2	17.7	0.30* (uncorticated)	2.50*	1.50*
Palm kernel cake	89.0	19.2	6.0	46.5	13.4	3.9	18.1	17.5	5.3	5.1	39.4	0.30	1.10	0.50
ditto double pressing (Malaya)	89.0	13.1	10.0	54.9	7.7	3.3	--	10.0	9.0	3.0	45.6	0.18	1.14	--
Palm cake (Congo)	87.7	27.1	14.3	32.0	10.1	3.6	--	--	--	--	--	--	--	--

Continued next page

Table 9 continued

Fodder	Average composition, %							Digestible nutrients, %				Mineral comp., %		
	Dry matter	Crude protein	Oil (ether extract)	Carbohydrate (nitrogen free extractives)	Crude fiber	Ash	True protein	Dig. crude protein	Dig. oil	Dig. fiber	Dig. carbohydrate (nitrogen free extractives)	C_aO	P_2O_s	K_2O
Miscellaneous and animal foods														
Urban pig swill (summer)	25.0	4.1	3.1	12.4	2.4	3.0	--	2.7	2.5	0.9	10.9	0.23	0.18	--
Mill swinepigs (Congo)	88.2	12.3	2.4	42.1	22.4	8.9	--	--	--	--	--	--	--	--
Soybean refuse, fresh (Malaya)	14.1	5.5	0.7	5.8	1.6	0.5	--	4.7	0.6	1.1	5.7	0.12	0.11	--
Brewers grains, dried	89.7	18.3	6.4	45.9	15.2	3.9	17.4	13.0	5.6	7.3	27.6	0.40	1.60	0.20
Padibran, coarse (Malaya)	90.5	6.2	2.7	37.8	33.1	10.7	--	4.3	2.2	8.6	28.0	0.10	0.90	--
Padibran, fine (Malaya)	89.2	11.4	6.8	45.4	14.1	11.5	--	7.9	5.6	3.7	33.6	0.09	0.83	--
Fresh fish (Malaya)	28.0	14.2	1.5	--	--	10.7	--	11.5	1.5	--	--	4.18	1.79	--
Fish meal, white	87.0	61.0	3.5	1.5	--	21.0	37.0	55.0	3.3	--	1.2	10.0	9.0	1.2
Blood meal	86.0	81.0	0.8	1.5	--	2.7	71.9	72.7	0.8	--	--	0.05	0.22	0.31
Silkworm pupae, fresh	35.4	19.1	12.8	2.3	--	1.2	--	--	--	--	--	--	--	--
ditto dried	90.0	35.9	24.5	6.6	--	1.9	--	--	--	--	--	--	--	--
ditto dried and defatted	91.1	75.4	1.8	8.4	--	5.6	--	--	--	--	--	--	--	--

[a] From Hickling 1962

Table 10a

COMPLETE DIET FORMULATIONS[a]

Ingredient	International feed no.	Amount in diet (%)
Forty-Percent-Protein Carp Grower		
Fish, meal mech extd, 65% protein[b]	5-01-982	46
Wheat, middlings, lt 9.5% fiber	4-05-205	28
Rice, bran w germ, meal solv extd	4-03-930	7
Wheat, bran	4-05-190	5
Soybean, seeds, meal solv extd, 44% protein[a]	5-20-637	5
Yeast, torula, dehy	7-05-534	4
Corn, gluten, meal	5-02-900	1.5
Vitamin premix[c]	--	0.5
Mineral premix[d]	--	0.5
Sodium chloride	--	0.5
Potassium phosphate	--	2.0
Twenty-Five Percent Protein[e] Catfish Pond Formula, Pelleted (Kansas Z-14)		
Wheat, bran	4-05-190	40.5
Sorghum, grain	4-04-383	17.5
Alfalfa, meal, s-c	1-00-025	10.0
Fish, meal mech extd	5-01-976	8.8
Soybean, meal, solv extd	5-04-604	8.5
Meat and bone meal (meat and bone scraps)	5-00-388	6.6
Corn, distillers solubles, dehy	5-02-147	5.0
Blood, meal	5-00-380	1.9
Dicalcium phosphate	6-01-080	0.57
Salt	6-04-152	0.5
Methionine, DL	--	0.09
Vitamin premix[b]	--	0.13

[a] From National Academy of Sciences 1977

[b] 6.25 x percent nitrogen

[c] Vitamin premix: vitamins added to cellulose powder to make 0.5% of diet (mg/kg):

Choline chloride	500	Riboflavin	25
Ascorbic acid	80	Pyridoxine	8
Inositol	80	Thiamine hydrochloride	5
Niacin	60	Biotin	0.05
Calcium pantothenate	80	Vitamin A	8,000 (IU/kg)
Vitamin E	45	Vitamin D_3	1,500 (IU/kg)

[d] Mineral premix: added to cellulose powder to make 0.5% of the diet (mg/kg):

Manganese	25	Magnesium	250
Iron	10	Cobalt	3
Zinc	25		

[e] 6.25 x nitrogen

Table 10b

RECOMMENDED ALLOWANCE FOR VITAMINS IN SUPPLEMENTAL AND COMPLETE DIETS FOR WARMWATER FISHES[a]

Vitamin	Amount/kg dry diet[b]	
	Supplemental	Complete
Vitamin A activity	2,000 IU	5,500 IU
Vitamin D_3 activity	220 IU	1,000 IU
VItamin E	11 IU	50 IU
Vitamin K	5 mg	10 mg
Choline	440 mg	550 mg
Niacin	17-28 mg[c]	100 mg
Riboflavin	2-7 mg[c]	20 mg
Pyridoxine	11 mg	20 mg
Thiamin	0	0.1 mg
D-Calcium pantothenate	7-11 mg[c]	50 mg
Biotin	0	0.1 mg
Folacin	0	5 mg
Vitamin B_{12}	1-20 µg	20 µg
Ascorbic acid	0-100 mg[c]	30-100 mg[c]
Inositol	0	100 mg

[a] From National Academy of Sciences 1977

[b] These amounts do not allow for processing or storage losses. Other amounts may be more appropriate for various species and under various environmental conditions.

[c] Highest amounts probably appropriate when "standing crop" of fish exceeds 500 kg/hectare of water surface.

method has yet been developed to assess the natural food produced in a pond and available to the fish, the formulation of a supplemental diet is even further complicated. Instead, most studies have concentrated on the standing stocks of natural foods, and not their rate of production or consumption by fish.

Hepher (1975) has pointed out that the three most important interrelated factors affecting fish production are stocking rates, fertilization, and supplemental feeding. The standing stocks of fish reach equilibrium at a pond's carrying capacity. Production increase can be realized only through enhanced feeding, either directly with supplemental feeding or indirectly through fertilization, or by reducing the stocking rate. In Israel, monoculture of common carp in unfertilized ponds yielded about 100 kg/ha; fertilization increased the carrying capacity to 460 kg/ha. In polyculture with Tilapia aurea, it was further boosted to 600 kg/ha (Hepher 1975).

Natural foods in intensive pond culture must be supplemented both quantitatively and qualitatively. The protein-to-energy ratio must be balanced, and a sufficient supply of vitamins and minerals added. The culturist must always be aware that diet composition changes should accompany an increase in the standing stocks. For example, protein level requirements will increase with increasing standing stocks (Hepher 1975). The deficit of growth factors in high densities is expressed by growth inhibition and lower feed utilization. For example, Hepher et al. (1971) found that the conversion rate of dietary protein to body protein in common carp increased from 4.5 at a density of 226-372 kg/ha to 12.2 at 372-461 kg/ha.

A considerable diversity of supplemental feeding patterns have evolved worldwide as a result of specific conditions, experience, and traditions. Locally available artificial feeds of plant or animal origin can often be obtained at nominal cost and efficiently utilized. China provides a good example. Tapiador et al. (1977) report that 60-70 kg of grass and vegetable tops can produce 1 kg of grass carp; 50 kg of snails and clams produce 1 kg or black carp; 100 kg of water fertilized with 77 percent bean curd residues and 23 percent fermented products residues produce 1 kg of silver carp; 500 g of fish waste produce 0.8 kg of silver or bighead carp; and 25 kg of animal manure produce 500 g of silver and bighead carp. Since grass and vegetables provide most of the supplemental feeds for fish, the pond dikes are used for plant cultivation. Experience indicates that 66.7 square meters of land is needed to provide 667 square meters of fish pond with plants as feed (i.e., 1:10).

Preparation of Foods

There are four basic methods of preparing foods for introduction into ponds:

1. The mechanical preparation (e.g., milling) of foods usually makes them easier to chew and swallow, except when the particle size becomes so reduced that it is unavailable for consumption by big fish (Woynarovich 1975). It is thus recommended these foods be kneaded with water into a dough. There is some nutritional loss in mechanically prepared foods because they dissolve in pond waters more readily than whole grains. Another disadvantage is that small wild fish may eat a significant portion of the small food particles. Therefore, the necessity for mechanical preparation depends on the size of the fish to be fed.

2. Some foods, especially grains, are soaked (usually for several hours) before presentation to fish so that the food may swell, soften, and sink. Hickling (1962) recommends feeding in designated areas so that fish can learn where food will be introduced and so that the fish farmer can inspect the pond bottom to evaluate the amount of food consumption.

It is not always necessary to soak hard food items. For example, whole lupin seeds, maize, and wheat can be fed to common carp during midsummer when the fish are large and hungry. However, Tilapia will not eat maize until it is well soaked. Smaller fish also require that ground or milled grains be soaked for optimal utilization.

3. Cooking or steaming of foods is an expensive method of preparation because of fuel costs and the labor involved. However, it is often unavoidable, especially with peas, beans, potatoes, and eggs. Cooked food is recommended for the feeding of fingerlings (Woynarovich 1975).

4. Mixing food with other materials not only increases palatability and FCR (see Table 9), but may be necessary to complete the nutritional requirement of fish. In some cases, food is mixed with minerals (lime) or yellow clay to help prevent body defects like osteomalacia (Woynarovich 1975).

Natural Feeds and the Role of Fertilization

There is no doubt that fertilization increases pond productivity (refer to the chapter on "Pond Fertilization Practices"). In the Congo, Hickling (1962) found that fish production declines after the initial application of fertilizers; if the practice is not regularly continued, feeding with leaves and household scraps will usually not improve the situation. The conclusion is that growth substances found in natural foods become depleted. For example, the addition of chironomid larvae and Daphnia to an experimental diet of Quaker oats and casein being fed to common carp noticeably increase their growth rates in excess of the nutritive value of these live foods (Yashouv 1956). Hickling (1962) estimated that 50 percent of a common carp diet must come from natural foods versus 10 percent for Tilapia.

The promotion of natural foods through fertilization is quite evident in brackishwater ponds where benthic algal growth is desired for milkfish culture. In Taiwan, supplemental feeding is unnecessary when abundant algal pasture exists (Chen 1970). However, when there is a shortage, especially during the rainy season when salinity decreases, rice bran, peanut meal, soybean meal, or flax seed cake are added. Generally the algal pastures consist of two groups of microscopic algae, the filamentous blue-green Cyanophyceae and the diatoms, Bacillariophyceae. In Table 11, Tang and Hwang (1967) have determined the nutritional information on four major groups of algae used as milkfish food in Taiwan.

It is estimated that the total amount of algal pasture grazed during the rearing season (April-October) for one fish is approximately 25,000 kg/ha (Tang and Chen 1967).

In the Philippines, milkfish farmers collect algae from other water areas and transfer them into ponds when benthic growth is insufficient. Although green algae, Chlorophyta, are generally less nutritious than blue-greens, the red seaweed Gracilaria appears very suitable for rearing fingerlings and adults. Many Philippine ponds grow milkfish to market

Table 11

NUTRITIONAL VALUE OF FOUR ALGAE GROUPS[a]

(Composition of Four Major Groups of Algae and Their Relative Nutritive Value as Milkfish Food)

		Total Composition						Digestive coefficient[b,c]						
Group of algae	Number of samples	Total dry matter, %	Crude protein, %	Crude fat, %	Nitrogen-free extract, %	Fiber, %	Mineral matter, %	Crude protein, %	Crude fat %	Nitrogen-free extract, %	Fiber, %	Digestible[d] protein, %	Total digestible nutrients[e] %	Nutritive ratio[f]
Chaetomorpha														
Fresh form	15	8.54:	2.82	0.91	1.50	1.22	2.09	3	72	87	21	0.09	3.12	1 33.44
Detritial form	15	10.72:	3.46	0.38	3.21	0.98	2.69	66	89	85	37	2.28	6.13	1 1.66
Phytoflagellates[g]														
Fresh form	5	11.98:	3.91	1.32	5.61	0.42	0.72	81	91	78	23	3.17	10.41	1 2.37
Diatoms[h]														
Fresh form	15	12.87:	2.89	0.94	2.25	0.27	6.52	87	96	84	19	2.51	6.48	1 1.54
Filamentous blue-green algae[i]														
Fresh form	15	9.86:	2.32	0.21	1.52	0.70	5.11	69	86	81	38	1.60	3.49	1 1.18

[a] From Tang and Hwang 1967

[b] Digestive coefficient: $\frac{\text{The amount of a class of organic nutrient in the feed - the amount of that class of organic nutrient in feces}}{\text{The amount of that class of organic nutrient in the food}} \times 100$

[c] The water temperature during digestion experiments ranged from 29° to 33°C and the salinity from 24 to 27 ppt

[d] Digestible protein: The percentage of protein in the food x digestion coefficient of protein

[e] Total digestible nutrients: The sum of digestible protein, fibre, nitrogen-free extract and fact x 2.25

[f] Nutritive ratio: $\frac{\text{The percentage of total digestible organic nutrients - the percentage of digestible protein}}{\text{The percentage of digestible protein}}$

[g] Centrifuged from the pond water where Chlamydomonas and Chilomonas flagellates bloomed predominantly

[h] Furnished as the diatom sludge

[i] Collected from the pond bottom where the dominant genera, Oscillatoria and Lyngbya, grew

size on a Gracilaria diet alone (Hora and Pillay 1962). The nutritional information for some food algae and weeds are shown in Table 12.

In Malaysian freshwater ponds, Hickling (1962) estimated that 150 fingerling grass carp ate about 8600 kg of Napier grass (57 kg of Napier grass/fish) during a two-month period; the crude conversion ratio was about 48:1. One advantage of culturing a herbivorous fish of this nature is the tremendous quantity of soft, partially digested feces it evacuates. This fecal material may be directly consumed by other fish species or serve as manure for the stimulation of plankton growth. The estimated amount of phosphorus recycled from the above Napier grass was 6.3 kg.

Larval Fish Diets

Larval fishes are usually cultured in rearing tanks rather than ponds and are often fed Artemia (brine shrimp). The increasing demand for Artemia has prompted efforts to find a suitable synthetic diet substitute that is not subject to seasonal and nutritional variation (Meyers 1979). Two main techniques exist for the preparation of these diets (van Limborgh 1979):

1. Preparation of a water stable matrix of dry ingredients followed by suitable grinding and sieving to proper particle size (e.g., pellets, flakes).

2. Incorporation of a solid, liquid, or finely suspended dietary component into properly sized microencapsulation.

It is important that nutritional loss be minimized during diet fabrication. Furthermore, as with fry and fingerling diets, the properties of water stability, density, size, color, taste, physical form, and attractants must be compatible with the feeding habits of larval fish.

ALTERNATIVE PROTEIN SOURCES FOR FISH MEAL

As mentioned above, the amino acid profiles of freshwater and marine fishes are quite similar. Thus fish meal is recognized as the best source of animal protein for most fish species. However, the increased costs and shortages of high quality fish meal impose very real constraints in the formulation of optimal diets. To emphasize the importance of finding a suitable fish meal substitute, Spinelli (1980) states that protein sources have been examined more intensely in the last 7 years than in the previous 50 years. Great differences exist in fish growth obtained from different sources of dietary proteins because of the different biological values in these proteins (Hepher 1979).

Table 12

NUTRITIONAL VALUE OF CERTAIN ALGAE AND SEAWEEDS[a]

Name	Moisture, %	Ash, %		Fat, %		Protein, %		Carbohydrates, %	
		F[b]	D[c]	F	D	F	D	F	D
Gracilaria confervoides	6.92	15.31	16.48	0.4	--	11.98	12.89	65.39	70.63
Enteromorpha intestinalis	81.35	6.02	32.27	0.48	2.57	3.66	19.61	8.49	45.55
Chaetomorpha spp.	85.50	2.82	19.50	0.27	0.71	3.72	27.66	7.87	52.13
Cladophora spp.	57.20	9.90	23.16	0.84	1.96	5.16	12.07	26.90	62.81
Eichhornia crassipes	89.81	1.34	13.15	--	--	2.19	21.49	6.66	65.36

[a] From Hora and Pillay 1962

[b] F - Fresh

[c] D - Dry

Soybean Meal

Considerable work has been conducted with soybean meal, and it has met with limited success. Soybean meal presents fewer problems with herbivorous than with carnivorous fishes and in supplemental diets rather than complete diets.

Soybean contains about 47-50 percent protein, 5-6 percent ash, 1 percent lipids, and about 40 percent carbohydrates (Spinelli 1980). It is deficient in tryptophan and the sulphur-containing amino acids. Although coldwater rainbow trout can thrive on soybean diets enriched with free amino acids, the utilization of supplemental amino acids by the warmwater fish (channel catfish and carp) remains obscure (Nose 1979). Either soybeans contain antigrowth factors, or certain warmwater fishes cannot utilize free amino acids.

Table 13 compares the amino acid profiles of fish meal and soybean meal.

Soybean meal is low in phosphorus and trace metals and contains factors deleterious to fish and farm animals. Heat treatment detoxifies these factors, but optimal guidelines for heat treatment have yet to be established (Nose 1979). In addition, heat increases the acceptability of soybean meal to fish and improves the availability of nutrients. This is accomplished by deactivating trypsin inhibitors and by denaturing the proteins for better digestibility.

Single Cell Proteins (SCP)

Single cell proteins are usually derived from unicellular yeast and bacteria, but may also come from fungi and algae (e.g., *Chlorella*, *Scenedesmus*,

Table 13

AMINO ACID PROFILES OF ALTERNATE PROTEIN SOURCES[a]

Amino acid	Fish meal	SCP, yeast	SCP, bacteria	Fly larva	Soy meal, 47%
Alanine	6.34	5.28	6.32	6.15	4.33
Arginine	5.82	7.79	8.01	5.42	7.15
Aspartic	9.35	6.26	6.31	10.8	10.90
Cystine	0.70	0.54	0.50	0.82	0.61
Glutamic	13.3	8.79	6.21	12.2	17.5
Glycine	5.90	4.63	5.84	5.40	4.34
Histidine	2.22	2.34	3.26	3.50	2.90
Isoleucine	4.85	2.73	3.25	4.13	4.34
Leucine	7.35	4.48	4.66	6.95	7.33
Lysine	7.85	6.80	7.19	7.37	6.0
Methioniue	2.84	0.70	1.45	2.24	1.21
Phenylalanine	4.35	2.01	1.30	6.95	4.90
Proline	4.35	2.27	3.4	3.66	4.53
Serine	4.55	.97	2.73	4.51	5.22
Threonine	4.55	3.59	3.65	4.53	5.0
Tryptophane	1.33	0.60	1.00	1.45	1.21
Tyrosine	3.45	1.11	1.05	8.10	3.94
Valine	5.65	3.31	3.37	5.60	4.33

[a] From Spinelli et al. 1979

Spirulina). They have a reasonably well-balanced amino acid profile (Table 13) and digestible lipids and carbohydrates (Spinelli 1980). In addition they are an excellent source of vitamins and minerals.

Although many studies have been conducted with SCP (Spinelli et al. 1979), it appears only algal SCP has been evaluated for warmwater pond fish culture. Hepher et al. (1979) tested three different diet treatments, using three polycultured ponds for each treatment. All diets contained 25 percent crude protein; one contained fish meal, the second soybean meal, and the third algae meal (SCP). The yield from the algae meal diet was more than 10 percent higher than from the fish meal diet; the soybean meal diet showed the lowest yield (Table 14). The greatest differences occurred in fish which accepted supplemental feeds (common carp and Tilapia), whereas almost no growth differences were noted in the silver carp, which prefers only natural food. In effect, it appears that algae meal is the only known protein source that can replace fish meal in diets for the common carp and Tilapia. The obvious obstacle to full employment of SCP is the cost of producing, harvesting, and processing it.

Other Protein Sources

Table 15 summarizes alternative sources of protein that have been examined as partial or complete replacements for fish meal in aquaculture diets.

INFLUENCE OF FEED ON PRODUCT QUALITY

Few systematic studies have been made relating diet compositions to the organoleptic quality of fishes. However, observations show that fish can quickly concentrate organoleptically active compounds in their tissues. These produce off-flavors that later pose serious marketing problems. The organoleptic properties of pond-cultured fish are influenced by many factors, as shown in Figure 1.

Most of the knowledge in this area has been derived from salmonids. There is very little basic and practical information available on pond-cultured warmwater fishes. Lovell and Sackey (1973) found that blue-green algae can synthesize compounds that are readily absorbed by channel catfish within two days. These compounds produce an earthy, musty flesh taste. The Actinomyces are also responsible for muddy taints in fish. Vale et al. (1970) compared kerosene-tasting mullet (Mugil cephalus) with untainted mullet and found the flesh lipid content to be 15 and 7 percent respectively. A related finding was reported for common carp grown in ponds fertilized with liquid cow manure compared to those grown on grain or high protein pellets. The carp grown in fertilized waters had better flesh color than those fed on prepared diets; the intramuscular fat levels were 6 and 15-20 percent respectively (Moav et al. 1977). Silkworm pupae fed to common carp can produce an undesirable flavor that can be purged from the flesh if pupae are eliminated from the diet for 30 days (Spinelli 1979). As more is learned about the effects of nutrition, feeding, and environment upon the quality of fish flesh, it may be possible to develop supplemental rations that will favorably alter the organoleptic characteristics of the fish (Spinelli 1979).

FEEDING PRACTICES

Feeding Behavior

The normal feeding habits of some warmwater cultured fish species are shown in Table 16. When fish feeding programs are developed, nutritional, economic, and feed preference aspects must be considered. For example, Cremer and Smitherman (1980) found that artificial feeds added to ponds were not consumed by silver carp, yet bighead carp enjoyed substantial growth; silver carp appear to accept only natural food (i.e., phytoplankton). Consequently, three types of feeding behavior in pond cultured fishes are recognized (Hora and Pillay 1962):

1. Those that thrive on artificial food (e.g., grass carp, black carp).

2. Those that thrive best on a diet of both artificial and natural food (e.g., common carp, mud carp).

3. Those that take only natural food and not artificial food (e.g., silver carp).

Rates and Amounts of Feeding

Feeding strategies differ with many factors, which include fish species, type of feed, and intensive or extensive culture. Considerable care must be practiced, for excessive feeding is not only wasteful but also deleterious to water quality. Conversely, underfeeding limits the capacity for full growth. Nose (1979) recommends the floating type of feed to prevent excessive feeding. Knowing the rates of digestion or evacuation will lead to the establishment of feeding schedules similar to those used for trout, which utilize feeding tables based on water temperatures and flow rates. There are several reports (Nose

Table 14

GROWTH ON FISH MEAL, SOYBEAN MEAL, AND ALGAE MEAL (SCP) DIETS[a]

Fish species	Fish meal diet		Soybean meal diet		Algae meal diet	
	Avg. harvest weight, g	Yield, kg/ha	Avg. harvest weight, g	Yield, kg/ha	Avg. harvest weight, g	Yield, kg/ha
Common carp	869.3	3196	1126.2	1921	992.8	3218
Tilapia aurea	433.8	614	402.8	621	431.6	700
Silver carp	1100.6	797	1051.9	739	1058.7	796
Bighead carp	878.3	119	918.2	140	1014.3	160
Total		4726		3421		4802
FCR[b]		2.1		3.6		2.1
C. carp (large)	398.8	1092	334.3	586	426.9	1556
C. carp (small)	159.9	589	156.4	566	190.6	631
Tilapia aurea	143.8	272	134.5	240	136.6	328
S. carp (large)	450.0	202	420.0	171	448.3	201
S. carp (small)	36.0	14	32.4	18	37.1	21
Grass carp	25.2	4	19.2	1	25.5	5
Total		2173		1582		2742
FCR[b]		2.0		2.8		1.7
Common carp		4877		3073		5405
Tilapia aurea		886		861		1028
Silver carp		1013		928		1018
Bighead carp		119		140		168
Grass carp		4		1		5
Annual Total		6899		5003		7624

[a] From Hepher et al. 1979

[b] $\text{Feed conversion ratio} = \frac{\text{Total feed}}{\text{Yield of C. carp and Tilapia}}$

Table 15

POTENTIAL ALTERNATE PROTEIN SOURCES[a]

Commercialized		
Vegetable	**Animal**	**Not commercialized**
Soy meal	Poultry byproducts	Insect larvae
Rapeseed meal	Feather meal	Single cell protein
Sunflower meal	Shrimp and crab meal	Grasses
Oat groats	Blood flour	Leaf protein
Cottonseed meal	Fish silage	Vegetable silage
Wheat middlings	Meat meal	Zooplankton (krill, etc.)
		Recycled wastes
		Yeast
		Phytoplankton
		Bacteria
		Algae
		Higher plants
Protein (range), %		
15-50	50-85	4-85

[a] From Spinelli 1980

Table 16

FEEDING HABITS OF SOME FISH SPECIES[a]

Feeding habit	Species
Algae and plankton	Silver carp
	Milkfish (*Chanos chanos*)
	Sarotherodon galileus
	Sarotherodon niloticus
	Grey mullet (*Mugil cephalus*)
Filamentous algae	Milkfish
Zooplankton	Bighead carp (*Aristichthys nobilis*)
Macrophytes	Grass carp (*Ctenopharyngodon idella*)
Benthos	Common carp (*Cyprinus carpio*)
	Black carp (*Mylopharyngodon piceus*)
	Mud carp (*Cirrhina molitorella*)
Detritus	*Sarotherodon aureus*
	Common carp
	Milkfish
	Sarotherodon niloticus
	Grey mullet

[a] From Schroeder 1980

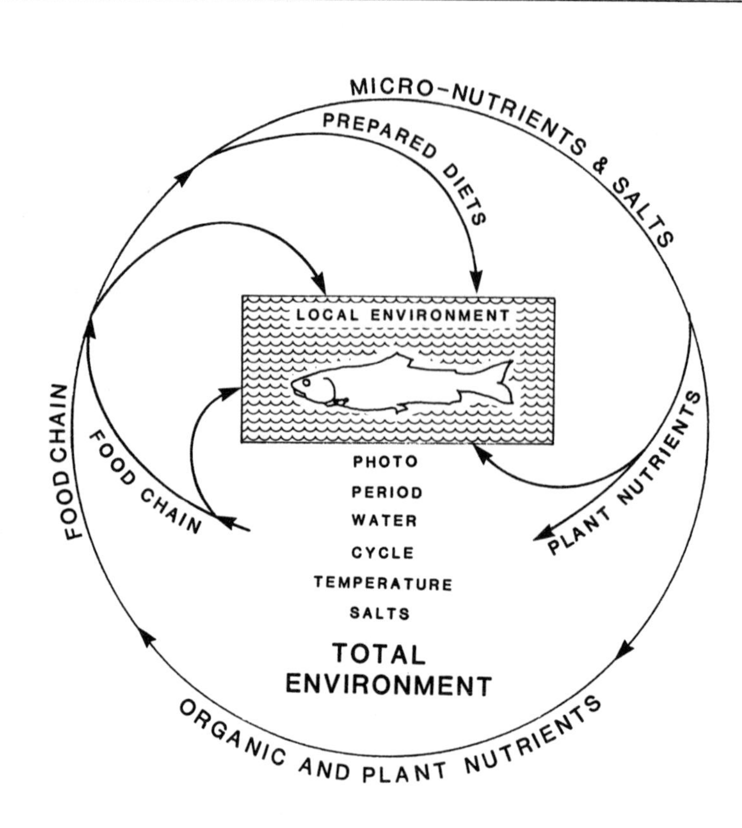

Figure 1

Factors affecting organoleptic qualities of fish. Figure from Spinelli 1979.

1979) indicating that feeding in excess of twice daily has no beneficial effect on food conversion and growth in fish with true stomachs (channel catfish and rainbow trout); however, Tilapia nilotica, which has a true stomach (Balarin and Hatton 1979), grew fastest when fed four to eight times daily (Lovell 1980). There is little information available for stomachless fishes like carp.

Shell (1967) found that the maximum growth rate of T. nilotica and T. mossambica fed 35 percent protein pellets does not necessarily accompany maximum feed utilization efficiency. These results possibly reflect differences in the quality, formulation, and physical consistency of the food as well as the mechanics and frequency of feeding. Although these findings were for Tilapia grown in troughs, the same results might occur in ponds.

Nose (1979) recommends that, ideally, feeding tables should be based on the dietary energy requirements of the fish and the digestible or metabolizable energy value of the fish feeds. However, since this information is usually not available, conventional practice is to feed a fixed percentage based on total fish body weight. A typical example for T. nilotica culture in the Philippines is (Guerrero 1980):

Fish Size	Feeding Rate
< 50 g	5 percent
50-100 g	4 percent
> 100 g	3 percent

According to Woynarovich (1975), when a pond population of common carp is unknown, the consumption of food supplied at specific feeding places within one hour indicates that about 1 percent of the total body weight was supplied or only the maintenance needs met; three to four hours is equivalent to 2-3 percent of the body weight; and six to eight hours is equivalent to 4-5 percent of the body weight.

For feeding methods, it is generally not advisable to broadcast feeds, because much will be lost in the bottom muds. Broadcasting is acceptable only when the pond is small, the stock crowded, and the fish very hungry. The best practice is to establish three to four fixed feeding places per hectare (Woynarovich 1975). If a boat is available, feeding in the interior of the pond is desirable; however, if only the shoreline is accessible, then the feeding area should not have soft mud and the water depth should range from 0.6-1.0 m. Hepher (1975) has described how feeding of cereal grains and pellets have become mechanized in many places. Container cars equipped with blowers blow the feed into the pond at fixed points. A less costly innovation has been the use of demand feeders originally developed for channel catfish culture in the United States, which presently enjoy widespread success in Israeli ponds.

FUTURE RESEARCH NEEDS

There are many variations in culturing fish, not only between and within fish species, but also between countries. Further study is required for optimizing the quality and quantity of both natural and artificial feeds needed for maintenance and growth by different species of pond fish at different life history stages, stocking intensities, and under varied environmental conditions. Besides defining the effects of individual constituents in a diet, it is equally important to balance their quantitative aspects for optimal performance in the interconnecting metabolic systems involved in growth. In addition, studies are needed of feed form, texture, size, smell, methods of feed distribution, and on the feeding behavior of fish at different life stages.

LITERATURE CITED

Andrews, S.W., and J.W. Page. 1974. Growth factors in the fishmeal component of catfish diets. J. Nutr. 104:1091-1096.

Aoe, H., I. Masuda, I. Abe, T. Saito, T. Toyoda, and S. Kitamura. 1970. Nutrition of protein in young carp. I. Nutritive value of free amino acids. Bull. Japn. Soc. Sci. Fish. 36:407-413.

Balarin, J.D., and J.P. Hatton. 1979. Tilapia: A Guide to Their Biology and Culture in Africa. University of Stirling, Scotland. 174 pp.

Boyd, C.E. 1979. Water Quality in Warmwater Fish Ponds. Auburn University Agricultural Experiment Station, Alabama. 359 pp.

Braaten, B.R. 1979. Bioenergetics - A review of methodology, pp. 461-504. In J.E. Halver and K. Tiews (eds), Finfish Nutrition and Fishfeed Technology. H. Heenemann GmbH and Co., Berlin. Vol. II.

Braekkan, O.R. 1977. Recent Advances in Animal Nutrition - 1977. Butterworth, London. 206 pp.

Castell, J.D. 1979. Review of lipid requirements of finfish, pp. 59-84. In Halver, J.E. and K. Tiews (eds), Finfish Nutrition and Fishfeed Technology. H. Heenemann GmbH and Co., Berlin. Vol. I.

Chen, T.P. 1970. Fertilization and feeding in coastal fish farms in Taiwan, pp. 410-416. In T.V.R. Pillay (ed), Coastal Aquaculture in the Indo-Pacific Region. FAO Fishing News Books Ltd., London, England.

Chow, K.W., and J.E. Halver. 1980. Carbohydrates, pp. 55-63. In Fish Feed Technology. UNDP/FAO, ADCP/Rep/80/11/.

__________, and W.R. Schell. 1980. The minerals, pp. 104-108. In Fish Feed Technology. UNDP/FAO, ADCP/Rep/80/11.

Collins, R.A., and M.V. Delmondo. 1976. Comparative economics of aquaculture in cages, raceways and enclosures, pp. 472-477. In FAO Technical Conference on Aquaculture, Kyoto. FIR/AQ/Conf/76/T.37.

Cowey, C.B. 1979. Protein and amino acid requirements of finfish, pp. 3-16. In J.E. Halver and K. Tiews (eds), Finfish Nutrition and Fishfeed Technology. H. Heenemann GmbH and Co., Berlin.

__________, and J.R. Sargent. 1972. Fish nutrition. Adv. Mar. Biol. 10:383-392.

Cremer, M.C. and R.O. Smitherman. 1980. Food habit and growth of silver and bighead carp in cages and ponds. Aquaculture 20:57-64.

Fischer, Z. 1979. Selected problems of fish bioenergetics, pp. 17-44. In J.E. Halver and K. Tiews (eds), Finfish Nutrition and Fishfeed Technology. H. Heenemann GmbH and Co., Berlin. Vol. I.

Guerrero, R.D. 1980. Cage culture of Nile tilapia. Asian Aquaculture 3:(12):4-6.

Halver, J.E. 1979. Vitamin requirements of finfish, pp. 45-58. In J.E. Halver and K. Tiews (eds), Finfish Nutrition and Fishfeed Technology. H. Heenemann GmbH and Co., Berlin. Vol. I.

__________. 1972. Fish Nutrition. Academic Press, New York. 713 pp.

__________. 1957. Nutrition of salmonid fishes. IV. An amino acid test for Chinook salmon. J. Nutr. 62:245-254.

__________, and K. Tiews. 1979. Finfish Nutrition and Fishfeed Technology. H. Heenemann GmbH and Co., Berlin. Volumes I and II.

Harris, L.E. 1980. Feedstuffs, pp. 111-170. In Fish Feed Technology. UNDP/FAO, ADCP/Rep/8011.

Hepher, B. 1979. Supplementary diets and related problems in fish culture, pp. 343-347. In J.E. Halver and K. Tiews (eds), Finfish Nutrition and Fishfeed Technology. H. Heenemann GmbH and Co., Berlin. Vol. I.

__________. 1975. Supplementary Feeding in Fish Culture. Proc. 9th Int. Congr. Nutrition, Mexico 1972, Vol. 3. 183-198. (Karger, Basel).

__________, J. Chervinski, and H. Tagari. 1971. Studies on carp nutrition. III. Experiments on the effect of fish yield on dietary protein source and concentration. Bamidgeh 23(1):11-37.

__________, E. Sandbank, and G. Shelef. 1979. Alternative protein sources for warmwater fish diets, pp. 327-341. In J.E. Halver and K. Tiews (eds), Finfish Nutrition and Fishfeed Technology. H. Heenemann GmbH and Co., Berlin. Vol. I.

Hickling, C.F. 1962. Fish Culture. Faber and Faber, London. 295 pp.

Hora, S.L., and T.V.R. Pillay. 1962. Handbook on Fish Culture in the Indo-Pacific Region. FAO Fisheries Biology Technical Paper No. 14. 204 pp.

Lall, S.P. 1979. Minerals in finfish nutrition, pp. 85-97. In J.E. Halver and K. Tiews (eds), Finfish Nutrition and Fishfeed Technology. H. Heenemann GmbH and Co., Berlin. Vol. I.

Ling, S.W. 1967. Feeds and feeding of warmwater fishes in ponds in Asia and the Far East. FAO Fish. Rep. 44(3):291-309.

Lovell, R.T. 1979a. Formulating diets for aquaculture species. Feedstuffs 51(27):29,32 (July 9).

__________. 1979b. Fish culture in the United States. Science 206:1368-1372.

__________, and L.A. Sackey. 1973. Absorption of channel catfish of earthy musty flavor compounds synthesized by cultures of blue-green algae. Trans. Amer. Fish. Soc. 102:774-777.

Lovell, T. 1980. Feeding Tilapia. Aquaculture Magazine 7(1):42-43.

Meyers, S.P. 1979. Formulation of water-stable diets for larval fishes, pp. 13-20. In J.E. Halver and K. Tiews (eds), Finfish Nutrition and Fishfeed Technology. H. Heenemann GmbH and Co., Berlin.

Moav, R., G. Wohlfarth, G.L. Schroeder, G. Hulat, and H. Barash. 1977. Intensive polyculture of fish in freshwater ponds. I. Substitution of expensive feeds by liquid cow manure. Aquaculture 10:25-43.

National Academy of Sciences. 1973. National Research Council, Subcommittee on warmwater Fishes, Nutrient Requirements of Warmwater Fishes. Washington, D.C. (Nutrient requirements of domestic animals). 1-78.

____________________. 1977. National Research Council, Subcommittee on Fish Nutrition, Nutrient Requirements of Trout, Salmon and Catfish. Washington, D.C. (Nutrient requirements of domestic animals). 11:1-57.

Noriega-Curtis, P. 1979. Primary productivity and related fish yield in intensely manured fish ponds. Aquaculture 17:335-334.

Nose, T. 1979. Diet composition and feeding techniques in fish culture with complete diets, pp. 283-296. In J.E. Halver and K. Tiews (eds), Finfish Nutrition and Fishfeed Technology. H. Heenemann GmbH and Co., Berlin. Vol. I.

Reay, P.J. 1979. Aquaculture. (The Institute of Biology's Studies in Biology: No. 106). University Park Press, Baltimore. 60 pp.

Schroeder, G.L. 1980. Fish farming in manure-loaded ponds, pp. 73-86. In S.V. Pullin and Z.H. Shehadeh (eds), Integrated Agriculture-Aquaculture Farming Systems. International Center for Living Aquatic Resources Management (ICLARM) Proceedings 4, Manila, Philippines.

______. 1978. Autotrophic and heterotrophic production of microorganisms in intensely manured fish ponds, and related fish yields. Aquaculture 14:303-325.

Schroeder, G. 1973. Factors affecting feed conversion ratio in fish ponds. Bamidgeh 25:104-113.

Shell, E.W. 1967. Relationship between rate of feeding, rate of growth and rate of conversion in feeding trials with two species of Tilapia, Tilapia mossambica Peters and Tilapia nilotica Linnaeus. FAO Fish Rep. 44(3):441-415.

Smith, R.R. 1980. Nutritional bioenergetics in fish, pp. 21-27. In Fish Feed Technology, UNDP/FAO, ADCP/Rep/80/11.

Spinelli, J. 1979. Influence of feed on finfish quality, pp. 345-352. In J.E. Halver and K. Tiews (eds), Finfish Nutrition and Fishfeed Technology. H. Heenemann GmbH and Co., Berlin. Vol. II.

______. 1980. Unconventional feed ingredients for fish feed, pp. 187-214. In Fish Feed Technology, UNDP/FAO, ADCP/Rep/80/11.

______, C. Mahnken, and M. Steinberg. 1979. Alternate sources of proteins for fish meal in salmonid diets, pp. 131-142. In J.E. Halver and K. Tiews (eds), Finfish Nutrition and Fishfeed Technology. H. Heenemann GmbH and Co., Berlin. Vol. II.

Swingle, H.S. 1968. Estimation of standing crops and rates of feeding fish in ponds. FAO Fish. Rep. 44(3):416-423.

Tal, S. and B. Hepher. 1967. Economic aspects of fish feeding in the Near East. FAO Fish. Rep. 44(3):285-290.

Tang, Y.A., and S.H. Chen. 1967. A survey of the algal pasture soils of milkfish ponds in Taiwan. FAO Fish. Rep. 44(3):365-372.

______, and T.L. Hwang. 1967. Evaluation of the relative suitability of various groups of algae as food of milkfish in brackishwater ponds. FAO Fish. Rep. 44(3):365-372.

Tapiador, D.D., H.F. Henderson, M.H. Delmendo, and H. Tsutsui. 1977. Freshwater Fisheries and Aquaculture in China. FAO Fish. Tech. Paper No. 168, FIT/T168. 85 pp.

Utne, F. 1979. Standard methods and terminology in finfish nutrition, pp. 437-444. In J.E. Halver and K. Tiews (eds), Finfish Nutrition and Fishfeed Technology. H. Heenemann GmbH and Co., Berlin. Vol. II.

Vale, G.W., G.S. Sidhu, W.A. Montgomery, and A.R. Johnson. 1970. Studies on kerosene-like taint in mullet (Mugil cephalus). I. General nature of the taint. J. Sci. Food Agric. 21(8):429-432.

van Limborgh, C.L. 1979. Industrial production of ready to use feeds for mass rearing of fish larvae, pp. 3-11. In J.E. Halver and K. Tiews (eds), Finfish Nutrition and Fishfeed Technology. H. Heenemann GmbH and Co., Berlin. Vol. II.

Webb, P.W. 1978. Partitioning of energy into metabolism and growth, pp. 184-214. In S.D. Gerking (ed), Ecology of Freshwater Fish Production. Blackwell Scientific, London.

Woynarovich, E. 1975. Elementary Guide to Fish Culture in Nepal. FAO. 131 pp.

Yashouv, A. 1956. Problems in carp nutrition. Bamidgeh 8(5):79-87.

Zeitoun, I.H., J.E. Halver, D.E. Ullrey, and P.I. Tack. 1973. Influence of salinity on protein requirements of rainbow trout (Salmo gairdneri) fingerlings. J. Fish. Res. Bd. Can. 30:1867-1873.

______, D.E. Ullrey, J.E. Halver, P.I. Tack, and W.T. Magee. 1974. Influence of salinity on protein requirements of coho salmon (Oncorhynchus kisutch) smolts. J. Fish Res. Bd. Can. 31:1145-1148.

POND PRODUCTION SYSTEMS: WATER QUALITY MANAGEMENT PRACTICES

by

Arlo W. Fast

INTRODUCTION

Water quality for aquaculture is defined here simply as "the degree of excellence that a given water possesses for the propagation of desirable aquatic organisms." The required quality is a function of the specific culture organisms and it has many components that are complexly interwoven. Sometimes a component of water quality can be dealt with individually, but, because of the complex interactions of components, one must usually view the total array and attempt to determine which components are critically limiting the culture organisms. Often, several components must be manipulated simultaneously to achieve a stable improvement in the "degree of excellence."

Water quality is one of the most important factors affecting successful pond fish culture. If water quality is excellent, then survival, growth, and reproduction can achieve high values; otherwise fish production will be reduced or impossible.

Some of the more commonly cultured warmwater pond fish species are relatively tolerant of poor water quality. They can exist and grow over a wide range of salinity and temperature, and they can tolerate low oxygen concentrations for brief periods (Table 1). Those species were undoubtedly selected by early aquaculturists for their hardiness. The milkfish, for example, can be reared over a salinity range of less than 10°/oo to more than 150°/oo (more than 4 times the salinity of seawater). Mullet and tilapia can be reared in fresh or marine waters, and all three are tolerant of low oxygen conditions. All of the other commonly cultured species, especially the carps and catfish, are very tolerant of low oxygen. Most are freshwater species, although they are also reared in brackish-water.

Maximum fish growth occurs with warm waters, high rates of photosynthesis (or feeding rates), a large range in oxygen fluctuations, and with water quality conditions far from pristine. High quality water for fish culture would not meet water quality standards for domestic drinking water. We must therefore keep in mind the intended use of the water and the water quality necessary to meet the intended use.

Many potential water quality problems can be avoided by proper pond siting. Ponds sited on unsuitable soil types, in areas with water of inadequate quantity or quality, or where temperature or salinity are too high or low are doomed to failure or marginal success. The time and money required to evaluate a potential site thoroughly are worth the investment. This evaluation will usually detect potential water quality problems before they occur and thus allow the aquaculturist to select a better alternative site if there is one.

There are a very large number of possible pollutants (used here to include both natural and man-induced situations) that can deleteriously affect aquacultural operations. Many authors credit industrial pollution and an uncontrolled world population increase for the substantial reduction in both wild fish stocks and aquaculture potential (Bardach 1978; Borgstrom 1978; Chen 1976). The list of these pollutants is indeed long. Some of the better known pollutants are reviewed by McKee and Wolf (1963), Environmental Protection Agency (1971, 1976), National Academy of Science and National Academy of Engineering (1972), Thurston et al. (1979), Colt et al. (1975), and Friedman and Shibko (1972).

In this chapter I will describe some of the more important components of water quality and discuss their management practices. Although we discuss each component individually, bear in mind that these are not mutually exclusive aspects and that the management of one aspect will affect others.

For further information on water quality principles and practices, one should refer to the publications cited in this review. Water Quality in Warmwater Fish Ponds by Dr. Claude E. Boyd (1979) is particularly well written and an excellent resource publication. This review relies heavily on that publication for illustrations.

Sampling practices or analytical procedures are not covered in this treatise. These are covered thoroughly elsewhere, especially in "Standard Methods for the Examination of Water and Wastewater," American Public Health Assoc. et al. (1971,1976); Boyd (1979); and Limnologist Methods, Welch (1948). Limnological concepts and principles are also important for the understanding of

Table 1

SALINITY, TEMPERATURE AND DISSOLVED OXYGEN VALUES FOR SELECTED WARMWATER POND FISHES[a]

	Salinity,[b] %			Temperature, °C			Min. Oxygen, mg/l	References
	Min.	Optimum	Max.	Min.	Optimum	Max.		
Chanos Milkfish	F	10-50	157	12-15	25-36	39-41	1.5	Schuster 1960; Bardach et al. 1972; Helfrich 1981; Crear 1981; Crear 1980
Mugil Mullet	1	18	50	5	21-33	35-40	2.5-7.0	Abbott 1981; Kuo 1979; Kuo et al. 1975;
Chlarias Catfish	F	F	S	20	29-32	37	5 (fry) 0.0 (adult)	Duodoroff and Shumway 1970; Bell and Canterbery 1976; Hora and Pillay 1962; Collins 1977
Ictalurus punctatus Catfish	F		8-12	8-20	25-30	39	0.8-2.0	Duodoroff and Shumway 1970; Colt et al. 1975
Tilapia and Sarotherodon Tilapia	--[c]	0-35	69	10-15	25-33	35-42	0.2-2.0	Balarin and Hatton 1979
Cyprinus carpio European carp	F		11	0-3	20	35	0.2-2.8	Duodoroff and Shumway 1970; Nakamura 1948; Black 1953; Sigler 1958; La Rivers 1962; Needham 1950; Burns 1966
Ctenopharyngodon idella Grass carp	F		7-10	12.8	18.3-29.4	35.5	0.2-0.6	Hickling, 1962; Duodoroff and Shumway 1970; Kilambi and Robinson 1979; Hora and Pillay 1962;
Hypophtalmichthys molitrix Silver carp	F		7-10	15	20-28	30	0.3-1.1	Hickling 1962; Duodoroff and Shumway 1970; Hora and Pillay 1962; Ling 1977; Chen 1976
Carp polyculture	F		--[c]	21.5	--[c]	--[c]	4	Dimitrov 1974
Common carp, Grass carp, Bighead carp and Silver carp	--[c]	--[c]	--[c]	21	27-29	32	0.5	Bortz et al. 1977
Catla catla Indian carp	F		13	14.4	26-29	34	0.7-1.0	Hickling 1962 Duodoroff and Shumway 1970; Bell and Canterbery 1976; Hora and Pillay 1962

[a] Salinity and temperature values are generally for long-term periods (e.g. months), whereas minimum oxygen values are generally for less than one day.

[b] F = freshwater, S = slightly brackish, c = information not available to author, or not determined.

pond ecology, or how the pond functions. These are covered in more detail in Welch (1952); Ruttner (1953); Hutchinson (1957, 1967, 1975); Cole (1975), Wetzel (1975); and Moss (1980).

Through the remainder of this chapter, discussion of some of the water-quality parameters and practices will be restricted to those that are considered particularly important and useful. Lastly, a few problem areas worthy of further research are listed.

DISSOLVED OXYGEN (DO)

Dissolved oxygen concentration is one of the most important water-quality parameters. Oxygen depletion is usually the principal cause of sudden, massive fish kills. Maintaining a "normal" or desirable oxygen regime in a pond not only helps assure the fish's health, but also indicates that the pond system is functioning suitably.

A productive pond will typically have supersaturated DO during the late afternoon, and undersaturated DO at dawn (Fig. 1). This daily cycle may range between 200 percent and 25 percent of saturation. Most warmwater pond fishes are quite tolerant of temporary low oxygen concentrations (Table 1). Maximum fish production rates are possible under these conditions provided that the night-time DO does not fall below 1 to 2 mg/l. Oxygen concentrations below this range indicate that undesirable conditions are developing in the pond and that corrective actions may be necessary.

Supersaturated DO has been implicated in fish kills (Weitkamp and Katz 1980), but supersaturation is not normally lethal. However, highly supersaturated DO during the day may indicate depletion to near zero at night due to the respiration of the dense plant population necessary to produce the supersaturation.

We can better understand how DO depletion occurs in a fish pond if we

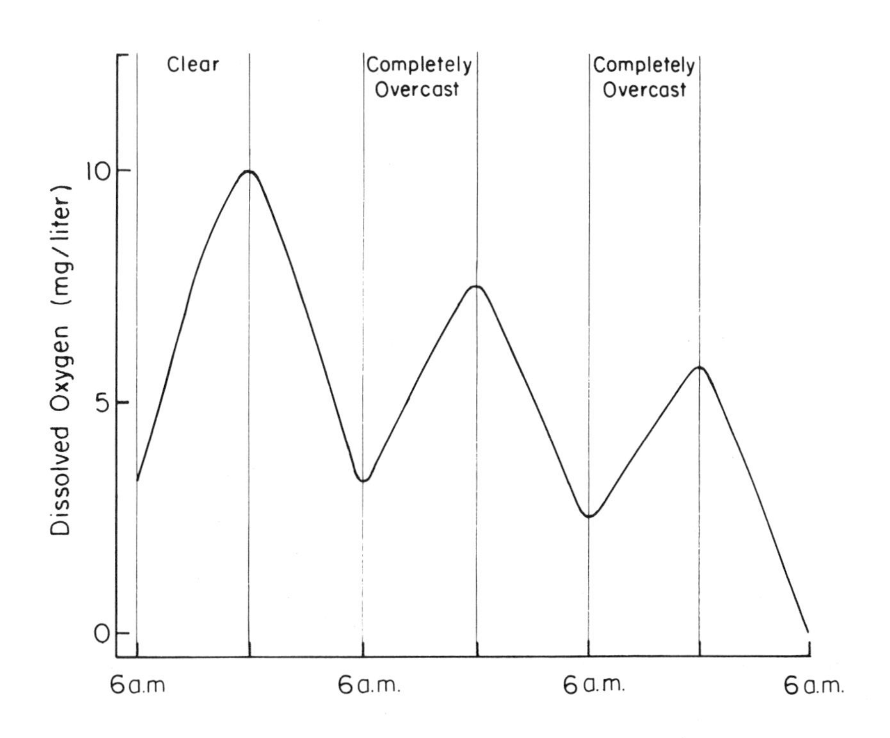

Figure 1

Characteristic dissolved oxygen cycle in a fish pond. Maximum DO occurs in the afternoon due to photosynthetic oxygen production, while DO minimum occurs at dawn due to nighttime respiration. Overcast weather can reduce photosynthesis and lead to oxygen depletion during the night. Figure from Boyd (1979).

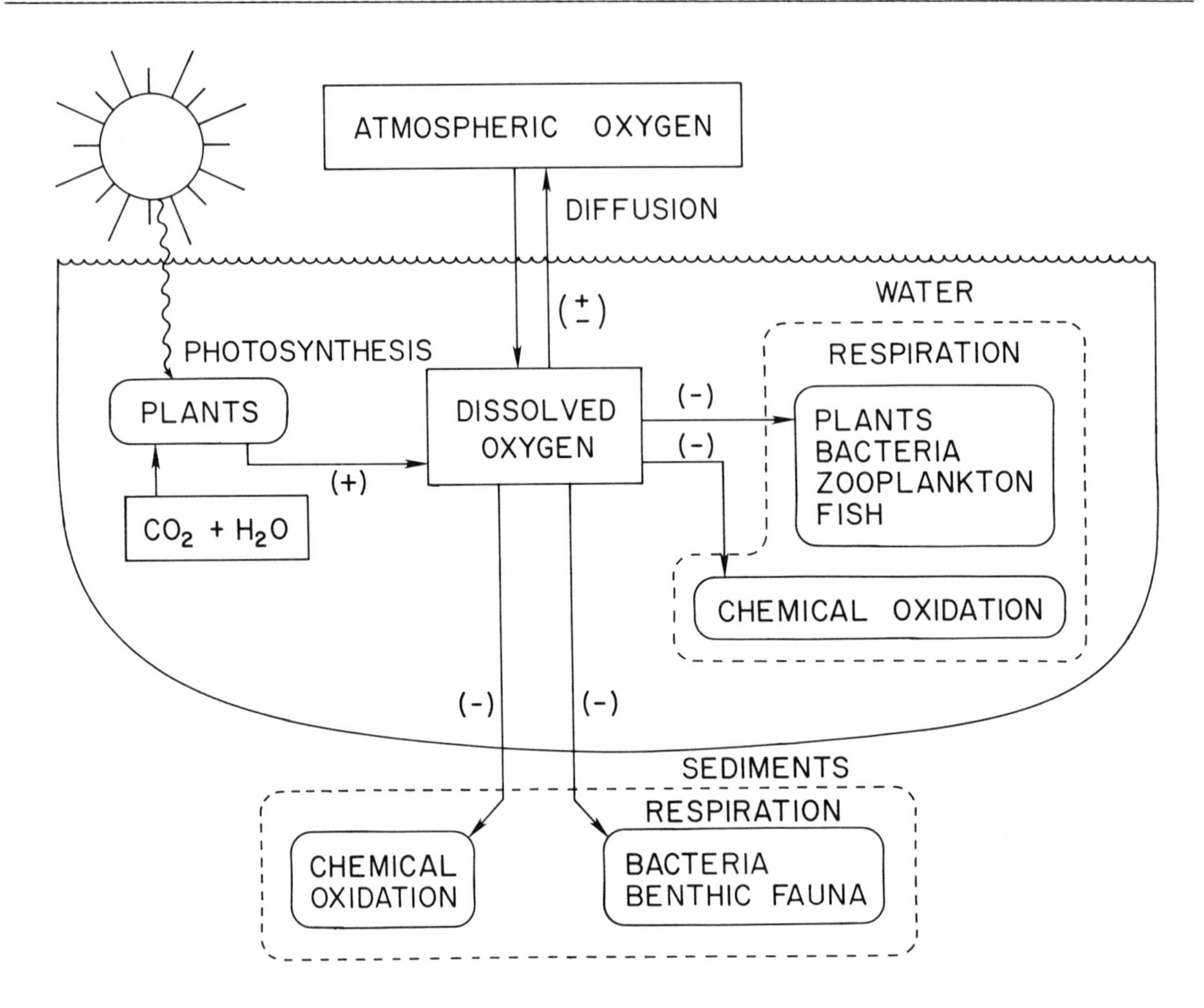

Figure 2

Principal sources and sinks for dissolved oxygen in a fish pond.

describe the interactions between the main variables which result in the daily DO cycle (Fig. 2). These variables include: <u>Photosynthesis</u>, <u>Diffusion</u> and <u>Respiration</u>.

A healthy, dense plant population will produce large amounts of DO by photosynthesis, and thus cause DO supersaturation during the day. Photosynthesis is often the principal source of oxygen, and in a healthy pond it exceeds the rate of oxygen consumption or losses during a daily cycle (Fig. 2).

Oxygen may be either gained or lost through diffusion into or out of a pond. Oxygen diffuses into the pond (gain) from the atmosphere when the pond water is undersaturated with oxygen. DO saturation values are largely a function of water temperature, elevation (barometric pressure), and salinity (Table 2). Oxygen diffuses out of a pond (a loss) when the DO in the pond is supersaturated. The rate of diffusion in or out increases substantially the greater the DO deviates from 100 percent saturation. This rate of diffusion (and direction) is shown in Table 3 (Schroeder 1975; Boyd 1979). For example, a pond with only 50 percent DO saturation at dusk will gain 1.69 mg/l during the night, whereas a pond with 250 percent DO saturation at dusk will lose 2.9 mg/l of DO during the night.

The respiratory consumption of DO by fish, plankton, and other organisms living in the water or on or in the sediments accounts for the greatest DO loss from the pond water. This oxygen is used to maintain aerobic metabolism and is converted to CO_2 and other products.

Table 2

SOLUBILITY OF OXYGEN (mg/l) IN WATER AT DIFFERENT TEMPERATURES, ELEVATIONS, AND SALINITIES (Dissolved Oxygen Values Are for 100% Saturation in Water Exposed to Water-Saturated Air)[a]

Temperature, °C	Elevation, ft (m)									
	0 (0)	500 (152)	1000 (305)	1500 (457)	2000 (610)	2500 (762)	3000 (915)	3500 (1067)	4000 (1220)	4500 (1372)
0	14.6	14.3	14.1	13.8	13.6	13.4	13.1	12.9	12.6	12.4
5	12.8	12.5	12.3	12.1	11.9	11.7	11.5	11.2	11.0	10.8
10	11.3	11.1	10.9	10.7	10.5	10.3	10.1	9.9	9.8	9.6
15	10.1	9.9	9.7	9.5	9.4	9.2	9.0	8.9	8.7	8.6
20	9.1	8.9	8.8	8.6	8.4	8.3	8 1	8.0	7.8	7.7
25	8.2	8.1	8.0	7.8	7.7	7.5	7.4	7.2	7.1	7.0
30	7.5	7.4	7.3	7.1	7.0	6.9	6.7	6.6	6.5	6.4
35	6.9	6.8	6.7	6.6	6.4	6.3	6.2	6.1	6.0	5.8
40	6.4	6.3	6.2	6.0	5.9	5.8	5.7	5.6	5.5	5.4
	Salinity, parts per thousand (saturation values at sea level)									
	0	5	10	15	20	25	30	35	40	45
0	14.6	14.1	13.6	13.2	12.7	12.3	11.9	11.5	11.1	10.7
5	12.8	12.3	11.9	11.6	11.2	10.8	10.5	10.1	9.8	9.5
10	11.3	10.9	10.6	10.2	9.9	9.6	9.3	9.0	8.8	8.5
15	10.1	9.8	9.5	9.2	8.9	8.6	8.4	8.1	7.9	7.6
20	9.1	8.8	8.6	8.3	8.1	7.8	7.6	7.4	7.2	7.0
25	8.2	8.0	8.8	7.6	7.4	7.2	7.0	6.8	6.6	6.4
30	7.5	7.3	7.1	6.9	6.8	6.6	6.4	6.2	6.0	5.9
35	6.9	6.8	6.6	6.4	6.2	6.1	5.9	5.8	5.6	5.5
40	6.4	6.2	6.1	5.9	5.8	5.6	5.5	5.4	5.2	5.1

[a] Data from Colt 1980; Weiss 1970

Mud respiration may range between 8 and 125 mg $O_2/m^2/hr$ (Mezainis 1977; Schroeder 1975). For a 1-meter deep pond, these values correspond to oxygen consumption rates from the water of 0.1 and 1.5 mg O_2/l per 12 hours. Much higher oxygen consumption rates by the mud are possible if the mud is suspended in the water, or if excessive amounts of organic materials (e.g., manure or feed) are added to the pond.

Fish respiration is primarily related to fish density (kg/hectare) and water temperature. Respiration rates typically range between 65 and 888 mg O_2/kg of fish/hour (Boyd 1975). Thus if a 1-meter deep pond stocked with 4,000 kg of fish/hectare contains fish respiring at 400 mg O_2/kg/hr, the oxygen consumption in the pond from fish respiration would equal 1.9 mg O_2/l/12 hrs.

From the foregoing, we see that nighttime oxygen depletion from respiration could equal:

O_2 consumption, mg/l/12 hrs

	Min.	Max.
Mud	0.1	1.5
Fish	0.3	4.2
BOD	0.2	8.4
	0.7	15.2

If the rate of respiration is near the higher value, and if DO at dusk is less than 16 mg/l, then oxygen depletion to near zero could occur before dawn.

From the above values, water BOD has the greatest potential for depleting DO. Most often high BOD corresponds to the decay of a dense growth of phytoplankton or macrophytes. The plants' death not only deprives the pond of its principal source of oxygen (photosynthesis), but it also creates a greatly increased water BOD due to the decay and the bacterial respiration on the dead plants. Mass death of plants often occurs for unknown reasons. Other times their death or moribund condition is caused by inclement weather or by toxic substances entering the pond. Substances (such as herbicides or algicides) may be nontoxic to the animals at certain concentrations but very toxic to the plants.

Dissolved Oxygen Management Techniques

There are a number of possible means to avert excessive oxygen depletion or to correct low DO when it does occur.

Table 3

PREDICTED GAINS (+) AND LOSSES (-) OF DISSOLVED OXYGEN FROM A FISH POND DURING THE NIGHT BASED ON OBSERVED OXYGEN SATURATION AT DUSK (These values are for a 1-m deep pond, for a 12-hour period)[a]

DO concentrations at dusk, % of air saturation	Gain or loss of DO during the night, mg/L
50	+1.69
60	+1.49
70	+1.18
80	+1.00
90	+0.77
100	0.44
110	0.16
120	-0.18
130	-0.55
140	-0.94
150	-1.48
160	-1.64
170	-1.82
180	-1.98
190	-2.11
200	-2.37
210	-2.42
220	-2.54
230	-2.67
240	-2.76
250	-2.91

[a]Table from Boyd 1979, modified from Schroeder 1985

Predicting a low DO level is highly desirable and should minimize fish losses. Romaire et al. (1978) and Boyd (1979) describe a predictive procedure for catfish and tilapia ponds based on water temperature, fish density (lb/acre) and DO concentrations at dusk. Pond Secchi disc or chemical oxygen demand (COD) values are used to predict safe or unsafe DO conditions. One example of this technique is shown in Table 4 for 1,000 lb/acre of channel catfish, and for safe and unsafe Secchi disc values. This predictive procedure is applied at dusk each day.

Another predictive technique involves measuring DO concentrations during the night and then projecting minimum DO values at dawn (Fig. 3, Boyd 1979). This procedure gives less lead time than the previous procedure, but it can be useful in some cases, and it does not require much knowledge of pond-specific conditions like fish standing crop.

Familiarity with the pond is equally important. An aquaculturist with intimate knowledge of the pond's behavior and history can often predict oxygen depletions, fish diseases, or other problems before they strike. Water color and fish behavior are clues. This knack is difficult to quantify, but is an essential part of most successful aquaculture projects.

Continuous aeration or circulation of pond water is another technique for preventing DO depletion, while at the same time increasing fish production. Loyacano (9174) increased average oxygen concentrations from 3.1 mg/l in non-aerated ponds to 4.5 mg/l in ponds continuously receiving 10.4 m^3 min/ha of compressed air (Table 5). Fish production was respectively increased from 2,736 kg/ha to 5,510 kg/ha. Busch and Goodman (1981) artificially circulated a catfish pond 18 hours each day with a low energy paddlewheel device. Compared with a noncirculated pond, oxygen concentrations were higher in the circulated pond, and fish production was 4.540 kg/ha in the circulated pond vs. only 3,450 kg/ha in the noncirculated pond. Emergency aeration was required on several occasions in the noncirculated ponds to prevent serious DO depletion and fish kill. The net result was that continuous circulation required less total energy and was less expensive than the emergency aeration in the noncirculated pond ($54.89/ha vs. $91.60/ha).

The mechanism whereby continuous aeration or circulation causes higher DO and greater fish production is not well understood. However, it could be related to two factors:

1. Continuous circulation reduces thermal stratification and thereby helps maintain aerobic conditions throughout

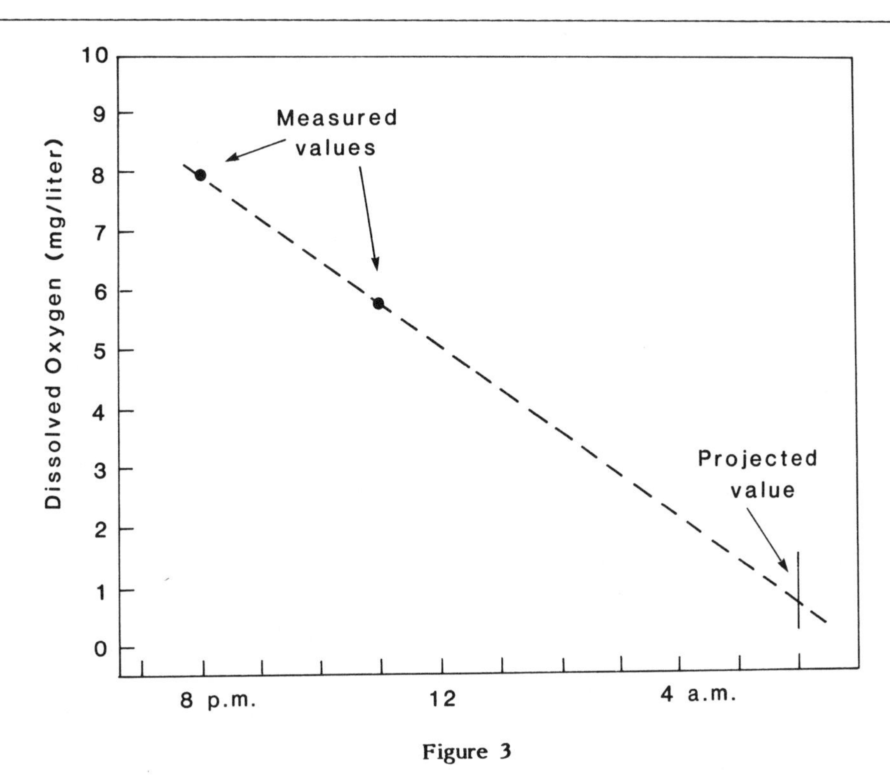

Figure 3

A method for predicting minimum dissolved oxygen during the night. This method requires at least two DO observations during the night. Figure from Boyd 1979.

the pond. This should result in a more constant decomposition of organic matter and even distribution of respiratory load.

2. Continuous circulation may also result in more oxygen diffusion into the pond and may help prevent phytoplankton die-offs. These die-offs greatly reduce the photosynthetic input of DO to the pond and at the same time greatly increase the respiratory load due to decomposition of the algae. Boyd (1975, 1979) found that these die-offs were invariably associated with dense growths of blue-green algae and calm, clear, warm weather. The calm weather allowed the blue-greens to float to the water's surface where intense sunlight killed them. A system of continuous circulation could prevent this occurrence.

Excessive fertilization from inorganic fertilizers, manures, feeds, or watershed run-off is another cause of oxygen depletion. Excessive fertilization normally leads indirectly to oxygen depletion by causing excessive algal growth. Excessive algal growth in turn can lead to DO depletions should the algae die suddenly. Controlled fertilization is a means of preventing DO depletion.

Emergency Aeration

Once the DO drops to a dangerously low value (e.g. 1-2 mg/l) and the fish show signs of distress, quick action must be taken to prevent fish loss. There are several emergency-aeration techniques.

Flushing Flushing of high DO water into low DO pond water is effective and inexpensive, provided an adequate supply of high DO water is available. Sources of such water include nearby streams, wells, ponds, or coastal waters. Gravity flow is the least expensive, but not often available. The water normally must be pumped, and most often from adjoining fish ponds.

Mechanical Aeration Emergency aeration may be achieved by either injecting air into water or by spraying the water into the air. The latter technique is more efficient in terms of kg O_2 dissolved per kw-hr (Rappaport et al. 1976; Boyd and Tucker 1979; Busch et al. 1974). Emergency aeration devices include: paddlewheel aerators that circulate and splash water into the air; floating sprayer types that pump water from below the surface and spray it into the air; air blowers that inject air either at one point in the pond or through a perforated

Table 4

CRITICAL SECCHI DISC DEPTHS (in cm) FOR A 1-METER DEEP FISH POND AS A FUNCTION OF WATER TEMPERATURE AND DISSOLVED OXYGEN CONCENTRATION AT DUSK[a,b]

Temperature, °C	DO concentration at dusk, mg/L										
	2	3	4	5	6	7	8	9	10	11	12
20	37	S	S	S	S	S	S	S	S	S	S
21	58	26	S	S	S	S	S	S	S	S	S
22	79	42	21	S	S	S	S	S	S	S	
23	90	58	32	16	S	S	S	S	S	S	S
24	100	69	42	26	S	S	S	S	S	S	
25	100	79	53	37	21	S	S	S	S	S	S
26	100	85	63	48	32	16	S	S	S	S	S
27	100	90	69	53	37	26	S	S	S	S	S
28	100	95	74	58	45	32	21	S	S	S	S
29	100	95	79	63	53	40	29	18	S	S	S
30	100	100	85	69	58	45	34	26	16	S	S
31	100	100	87	74	63	50	40	32	21	S	S
32	100	100	90	79	66	55	45	37	29	18	S

[a] Table from Boyd 1979.

[b] In the example, the pond contained 1,120 kg/ha (1,000 lb/acre) of channel catfish. Observed Secchi disc values of less than the table values indicate that DO concentrations will drop below 2 mg/l by dawn. An (S) indicates that DO will not drop below 2 mg/l regardless of observed Secchi disc values.

pipe; large-volume water pumps that pump water from the pond (or another source) and spray into the air above the treatment pond; and venturi aerators that suck air into a pipe through which the low DO water is pumped. Of these systems the paddlewheel design is generally considered the most effective.

Two other mechanical aeration devices that utilize pumped water (or gravity flow) are the inclined plane and the packed column. Wirth (1981) described the use of an inclined plane, or cascade aerator, to prevent winterkill in Wisconsin lakes. This aerator is trailer mounted and may be towed to location. It measures 2.4 m (8 ft) wide by 11 m (36 feet) long and is basically a rectangular box with baffles. In operation, the box is tilted so that the upper end is about 2.4 m (8 ft) above the ground. Wirth pumped 38 m^3/min (10,000 gpm) through this aerator and brought the DO from 1 to 11 mg/l on one pass. Owsley (1978) describes a packed-column aeration device which effectively increased DO to saturation while at the same time

Table 5

AVERAGE FISH PRODUCTION RATES, AVERAGE FISH SIZE, AND AVERAGE OXYGEN CONCENTRATIONS IN WHITE CATFISH PONDS RECEIVING THREE LEVELS OF ARTIFICIAL AERATION[a,b]

Air injection volume, m^3/min•ha	Average fish production rate, kg/ha	Average fish size, g	Average oxygen concentration, mg/l
0	2,736	194	3.1
6.9	4,562	287	3.9
10.4	5,510	317	4.5

[a] Data from Loyacano 1974.

[b] Air injection was continuous in the aerated ponds. Twelve ponds (four for each aeration level) were used.

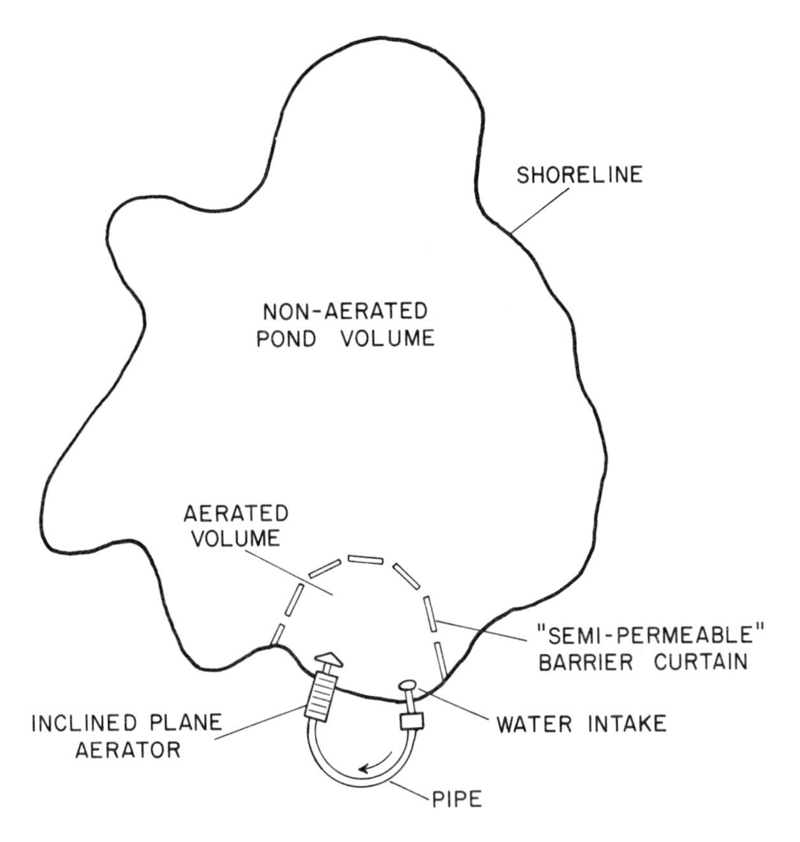

Figure 4

A proposed technique for partial aeration of a fish pond. The aerated portion is separated from the main pond area by a semipermeable barrier. An inclined plane aeration system is shown in the aerated volume, although other aeration devices could be used.

reducing the concentration of other gases like nitrogen (N_2). The packed column consists of a pipe packed with plastic elements. Water cascades through the aerator and forms a thin surface film on the plastic elements. This increases oxygen transfer between the air in the packed column and the water. Both the inclined plane and the packed column aerators may have special uses for emergency pond aeration.

Partial Aeration Most emergency aeration devices are intended or designed to aerate the entire pond. This may not be necessary. Aeration of only a portion of the pond is an alternative. The major advantage of partial aeration is that a much smaller aeration system is needed.

To the best of my knowledge, partial aeration has not yet been attempted. It could be used if a removable partition could be quickly installed when aeration is needed (Fig. 4). A plastic sheet suspended from a float line should suffice. Holes in the plastic would permit entrance of the fish. The fish will readily home on the higher DO within the sheet. They would remain within the aerated portion until the DO was again suitable in the larger pond area, at which time the sheet could be removed and the fish dispersed. Partial aeration could be used with any of the aforementioned aeration devices. The device should be portable so that it could be moved to the affected ponds. This technique may have special applications in very large ponds that would be difficult and expensive to aerate otherwise.

TEMPERATURE

Effects on Fish Growth

Water temperature has a profound influence on the species of fish that can be cultured, growth rates, the quality of the fishes' flesh, food conversion efficiency, and the economics of a fish culture operation.

Fish typically have a series of growth curves related to water temperature and feeding rate (Fig. 5). These curves show a maximum growth rate for a given temperature and feeds rate. For example, when sockeye salmon are fed in excess they have a maximum growth rate of 1.4 percent of their body weight per day, at a temperature of about 15°C. At temperatures above or below 15°C, or at lower feed rates, growth is reduced. These growth data can be further refined to establish the most efficient or economical set of conditions for a given species (Brett et al. 1969). Although the curves shown in Fig. 5 are for sockeye salmon, other species have similar curves with different values.

Colt et al. (1975) reviewed temperature effects on channel catfish. They report that channel catfish: cease feeding below 8 to 10°C; have optimum secretion of digestive enzymes at 23.9°C and maximum digestion rates between 26 and 30°C; show optimum growth near 30°C and decreasing growth above 32°C even with unlimited feed); and have an upper lethal temperature of 39°C. In addition, the "quality" of the catfish also changes as a function of temperature. They report a linear increase in the whole carcass fat content of channel catfish from 23.8 percent to 43.6 percent in fish reared over the temperature range of 18 to 34°C.

Although such details are largely unknown for most commonly cultured warmwater fish species, it is generally known over what temperature range a fish species performs best. A fish culturist

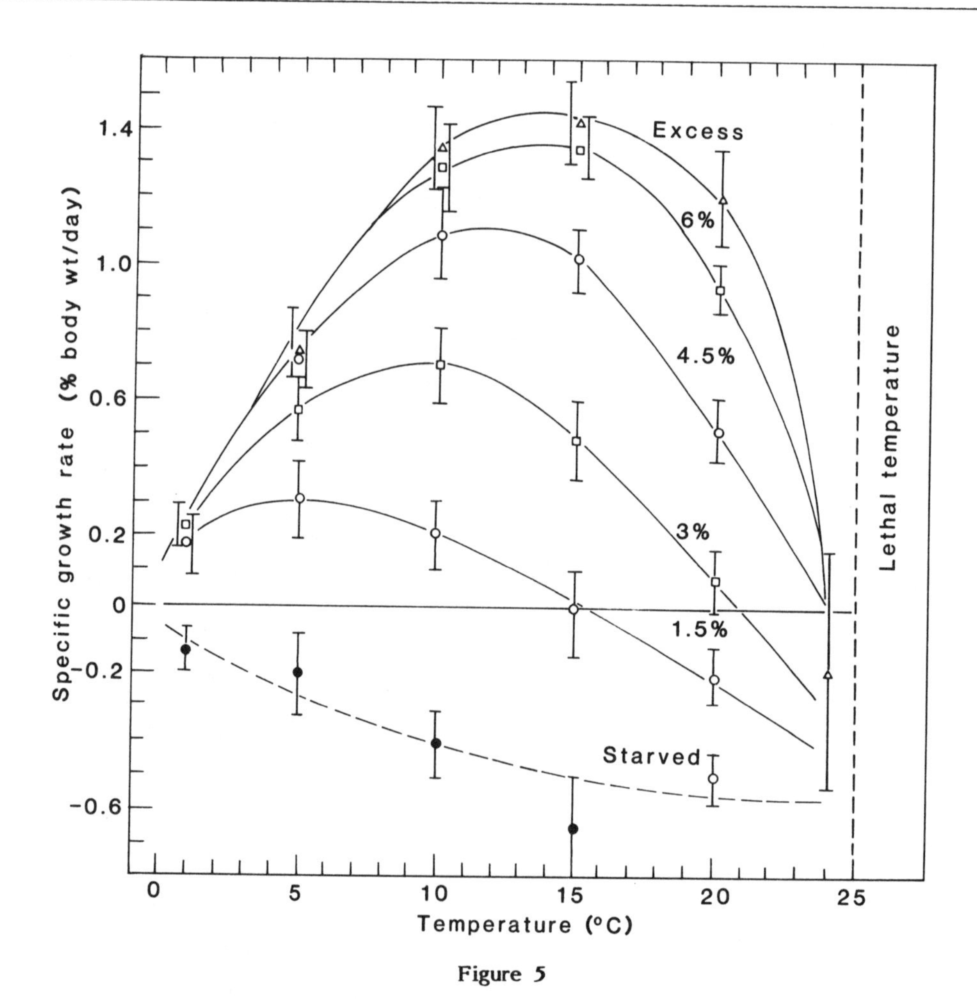

Figure 5

Growth rates of sockeye salmon as a function of water temperature and feeding rates. Figure from Brett et al. 1969.

should try to site ponds in a location that will maximize the growth potential for the target species. Conversely, if the ponds are already existent or must be sited at a given location, then species that are best suited for these conditions should be selected. Application of this advice may require the accumulation of site-specific temperature data for a year or more if the data does not already exist for comparable sites. This can save the considerable expense of an unsuitable site.

It is usually impractical to heat or cool water artificially. The energy requirement is great, especially if the water has much flow-through. However, some unique situations exist whereby the aquaculturist can capitalize on natural or man-made conditions to optimize his water temperatures for fish culture. Springs or wells often contain water of an ideal and uniform temperature. This has led to the establishment of a large trout industry in Idaho (Klontz and King 1974). Power plant cooling waters can yield temperatures nearly ideal for the growth of warmwater fish (Guerra et al. 1979; Ford et al. 1975). Fast (1977, 1979) has proposed a system of fish culture that capitalizes on natural thermal stratification in lakes, quarries, reservoirs, or the ocean. By selective depth withdrawal within a thermally stratified water body, optimum water temperatures and natural feed items may be provided nearly year-round. Lastly, many tropical or subtropical locations like Hawaii have nearly ideal water temperatures year-round for many warmwater species. Not only is the temperature near the optimum for feed conversion and growth, but there is little seasonal fluctuation in temperature, light, or other conditions required to maintain a highly productive pond ecology.

Thermal Stratification

Thermal and chemical stratification can develop in even shallow ponds. Boyd (1979) observed diurnal thermal stratification in shallow ponds averaging 1-m deep at Auburn, Alabama, where they "stratify during daylight hours in warm months only to destratify at night when the upper layers cooled. Larger, deeper ponds (0.5 ha or more with average depths of 1.5 to 2.0 m) may remain stratified throughout the warm months." In Boyd's shallow ponds, oxygen depletion may occur over much of the pond bottoms during a part of each day. In the deeper ponds, the oxygen may become depleted over much of the pond bottoms for much of the summer (Fig. 6 and 7).

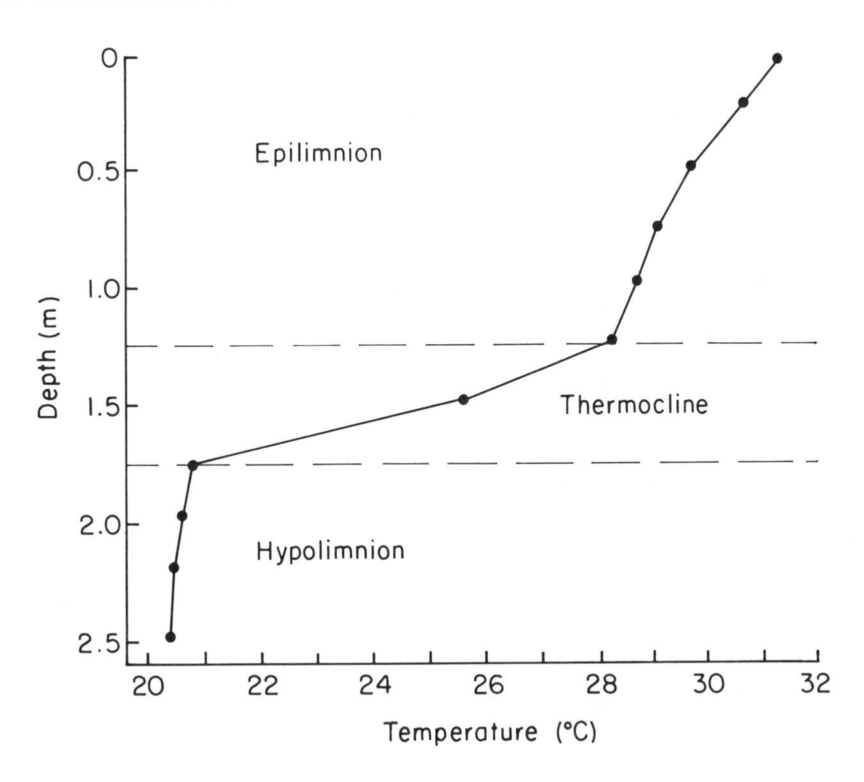

Figure 6

Thermal stratification in a fish pond. Figure from Boyd 1979.

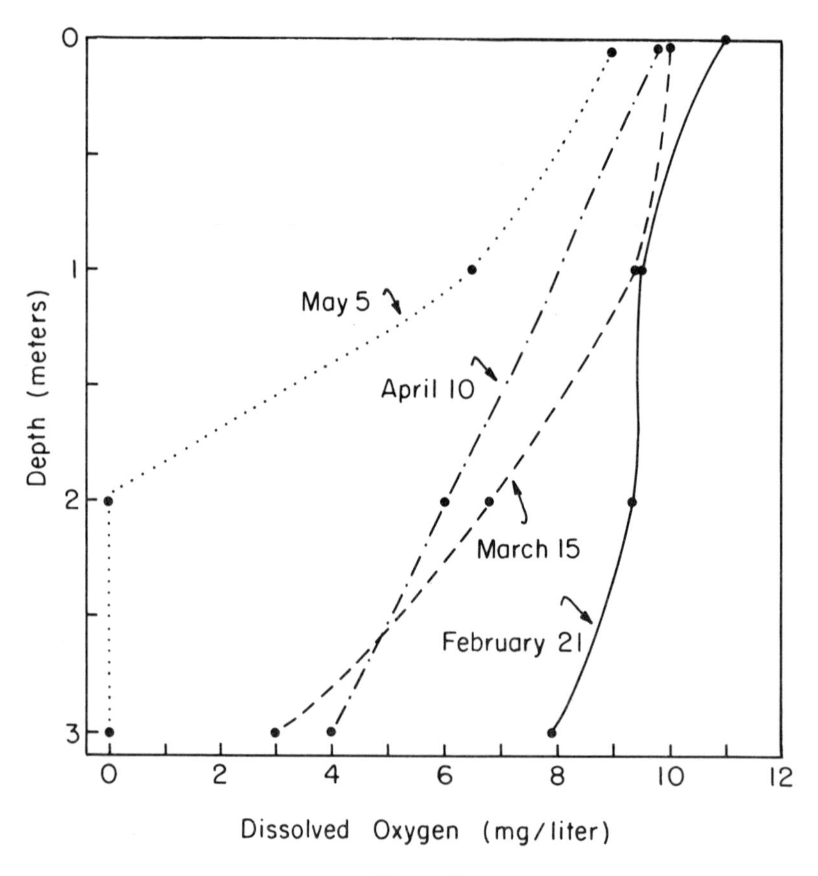

Figure 7

Oxygen depletion in a thermally stratified fish pond at Auburn, Alabama. Thermal stratification and high turbidity led to oxygen depletion below the 2 m depth. Figure from Boyd 1979.

The effect of this stratification on pond productivity is not well documented. However, it is reasonable to assume that such stratification could render these bottom areas uninhabitable to the fish, as it does in deeper water bodies (Miller and Fast 1981). It may also reduce the production of benthic forage organisms and/or make these organisms largely unavailable to the fish (Fast 1973). If this happens, then productivity could be substantially reduced, especially in cases where the fish obtain a considerable portion of their diet from food produced by the pond.

The few studies where continuous aeration or circulation of the pond was attempted indicate that artificial mixing will increase fish production (Loyacano 1974; Busch and Goodman 1981). These studies are also discussed in the section on dissolved oxygen. Although neither study thoroughly documented the effect of continuous mixing on thermal stratification, benthic oxygen concentrations, benthic fauna populations, and fishes' feeding behaviors, the implication is strong that continuous mixing greatly enhanced conditions for the fish.

SALINITY

Salinity is one of the major factors that determine the kind of fish that can be reared at a given location. Although some species like milkfish and mullet are euryhaline, most fishes tolerate a relatively narrow salinity range (Table 1).

Not only is it important to establish the salinity of a fish pond, but it is equally important to establish the salinity fluctuations that can occur. Some coastal ponds are subject to freshwater runoff, in which case salinities could vary from marine to freshwater. Some coastal ponds or ponds located in arid regions experience hypersaline conditions due to high evaporation rates and a low inflow of freshwater. Milkfish ponds on Christmas Island often exceed 150°/oo (Helfrich 1981), while Laguna Madre, Texas Lagoon, often has salinities in the rage of 50 to 80°/oo (Pearce and Gunter 1957). Table 6 shows one classification of water based on salinity.

Potential salinity problems should be evaluated before siting a pond. If salinities are unsuitable for the fish species under consideration, then the pond should be sited elsewhere, a different culture species evaluated, or corrective measures sought. Corrective measures, if any are possible, typically include providing water with salinities needed to maintain the desired salinity. For example, this could mean a good freshwater source in a situation where high evaporation and hypersaline problems exist, or conversely, diverting freshwater inflow and allowing marine-water influxes.

Table 6
A CLASSIFICATION OF WATER BASED ON SALINITY[a]

Classification	Salinity, %
Fresh water	<0.5
Oligohaline	0.5 - 3.0
Mesohaline	3.0 - 16.5
Polyhaline	16.5 - 30.0
Marine	30.0 - 60.0
Brines or Hypersaline	>40.0

[a]Based on Hedgpeth 1957

ALKALINITY, HARDNESS, AND pH

Alkalinity and pH conditions cause some of the most common water-quality problems. Usually the problem is with highly variable pH, and low alkalinity. The problem occurs in fresh or brackishwaters.

Alkalinity is defined as the capacity of water to accept protons and it is measured by adding acid until a pH value of about 4.5 is reached. The acid combines with base substances such as carbonate, bicarbonate, and hydroxides. The amount of acid used is a measure of the total alkalinity.

Hardness tests were originally developed as an indicator of the water's capacity to precipitate soap. Hard water readily precipitates soap, thus creating a heavy scum and reducing the soap's cleansing capacity. Soap is usually precipitated by calcium and magnesium ions, but other polyvalent ions such as aluminum, iron, and manganese will also create hardness. Sodium and potassium ions do not contribute to the precipitation of soap. In most freshwaters, calcium and magnesium carbonates and bicarbonates predominate. In these waters, hardness and alkalinity are nearly the same. However, in waters with high concentrations of sodium carbonate, alkalinity would be much greater than hardness. Conversely, waters with high concentrations of calcium sulfate could have much greater hardness than alkalinity. Of these two parameters, alkalinity is usually the more important for fish culture. However, because these values are usually very similar, alkalinity and hardness are often used synonymously. They are both relatively easy to measure, but hardness may be somewhat easier to determine.

The hydrogen ion concentration (pH) of water is a measure of the water's acid/base condition. It ranges from pH = 7 (neutral), to pH = 14 (extremely basic). The toxicity of many substances, such as ammonia and hydrogen sulfide, are greatly influenced by pH. The pH of most freshwaters ranges between 6 and 9, which is an adequate range for most fishes. Values below pH 6 are generally deleterious to fish, and pH 4.5 to 5.0 is the lower limit of fish survival. Likewise, values above 9 to 10 are usually lethal. The most desirable range for fish is from pH 7 to 8, with minimum variation. Waters with a high capacity to resist changes upon the addition of acids or bases are said to be well buffered. Waters with high alkalinities are generally well buffered. Ocean waters have a very high buffering capacity and a stable pH in the range of 7.8 to 8.3 (Sverdrup et al. 1942).

Low-alkalinity waters are usually unproductive (see fertilization section). They have low nutrient concentrations, little plant growth, large variations in pH, and low fish yields (Schaeperclaus 1933; Hickling 1962; Boyd 1979). Schaeperclaus was one of the first to classify the productivity of pond waters based on their alkalinity (Table 7). Although Schaeperclaus felt that optimal productivity occurred at alkalinities of 100 mg/l as $CaCO_3$ or greater in Germany, Boyd (1979) found that alkalinities of 25 to 30 mg/l as $CaCO_3$ (corresponds to total hardness of about 20 mg/l as $CaCO_3$) provided optimal fish production in the southeast U.S. Boyd states that "lime applications (to increase alkalinity) may not be worth the bother and expense in waters containing slightly less than 20 mg/liter total hardness."

Liming of ponds with low-alkalinity water and acid sediments to increase fish production is a well-established water-quality management practice (Schaeperclaus 1933; Ness 1946; Huet 1970). Hickling (1962) reported increased Tilapia production of 243 to 385 kg/ha

Table 7

POND PRODUCTIVITY RELATED TO TOTAL ALKALINITY[a]

Alkalinity, mg $CaCO_3$/l	Significance in pond culture
Zero	Water strongly acid, unusable for hatchery purposes, adding lime to the water unprofitable in most cases.
5 to 25	Alkalinity very low. Danger of fish dying, pH variable, carbon dioxide supply poor, water not very productive.
25 to 100	pH variable, carbon dioxide supply medium, productivity medium.
100 to 250	pH varies only between narrow limits, carbon dioxide supply and productivity optimal.
>250	Rarely found. pH very constant, productivity alleged to decline but not proven so far. Health of fish not endangered.

[a] Data modified from Schaeperclaus 1933

through the application of 2,200 kg/ha of limestone and inorganic fertilizer. Arce and Boyd (1975) increased Tilapia production by about 25 percent in Alabama ponds limed at 3,836 to 4,371 kg/ha. Other researchers and aquaculturists have experienced similar results.

Liming and the corresponding increases in alkalinity cause increased fish production by several mechanisms. The lime increases the sediment pH and thus reduces the capacity of the sediments to bind plant nutrients like phosphorus (Ohle 1938; Bowling 1962). This not only releases these nutrients to the water where the phytoplankton can use them but also makes added nutrients more available for plant growth. Higher sediment pH may also create more favorable conditions for microbial growth and thus lead to a more efficient detrital food chain and the increased production of benthic fish-food organisms (Pamatmat 1960). Arce (1974) observed an average pH increase from 5.2 to 6.8 in five limed Alabama ponds, while five unlimed (control) ponds had average pH values of 5.4 and 5.5 during the same period. Liming also increases the available carbon dioxide for photosynthesis (Arce and Boyd 1975), creates a more desirable water pH range, and buffers against drastic daily pH changes.

Lime is commonly available as powdered limestone ($CaCO_3$, $MgCO_3$) hydrated or slaked lime ($Ca(OH)_2$), or quick lime or unslaked lime (CaO). It is applied to the pond water or bottom. The preferred procedure is to apply the lime to the soil after the pond has been drained and dried, although liming is often done without draining the water.

Liming application rates vary greatly from pond to pond and from region to region. Boyd (1979) developed and described a simplified procedure for estimating liming rates for most Alabama ponds. His procedure involves measuring sediment (mud) pH before and after the application of a p-nitrophenol buffer solution. Recommended lime application rates ranged from 91 to 7056 kg/ha of $CaCO_3$ (Table 8).

Using Boyd's techniques, alkalinity increased from about 17 mg/l to more than 30 mg/l in five Alabama ponds (Fig. 8). Hardness increased from about 10 mg/l to more than 25 mg/l. Alkalinity was typically 5 to 10 mg/l greater than hardness in these ponds, presumably due to nonalkaline earth carbonates.

Liming typically causes a marked increase in alkalinity and pH immediately after the lime application. This is

Table 8

ESTIMATED LIME REQUIREMENTS IN kg/ha NEEDED TO INCREASE THE TOTAL HARDNESS AND ALKALINITY TO 20 mg/l OR GREATER[a,b]

Mud pH in water	Calcium carbonate required according to mud pH in buffered solution									
	7.9	7.8	7.7	7.6	7.5	7.4	7.3	7.2	7.1	7.0
5.7	91	182	272	363	454	544	635	726	817	908
5.6	126	252	378	504	630	756	882	1,008	1,134	1,260
5.5	202	404	604	806	1,008	1,210	1,411	1,612	1,814	2,016
5.4	290	580	869	1,160	1,449	1,738	2,029	2,318	2,608	2,898
5.3	340	680	1,021	1,360	1,701	2,042	2,381	2,722	3,062	3,402
5.2	391	782	1,172	1,562	1,548	2,344	2,734	3,124	3,515	3,906
5.1	441	882	1,323	1,765	2,205	2,646	3,087	3,528	3,969	4,410
5.0	504	1,008	1,512	2,016	2,520	3,024	3,528	4,032	4,536	5,040
4.9	656	1,310	1,966	2,620	3,276	3,932	4,586	5,242	5,980	6,552
4.8	672	1,344	2,016	2,688	3,360	4,032	4,704	5,390	6,048	6,720
4.7	706	1,412	2,116	2,822	3,528	4,234	4,940	5,644	6,350	7,056

[a] Table from Boyd 1979

[b] The lime required (as calcium carbonate) is estimated from the pH of the pond muds before and after the addition of a buffer solution.

often followed by a reduction in alkalinity and pH and a leveling off of these values (Fig. 9). If a suitable amount of lime is applied and if water flushing rates are not excessive, the benefits from liming should last several years. High flushing rates, however (e.g., Grier's Pond, Fig. 9), can nullify the lime applications since the lime is rapidly flushed from the pond. Boyd (1979) found that an annual lime application of 25 percent the initial dose was adequate to maintain high fish productivity in Alabama fish ponds with limited flushing rates.

The above procedures for estimating lime application rates do not apply to acid sulfide soils. These ponds often have a sediment pH of 4.5 or less, high dissolved-metal content, and are often unsuitable for fish culture even with intensive treatment. Boyd (1979) recommends a procedure for estimating the lime requirement of these soils, but cautions that "the efficacy of the procedure in estimating the lime requirement of pond muds with acid sulfide problems has not been evaluated."

DEWATERING AND DRYING

It is highly desirable, if not essential, to be able to fully drain a fish pond. Draining allows for the complete harvest of the fish crop, but that is not the only purpose for draining a fish pond. Reasons for dewatering and drying fish ponds include (Hickling 1962, Huet 1970):

1. Nutrient Regeneration. The drying of pond soils causes an accelerated rate of organic matter mineralization. This aerobic process results in mineral forms that are more readily released to the water when the pond is refilled. These nutrients increase plant growth and, consequently, fish production.

2. Fish Population Control. Periodic dewatering allows good control of fish populations. Fish can be sorted and restocked at densities that will maximize the production of desirable-sized fish. Stunting and the excessive production of small fish can be avoided by sorting and thinning the fish stock. Undesirable fish species can be removed.

3. Reduce Oxygen Demand of Sediments. Ponds with large amounts of organic matter develop anaerobic bottom conditions and the decomposition of the organic matter is slowed. Dewatering and drying accelerates the aerobic oxidation and decomposition of this organic matter. This creates a better pond sediment

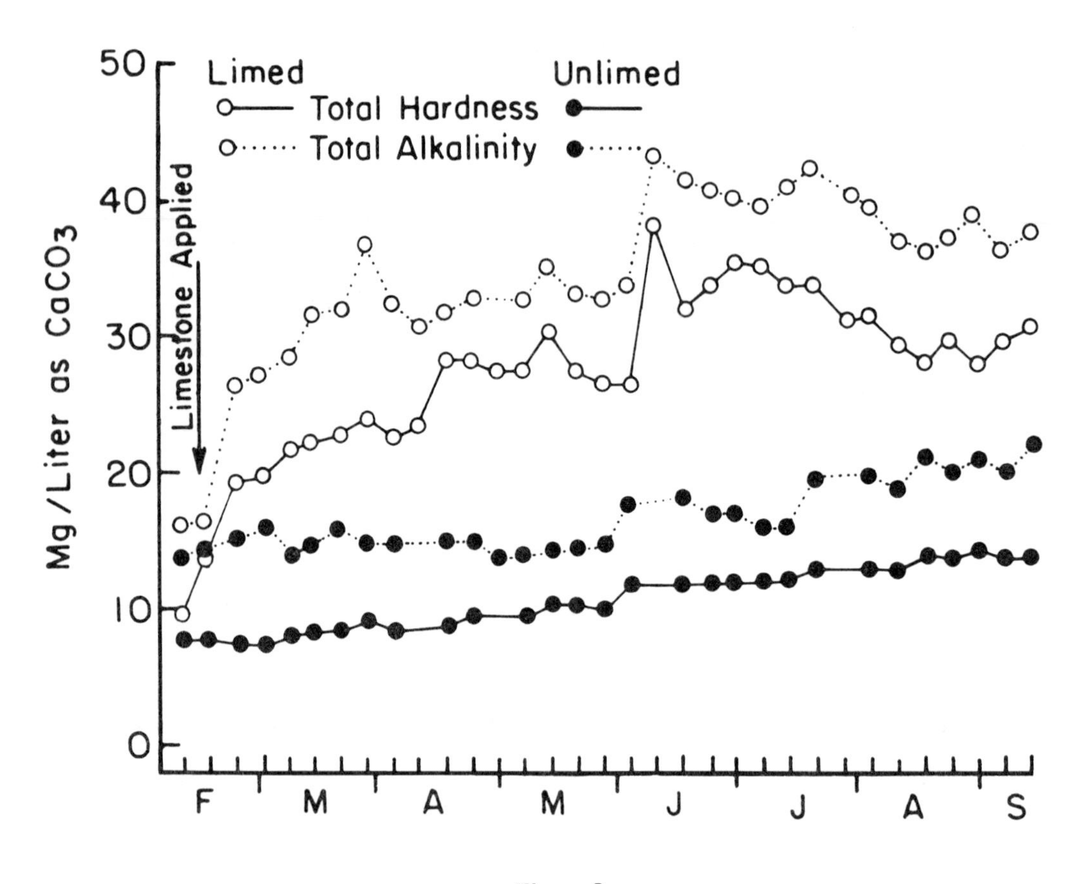

Figure 8

Total hardness and total alkalinity in limed and un-limed fish ponds. Figure from Boyd 1979, after Arce and Boyd 1975.

substrate for benthic fauna production and reduces the oxygen demand of the sediments when the pond is refilled.

4. Oxidation and Leaching of Acid Sulfate Soils. Dewatering and drying enchances the oxidation of sulfur in acid sulfate soils. The resultant acids can then be leached from the soil and the soil can be limed to create a favorable pH for fish culture (Singh 1980).

5. Control Vegetation. Vegetation can be controlled through dewatering and drying. If the vegetation also dries, it can be burned to remove the aboveground growth, and the roots can be plowed or disked under. If the vegetation does not dry sufficiently it may be plowed under or used as cattle fodder. Some plants are controlled by prolonged dessication.

6. Disease Control. Certain fish and human diseases are difficult to control in water-filled ponds (see diseases chapter). Malaria mosquitoes are a particular problem in weed-choked ponds. Routine drying helps control the weeds and allows larvivorous fish (such as Gambusia) to control the mosquitoes when the ponds are refilled. The Schistosoma disease (Bilharzia) has its intermediate stage in snails and its primary stage in the blood of man. The snail is effectively controlled by periodic drying and liming of the soil. This also helps control weed growth, thus reducing the habitat available to the snail upon refilling. Stocking herbivorous fish further reduces weed growth and makes the snail more vulnerable to predation by carnivorous fish.

Two fish parasites, the fish louse Argulus and the parasitic copepod Lernaea, can be controlled by periodic dewatering and drying. Lime application during the dry period greatly enhances treatment of these and other parasites.

7. Pond Maintenance. Dewatering and drying simplifies pond maintenance.

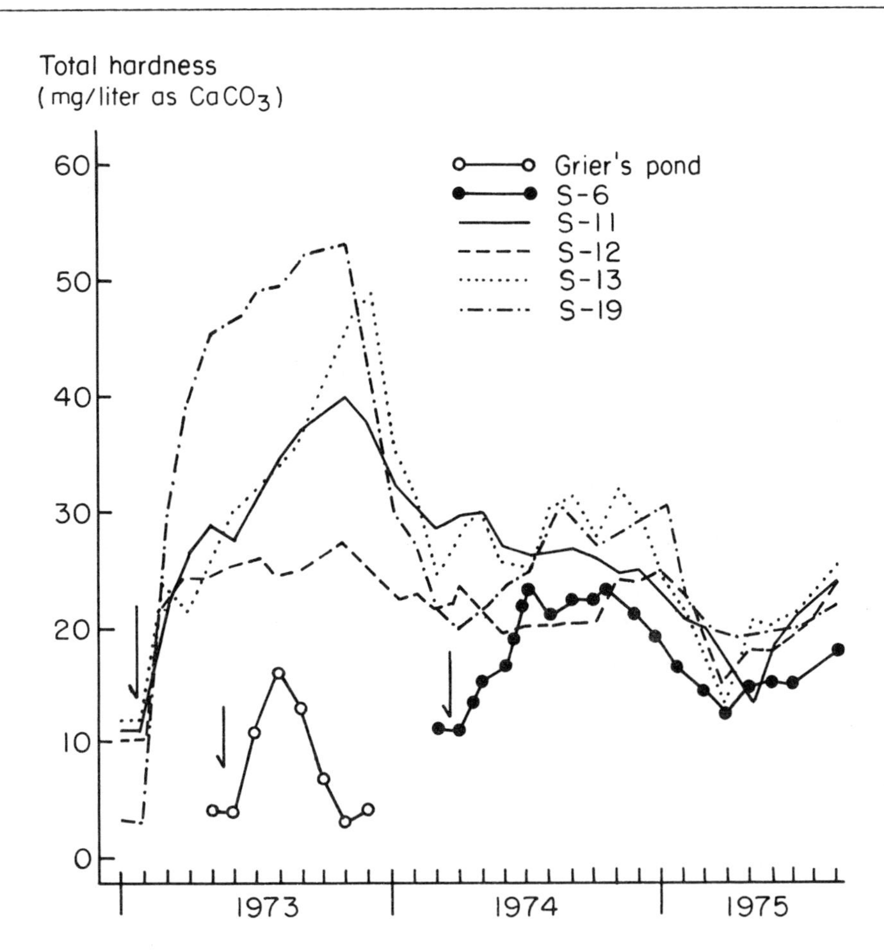

Figure 9

Long term effects of liming on six Alabama ponds. Grier's Pond had the highest water flow-through rate (3 weeks) and showed the least benefit from liming. The arrows indicate lime applications. Figure from Boyd 1979, after Boyd 1976.

Excess accumulations of sediment and debris can easily be removed by earthmoving equipment or by hand. This sediment is often a valuable addition to fields and gardens since it has a high nutrient content and good soil-conditioning properties. While they are dry, pond drainage channels and sluice boxes can be cleaned and restored. Fertilizers and lime can be added to the pond soil and disked in more easily when it is dry.

8. Crop Rotation. Often, crops such as rice, cotton, clover (and other legumes), or cereals are alternated with fish crops. These crops tend to increase the productivity of the fish crops, probably by (a) enhancing soil oxidation and thus making mineral nutrients more available when the pond is reflooded, (b) producing organic material that promotes the growth of fish food organisms, and/or by (c) fixing atmospheric nitrogen, which is released to the water upon reflooding.

When to dewater and dry, and for how long? There are no good rules for the frequency or duration of dewatering and drying. In cold climates, the common practice is to dewater and dry each winter, at least in the shallow pond. The fish are either harvested in the fall or moved to deeper overwintering ponds that are less subject to oxygen depletion and fish kills. Fish are less stressed by handling during cold periods. They have greatly reduced metabolic and food requirements during the winter and do not grow much. In China, a pond may be drained every one to two years (Hoffman 1934), while in Germany the practice was to rear fish for three years and then to grow dry land crops for three years (Wunder 1949). The latter program is probably excessive.

The appropriate frequency and duration of drying will largely depend on the nature of the pond soils, the climate, the vegetation in the pond, and other relevant factors. The appropriate period could be as short as a few days or as long as a year or more. Ponds with soils and construction that facilitate rapid dewatering, with a warm, dry climate, and with a minimum of vegetation cover may require only a few days of drying (Hickling 1962). However, pond soils that remain wet may not benefit greatly from much longer periods of "dewatering" and "drying."

TURBIDITY

Pond turbidity is caused by organic matter like phytoplankton and by suspended organic matter like silt and clay. Inorganic turbidity can be most troublesome, since it can reduce the light penetration required for photosynthesis and thus reduce oxygen generation and phytoplankton production. Inorganic turbidity may also reduce the benefits of artificial fertilization since phosphorus may absorb or adsorb on the sediment particles.

Persistent inorganic turbidity is often caused by negatively charged colloidal clay particles. They stay dispersed and in suspension due to their small size and electrical charge. In these cases, turbidity may be reduced by applying organic materials such as cut hay or manure to the ponds (Irwin and Stevenson 1951; Swingle and Smith 1947). Decomposition of these organic materials leads to increased CO_2 concentrations, decreased pH, and precipitation of the clay. Irwin and Stevenson recommend hay applications of 0.05 kg/m^3 of pond water with 25 mg/l turbidity and 0.4 kg/m^3 for 200 mg/l turbidity. Swingle and Smith recommend two or three applications of barnyard manure at the rate of 2,440 kg/ha.

Boyd (1979) tested alum (aluminum sulfate), gypsum ($CaSo_4 \cdot 2H_2O$), and slaked lime ($Ca(OH_2)$) on several soil types. He found alum to be the most effective. In six test ponds an average alum application of 20 mg/l resulted in an average turbidity decrease of 91 percent. In one pond, turbidity decreased from 830 mg/l before treatment with 20 mg/l of alum to 24 mg/l after treatment (97 percent reduction). Alum is relatively nontoxic to fish and other aquatic organisms, although it can cause a substantial drop in water pH.

DOMESTIC WASTES

Domestic wastes contain high concentrations of valuable nutrient materials, but they also contain potentially high concentrations of human pathogens. These wastes have been used for many centuries for both agricultural and aquaculture purposes. However, the threat of disease is ever-present when domestic waste is used to culture food products for human consumption. The threat is particularly acute for species that are sometimes eaten raw (such as oysters or mussels) or for species that are sometimes pickled or not thoroughly cooked.

The United States has strict laws regarding water-quality standards for shellfish-rearing areas, but infectious diseases (e.g. hepatitis) still are attributed to these sources. There are also strict laws regarding the use of domestic wastes for fish culture in the United States. Although there is much interest in the possible use of these waste waters for aquaculture (Devik 1976), it is unlikely that we will see large scale uses in the United States in the foreseeable future.

The use of domestic or agriculture wastes for aquaculture is an established practice in much of the world. It is relatively safe, providing precautions are taken to assure that the end-product is properly handled and cooked before consumption.

AGRICULTURAL POLLUTION

The careless use of pesticides and herbicides is probably agriculture's greatest threat to aquaculture. These toxicants may drift into fishponds as overspray from agricultural fields or may concentrate in water sources that drain the treated area. In addition, some pesticides and herbicides may contaminate the soil, so that rotating field crops with aquacultural crops may cause excessive contamination of the fish.

As with industrial wastes, the best approach is to site the aquacultural operation away from sources of potential agricultural pollution. Steps should be taken to alert the crop farmer to the damage he can cause the fish farmer through careless use of his chemicals.

In addition to agricultural toxicants, misused fertilizers, runoff from animal feed lots, and excessive erosion are other potential threats posed by agriculture to fish culturists. Fertilizers and animal wastes can be directly toxic, especially if ammonia concentrations and water pH values are high. These materials can also lead to excessive nutrient concentrations in the fish pond, excessive plant growth, and ultimately oxygen depletion. Erosion carries silt, sand, and other materials into ponds where they settle and lead to a filling in of the pond. This shortens the useful life of the pond, creates problems with macrophytes, reduces the productive volume, sometimes increases turbidity (and thus reduces phytoplankton growth), and reduces fish production.

INDUSTRIAL WASTES

There are a large number of industrial wastes that are potentially harmful to aquaculture. A recital of these wastes is unnecessary, and indeed beyond the scope of this review. They are, however, of paramount importance to the aquaculturists. Great care should be exercised when siting an aquaculture operation to avoid contamination from this source. Precautions must also be taken by industrial concerns to avoid contaminating established aquaculture operations with their discharges. Liability for damages that result from such discharges will almost always lie with the discharger.

AQUATIC PLANTS

Macrophytes

Macrophytes are large aquatic plants that are either rooted to the bottom or have connection with the bottom or some other stable substrate. The aquatic plants that cause problems for pond aquaculturists may be classified as follows:

1. Riparian, or shore, plants are plants that grow at or above the water's edge. They may be woody or soft stemmed (e.g. willow).

2. Emergent plants are those plants that are found below water, usually rooted to the bottom, and have substantial portions of the plant extending above the water's surface (e.g. cattails, rushes).

3. Submergent plants are those plants that are totally, or nearly always, underwater. They typically lack woody parts, and they may be rooted or not (e.g. *Elodea*, *Potamogeton* and filamentous algae).

4. Floating plants, where most of the plant grows on the water's surface (e.g. water hyacinth, water lilies, duckweed).

Proper identification of a nuisance macrophyte is usually necessary before it can be properly dealt with. This is especially true for chemical control methods, since many macrophyte toxins are specific for certain plants. Dosage rate and application methods may also vary for different plants. There are a number of useful publications for identifying aquatic plants, including Fassett (1960), Weldon et al. (1969); Klussman and Lowman (1975), and Applied Biochemists (1979).

Aquatic macrophytes may seriously impact fish ponds if they become excessively abundant. Excessive macrophyte growth may: tie up nutrients and biomass that could be more efficiently used by the fish through the phytoplankton foodweb; greatly restrict phytoplankton production; provide abundant shelter for small fishes and thus lead to their overabundance and stunting; greatly hinder harvest of the fish crop; restrict water circulation; contribute to accelerated sedimentation of the pond; restrict movement and living space for the fish; and contribute to oxygen depletion and fish kills when the plants die (Hickling 1962, Huet 1970; Stickney 1979). These problems can range from the minor to the very serious. In the latter case, fish production is often greatly reduced and the rearing operation becomes uneconomical. The amount of tolerable macrophyte cover is not easy to define, but Boyd (1979) suggests that 10 to 20 percent cover of the pond is the upper limit.

Troublesome macrophytes may not only quickly fill in or "incapacitate" a pond, but they also produce relatively low standing crops (Forsberg 1960; Westlake 1965; Boyd 1975). Although they may be used for animal feed or mulch, their value is minimal. Furthermore, very few fresh water macrophytes are of any value to humans. The more prominent ones are water spinach (*Ipomea*), watercress (*Nasturtium*), arrowhead (*Sagittaria*), wild rice (*Zizania*), and cattails (*Typha*) (Bardach et al. 1972). Except for watercress there is little interest in culturing these plants for human consumption and, with the exception of the cattails, they seldom cause serious problems in fish ponds.

Macrophytes are a particular problem in shallow ponds where light can penetrate to the bottom during the early growing season. Macrophytic growth may progress at a rapid rate and derive phytoplankton of the nutrients needed for their growth. Many macrophytes can draw nutrients from the sediments as well as from the water. This gives them an added advantage in their competition with phytoplankton. Deep ponds (greater than 1.5 m deep) with moderate-to-high turbidity seldom have a problem with macrophytes, except along their shorelines.

Macrophyte Control Methods

Mechanical Control Macrophyte control by mechanical or biological means is ecologically the most desirable. As with any plant-control technique, mechanical control is best achieved before dense growth of the plants occurs. Younger plants not only have less biomass, but they are often easier to cut (they are less fibrous) or detach.

Macrophytes may be cut by hand or machine. Hickling (1962) used hand-scythes and a motorized cutter on reeds, rushes, and sedges in Malaysia with good success. He obtained the best results when the plants were first cut early in the season and then once or twice later on. The regrowth of the plants exhausted their stored root reserves and led to longer-term control. His success was greatest in water of 1 m depth or deeper, especially when phytoplankton growth resulted in greater turbidity following the cuttings.

After mechanical cutting of the macrophytes, the cut weeds should be removed from the water to prevent oxygen depletion and excessive filling of the pond. Huet (1970) recommends that not more than 1,500 kg/ha of weeds remain.

A properly constructed pond is one of the best mechanical means of controlling plant growth. This usually means a minimum depth of 1 to 1.5 meters and relatively steep slopes near shore. The slope should not exceed 4:1 for safety reasons (Glenn 1980). A pond of this design will greatly reduce the potential for macrophyte growth since it creates favorable conditions for phytoplankton growth and for the shading of the bottom areas.

Many aquatic plants can be effectively controlled by periodic drying of the pond. Drying and mowing, drying and burning, and drying and disking are effective (Hickling 1962).

Biological Control Biological control often results in both effective control of nuisance macrophytes and increased fish production.

Biological control of macrophytes through the enhancement of phytoplankton is probably the most popular technique employed. Macrophytes are generally sparse or absent from ponds with moderate-to-heavy plankton blooms and water depths of 1.5 m or greater. Phytoplankton turbidity or inorganic turbidity effectively reduce light penetration to the pond bottom and thus limit macrophyte growth. Hutchinson (1975) found that macrophytes did not grow below 1.5 m when Secchi disc transparencies were 1 m, nor below 2.5 m when the Secchi disc depth was 2 m. Boyd (1975) found similar results and suggested that macrophytes could not grow at depths greater than twice the Secchi disc transparency. Since Secchi disc depths of about 0.5 are generally considered optimal for phytoplankton densities, it follows that not much macrophyte growth will occur below the 1.5 m depth.

Herbivorous animals such as grass carp, tilapia, ducks, geese, swans, cattle, and nutria, are often used to control macrophyte growth (Hickling 1962; Huet 1970). Cattle, nutria, and birds are most effective in controlling plants in shallow water and on the pond bank. Not only can they be very effective control measures, but they also contribute manure and nutrients to the pond and provide an additional protein crop for market. Grass carp and tilapia are particularly valued fish crops. However, they are less successful in areas with cool climates and their importation in some areas is also restricted.

The common, or Israeli, carp is sometimes used for weed control (Huet 1970). Although it derives some nutrition from the vegetation, it mainly controls the weeds by dislodging them through its "rutting" action on the pond bottom. It also increases the water turbidity and thus helps reduce light penetration to the pond bottom. The disadvantage of the common carp is that large numbers are often required for effective control (2,500 to 3,000/ha; Huet 1970). They also increase inorganic turbidity (thus reducing phytoplankton growth) and are not as desirable a market fish as certain other species.

Chemical Control Chemical control of macrophytes is often very cheap and effective. However, it can upset the ecological balances in a fish pond and cause reduced production (or death) of the fish and their forage organisms. Herbicides may be directly toxic to fish and their food organisms, or they may cause oxygen depletion through decomposition of the dead macrophytes.

Although there are a large number of herbicides, only a few are approved in the United States for use in food fish ponds. Meyer et al. (1976) list only copper sulfate, 2,4-D, Diquat dibromide, endothall, and simazine. In the United States these chemicals can only be sold to those persons licensed to apply them.

Boyd (1979) cautions that "in pond fish culture, herbicides should be used only to eliminate aquatic vegetation so that plankton blooms may develop." He further states that "unless a plankton bloom is encouraged, macrophytes will simply regrow as soon as herbicide treatments return to nontoxic levels."

There is sometimes no suitable alternative to chemical treatment. In these cases application should be made as directed by the manufacturer and the results closely monitored.

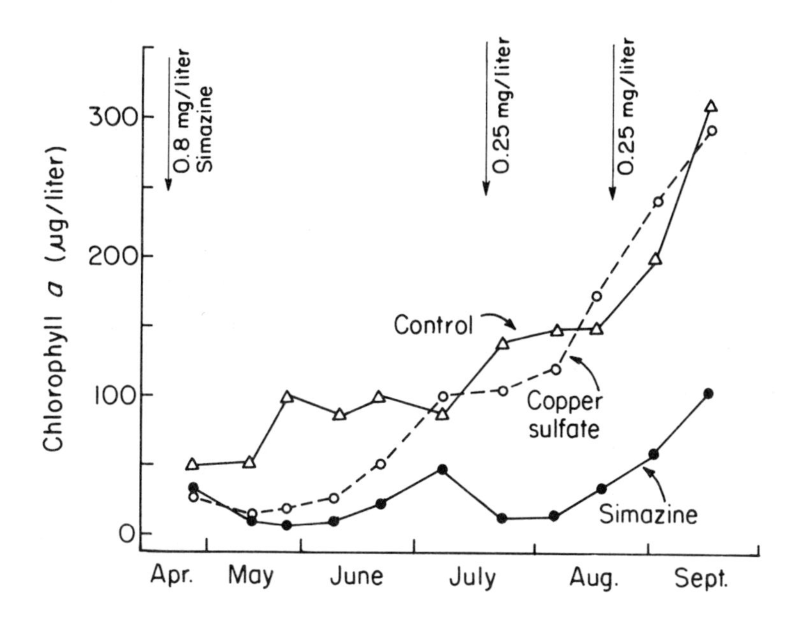

Figure 10

Effects of copper sulfate and simazine on chlorophyll a concentrations in fish ponds. The dates when the algacides were added are shown by arrows. Figure from Boyd 1979.

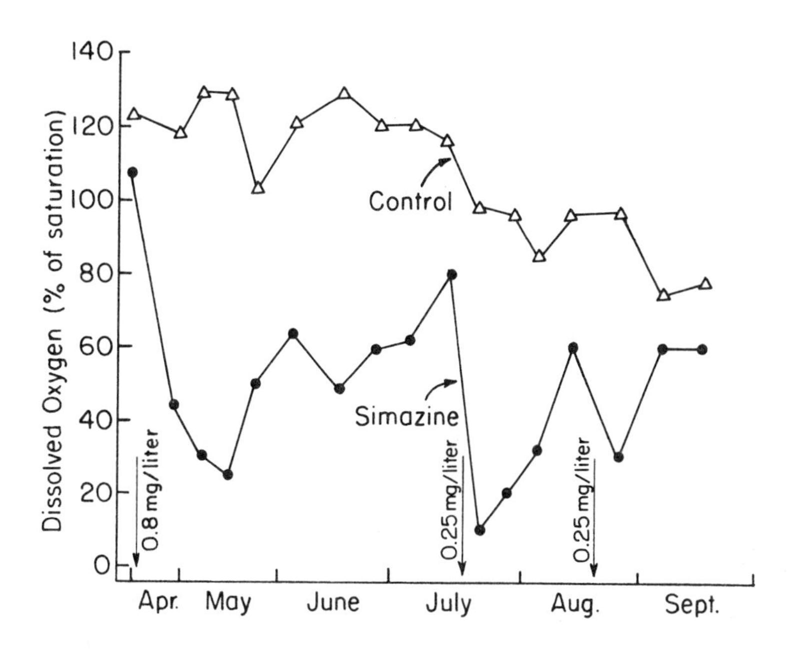

Figure 11

Effects of simazine on dissolved oxygen concentrations in fish ponds. The dates when the algacide was added are shown by arrows. Figure from Boyd 1979, after Tucker and Boyd 1978.

Phytoplankton

Phytoplanktons are small, usually microscopic plants that are carried about by water currents. They give the water its green appearance and they are an important link in the food web. Through photosynthesis they convert solar energy and inorganic compounds into organic compounds. In this process they also produce oxygen, which dissolves in the water and helps maintain aerobic conditions.

Excess fertilization may cause phytoplankton growth to become excessive or undesirable forms of phytoplankton to develop. This can lead to oxygen depletions of off-flavors in the fish flesh through the production of geosmin by the algae (Lovell and Sackey 1973).

Phytoplankton Control Methods Sometimes it is easier to treat consequences of excessive phytoplankton growth than to treat the phytoplankton directly. This is treating the symptom rather than the cause, but it is often as effective and more practical. The usual symptom is oxygen depletion which is discussed earlier in this chapter.

Long-term treatment usually requires some reduction of nutrient inputs. This is most easily achieved when the principal inputs are applied by the pond operator; reduction of nutrient inputs is more difficult when they are from watershed runoff or groundwater inflows.

Flushing clean water through the pond is one means of reducing nutrient and phytoplankton populations. Water can be exchanged between ponds or washed downstream. However, the latter remedy may require large amounts of source water.

Biological control of phytoplankton has not been thoroughly evaluated. Some results indicate that plankton-eating fishes will stimulate phytoplankton growth rather than retard it (Hurlbert et al. 1972; Perschbacher 1975). Primary production was 5.0 mg O_2/1/day in ponds with planktivorous fish, vs 3.5 mg O_2/1/day in ponds without such fish. Likewise, phytoplankton density was 7,500 cells/ml in the fish ponds, vs 3,100 cells/ml in the ponds without the fish. Malea (1976) and Boyd (1979) found that although phytoplankton densities may not be decreased by a polyculture program using planktivorous fish, the total fish production was substantially increased. In addition, there could be a change in the types of algae produced in these systems even though total algal density is not decreased. For example, if blue-green algae (which produce geosmin and often are associated with oxygen depletion) could be replaced by green algae or diatoms, there could be a net improvement in water quality even though the total algal biomass is the same. There has not yet been sufficient investigation of this management technique to assess its efficacy.

Copper sulfate is the common algacide used. It is usually broadcast into the water as a crystal or dissolved in the water by various means. More recently, its chelated form (with citric acid) has become popular. Simazine has also become popular recently for controlling both phytoplankton and certain macrophytes. Although both chemicals are very effective algacides, they can reduce both fish production and dissolved oxygen concentrations (Boyd 1979). Tucker and Boyd (1978) observed a substantial reduction in dissolved oxygen (and phytoplankton density) in a fed catfish pond which was treated with simazine (Figs. 10 and 11). The oxygen depression persisted all summer.

Copper sulfate may be a useful means of controlling scum-forming blue-green algae (Kessler 1960; Crance 1963). However, special care should be exercised not to cause toxicity to the fish or cause excessive oxygen depletion.

Boyd (1979) concluded that "the use of algacides to limit phytoplankton growth in ponds used for intensive fish culture is analogous to a human losing weight by taking periodic, sublethal doses of a toxin instead of reducing his food intake."

SELECTED FUTURE RESEARCH NEEDS

1. Thermal and Chemical Stratification. The effects of thermal stratification and oxygen depletion on the bottom waters should be further evaluated. In particular, the production of benthic fauna forage organisms, depth distribution and feeding habits of the fish, zooplankton depth distribution, and nutrient cycling should be studied.

Even shallow ponds of 1.5 to 2.0 meters depth may stratify periodically, while deeper ponds may have long periods of stratification. These conditions could cause a reduction in forage production and the availability of forage to the fish, and consequently a reduction in fish production.

The research design should include a variety of ponds, including shallow ponds with ephemeral stratification and deep ponds with sustained periods of stratification.

2. Continuous Circulation. In connection with the above, the effects of continuous circulation by artificial

means should be evaluated. Preliminary research results indicates that continuous circulation/aeration can result in a substantial increase in fish production and increased DO. However, the mechanism whereby this occurs is not known.

3. Emergency Aeration. Partial aeration of a fish pond using a semipermeable barrier should be evaluated and compared with aeration of the entire pond. In particular, the energy requirements, rate of oxygen increase, and movements of fish into the aerated zone should be studied.

4. Liming Procedures. Liming procedures should be developed and tested for a variety of soil types and regions. The methods developed by Boyd (1979) for certain Alabama soils are most useful, but their widespread application is untested. Possibly these techniques, with modifications, could be developed for general use.

5. Acid Soil Restoration. Some acid soils are not readily restored by existing reclamation practices. Alternative methods of restoring these soils should be evaluated.

6. Phytoplankton Manipulations. Although phytoplankton population densities are often at desirable levels, the species composition may not be optimal. Conditions favoring high phytoplankton production also tend to favor the growth of blue-green algae or other forms that are not preferred foods of many herbivores. It may be possible to artificially manipulate the species composition of phytoplankton by the use of selective algacides, nutrient additions, herbivorous fish, or artificial circulation.

LITERATURE CITED

Abbott, R. 1981. Personal communication. Oceanic Institute, Waimanalo, Hawaii.

American Public Health Association, American Water Works Association, and Water Pollution Control Federation. 1971. _Standard Methods for the Examination of Water and Wastewater_, 13th ed., APHA, Washington, D.C. 874 pp.

__________. 1976. _Standard Methods for the Examination of Water and Wastewater_, 14th ed., APHA, Washington, D.C. 1193 pp.

Applied Biochemists, Inc. 1979. How to Identify and Control Water Weeds and Algae. Applied Biochemists, Inc., Mequon, Wisconsin. 64 pp.

Arce, R.G. 1974. Effects of Applications of Agricultural Limestone on Water Chemistry, Phytoplankton Productivity, and Tilapia Production in Soft-Water Ponds. M.S. Thesis, Auburn Univ., Auburn, Alabama. 28 pp.

__________, and C.E. Boyd. 1975. Effects of agricultural limestone on water chemistry, phytoplankton productivity, and fish production in soft-water ponds. _Trans. Amer. Fish. Soc._ 104:308-312.

Balarin, J.D., and J.P. Hatton. 1979. _Tilapia. A Guide to Their Biology and Culture in Africa_. Unit of Aquatic Pathobiology, University of Stirling, Scotland.

Bardach, J.E. 1978. The growing science of aquaculture, pp. 424-446. _In_ S.D. Gerking (ed), _Ecology of Freshwater Fish Production_. John Wiley & Sons, Interscience, New York.

__________, J.H. Rhyther, and W.O. McLarney. 1972. _Aquaculture, the Farming and Husbandry of Freshwater and Marine Organisms_. Wiley-Interscience, New York.

Bell, F.W., and E.R. Canterbery. 1976. _Aquaculture for the Developing Countries_. Ballinger Publishing Co., Cambridge, Mass. pp. 29-31.

Black, E.C. 1953. Upper lethal temperatures of some British Columbia freshwater fishes. _Fish. Res. Bd. Canada, Jour._ 10:196.

Borgstrom, G. 1978. The contribution of freshwater fish to human consumption, pp. 469-491. _In_ S.D. Gerking (ed), _Ecology of Freshwater Fish Production_. John Wiley & Sons, New York.

Bortz, B., J. Ruttle, and M. Podens. 1977. Raising Fresh Fish in Your Home Waters. Rodal Resources Inc. No. 17. 33 pp.

Bowling, M.L. 1962. The effects of lime treatment on benthos production in Georgia farm ponds. _Proc. Annual Conf. S.E. Association Game and Fish Comm._ 16:418-424.

Boyd, C.E. 1975. Competition for Light by Aquatic Plants in Fish Ponds, Auburn Univ. (Ala.) Agr. Exp. Sta., Cir. 215. 19 pp.

Boyd, C.E. 1976. Lime requirement and application in fish ponds. In T.V.R. Pillay and W.A. Dill (eds), FAO Tech. Conf. on Aquaculture Advances in Aquaculture. Fishing News Books, Ltd., Farnham, England.

__________ 1979. Water Quality in Warmwater Fish Ponds. Auburn University, Agricultural Experiment Station, Auburn Alabama. 359 pp.

__________ and C.S. Tucker. 1979. Emergency aeration of fish ponds. Trans. Amer. Fish. Soc. 108:299-306.

Brett, J.R., J.E. Shelbourn, and C.T. Shoop. 1969. Growth rate and body composition of fingerling sockeye salmon, Oncorhynchus nerka, in relation to temperature and ration size. J. Fish. Res. Bd. Canada 26:2363-2394.

Burns, J.W. 1966. Carp, pp. 510-515. In A. Calhoun, (ed), Inland Fisheries Management. California Dept. Fish and Game, Sacramento, California.

Busch, C.D. and R.K. Goodman. 1981. Water circulation, an alternative to emergency aeration. Proc. of World Mariculture Soc. 12(1):13-19.

__________, J.L. Koon, and R. Allison. 1974. Aeration, water quality, and catfish production. Trans. Amer. Soc. Agr. Eng. 17:433-435.

Chen, T.P. 1976. Aquaculture Practices in Taiwan. Fishing New Books, Ltd., Surrey, England. 176 pp.

Cole, G.A. 1975. Textbook of Limnology. C.V. Mosby Co., St. Louis, MO. 283 pp.

Collins, Charles M. 1977. Catfish cage culture - fingerlings to food fish. The Ker Foundation, Inc. Publication No. 13.

Colt, J. 1980. Dissolved oxygen. Dept. of Civil Engineering, University of California, Davis, 2 pp. (mimeo).

________, G. Tchobanoglous, and B. Wong. 1975. The requirements and maintenance of environmental quality in the intensive culture of channel catfish. Civil Engineering Department, University of California, Davis, 119 pp.

Crance, J.H. 1963. The effects of copper sulfate on Microcystis and zooplankton in ponds. Prog. Fish-Cult. 25:198-202.

Crear, D. 1980. Observations on the reproductive state of milkfish populations (Chanos chanos) from hypersaline ponds on Christmas Island. Proc. of World Mariculture Soc. 11:548-556.

________. 1981. Personal communication. Aecos Inc., Kaneohe, Hawaii.

Devik, O. 1976. Harvesting Polluted Waters. Plenum Press, New York. 324 pp.

Dimitrov, M. 1974. Mineral fertilization of carp ponds in polyculture rearing. Aquaculture. 3:273-285.

Duodoroff, P. and D.L. Shumway. 1970. Dissolved oxygen requirements of freshwater fishes. FAO United Nations, Fish. Tech. Paper. 86. 291 pp.

Environmental Protection Agency. 1971. Water Quality Data Book - Vol. 3, Effects of Chemicals on Aquatic Life. U.S. Government Printing Office. 100 pp.

______________________________. 1976. Quality Criteria for Water. U.S. Government Printing Office. 256 pp.

Fassett, N.C. 1960. A Manual of Aquatic Plants. U. of Wisconsin Press, Madison. 405 pp.

Fast, A.W. 1973. Effects of artificial aeration on primary production and zoobenthos of El Capitan Reservoir, California. Water Resources Res. 9(3):607-623.

__________ 1977. Floating fish rearing system. U.S. patent No. 4,044,720.

__________ 1979. Artificial aeration as a lake restoration technique, pp. 121-131. In Lake Restoration, Proceedings of a National Conference. August 22-24, 1978, Minneapolis, MN. EPA 440/5-79-001.

Ford, R.F., J.C. Van Olst, J.M. Carlsberg, W.R. Dorgand, and R.L. Johnson. 1975. Beneficial use of thermal effluent in lobster culture. Proc. World Maricult. Soc. 6:509-519.

Forsberg, C. 1960. Subaquatic macrovegetation in Osbysjon, Djursholm. Oikos 11:183-199.

Friedman, L. and S.I. Shibko. 1972. Non-nutrient components of the diet, pp. 181-254. In J.E. Halver (ed), Fish Nutrition. Academic Press, New York.

Glenn, J.H. 1980. Personal Communication. J. Harlan Glenn & Assoc., Anaheim, California.

Guerra, C.R., R.E. Resh, B.L. Godriaux, and C.A. Stephens. 1979. Venture analysis for a proposed commercial waste heat aquaculture facility. Proc. World Maricult. Soc. 10:28-38.

Hedgpeth, J.W. 1957. Classification of marine environments, pp. 17-27. In J.W. Hedgpeth (ed), Treatise on Marine Ecology and Paleoecology. Vol. 1, Memoir 69, Geological Soc. Amer., New York.

Helfrich, P. 1981. Personal communication. Hawaii Institute of Marine Bioloby, Kaneohe, Hawaii.

Hickling, C.F. 1962. Fish Cultures. Faber and Faber, London. 295 pp.

Hoffman, W.E. 1934. Preliminary notes on the freshwater fish industry in South China, especially Kwangtung Province. Lingnan Univ. Sci. Bull. No. 5, Canton.

Hora, S. L. and T.V.R. Pillay. 1962. Handbook on fish culture in the Indo-Pacific Region. FAO Fisheries Biology Technical Paper No. 14. 204 pp.

Huet, M. 1970. Textbook of Fish Culture: Breeding and Cultivation of Fish. Fishing News Books Ltd., Surrey, England. 436 pp.

Hurlbert, S.H., J. Zedler, and D. Fairbanks. 1972. Ecosystem alternation by mosquitofish (Gambusia affinis). Predation Science 175:639-641.

Hutchinson, G.E. 1957. A Treatise on Limnology: Vol. I. Geography, Physics, and Chemistry. John Wiley and Sons, New York. 1015 pp.

_______________. 1967. A Treatise on Limnology: Vol. II. Introduction to Lake Biology and the Limnoplankton. John Wiley and Sons, New York. 1115 pp.

_______________. 1975. A Treatise on Limnology: Vol. III. Limnological Botany. John Wiley and Sons, New York. 660 pp.

Irwin, W.H. and Stevenson 1951. Physiochemical nature of clay turbidity with special reference to clarification and productivity of impounded waters. Okla. Agr. Mech. Coll., Bull. 48. 1-54 pp.

Kessler, S. 1960. Eradication of bluegreen algae with copper sulfate. Bamidgeh 12:17-19.

Kilambi, R.V. and W.R. Robinson. 1979. Effects of temperature and stocking density on food consumption and growth of grass carp Ctenopharyngodon idella. Val. Journal of Fish Biology 15:337-342.

Klontz, G.W. and J.G. King. 1974. Aquaculture in Idaho and nationwide. Research Tech. Compl. Report IDWR Project 45-080, Idaho Dept. of Water Resources.

Klussman, W.G. and F.G. Lowman. 1975. Common aquatic plants. Texas Agricultural Extension Service, College Station. 15 pp.

Kuo, C. 1979. Mass propagation of baitfish (grey mullet). A proposal to the Dept. of Land & Natural Resources, State of Hawaii.

_______. 1975. The induced breeding and and rearing of grey mullet (soundtrack text). The Oceanic Institute.

La Rivers, I. 1962. Fishes and Fisheries of Nevada. Nevada Fish and Game Comm. 782 pp.

Ling, S.W. 1977. Aquaculture in Southeast Asia - A Historical Overview. A Washington Sea Grant Publication. 108 pp.

Lovell, R.T. and L.A. Sackey. 1973. Absorption by channel catfish of earthy-musty flavor compounds synthesized by cultures of blue-green algae. Trans. Amer. Fish. Soc. 102:772-777.

Loyacano, H.A. 1974. Effects of aeration in earthen ponds on water quality and production of white catfish. Aquacult. 3:261-271.

Malea, R.P. 1976. Polyculture Systems with Channel Catfish as the Principal Species. Ph.D. diss., Auburn, Ala. 202 pp.

McKee, J.E. and H.W. Wolf, (eds). 1963. Water Quality Criteria, 2nd ed. State of California, State Water Quality Control Board, Publ. No. 3-A. Sacramento. 548 pp.

Meyer, F.B., R.A. Schnick, and K.B. Cumming. Registration status of fishery chemicals, February 1976. Prog. Fish-Cult. 38:3-7.

Mezainis, V.E. 1977. Metabolic Rates of Pond Ecosystems under Intensive Catfish Cultivation. M.S. Thesis, Auburn Univ., Auburn, Ala. 107 pp.

Miller, L.W. and A.W. Fast. 1981. The effects of artificial destratification on fish depth distribution in El Capitan Reservoir, pp. 498-514. In F.L. Burns and I.J. Powling, (eds), Destratification of Lakes and Reservoirs to Improve Water Quality. Australian Govern. Publ. Service, Canberra.

Moss, B. 1980. Ecology of Fresh Waters. John Wiley & Sons, New York. 332 pp.

Nakamura, N. 1948. On the Relation Between Salinity Contents of the Water and Living Condition and Productivity of Carps in Ponds near the Sea. Tokyo Univ., Physiograph Sci. Res. Inst., Bull., Vol. I, No. 51 (Biol. Abs., Vol. 27, 1953).

National Academy of Science and National Academy of Engineering. 1972. Water Quality Criteria 1972. U.S. Government Printing Office. 594 pp.

Needham, P.R. 1950. Fisheries Management Laboratory Syllabus. Berkeley, University of California. 71 pp. (mimeo).

Ness, J.C. 1946. Development and status of pond fertilization in Central Europe. Trans. Amer. Fish. Soc. 76:335-358.

Ohle, W. 1938. Die Bedeutung der Austauschvorgange Schlamm und Wasser fur den Stoffkreislauf der Gewasser. Vom Wasser. 13:87-97.

Owsley, D.E. 1978. Packed Columns for Nitrogen Gas Removal. Dworshak National Fish Hatchery, Ahsahka, Idaho. 22 pp. (mimeo).

Pamatmat, M.M. 1960. The Effects of Basic Slag and Agriculture Limestone on the Chemistry and Productivity of Fertilized Ponds. M.S. Thesis, Auburn Univ., Auburn, Ala. 113 pp.

Pearce, A.S. and G. Gunter. 1957. pp. 129-157. In J.W. Hedgpeth (ed), Treatise on Marine Ecology and Paleoecology. Vol. I, Memoir 69, Geological Soc. Amer., New York.

Perschbacher, P.W. 1975. The Effect of an Herbivorous Fish, Tilapia aurea (Steindachner), on the Phytoplankton Community of Fertilized Ponds. M.S. Thesis, Auburn Univ., Auburn, Ala. 54 pp.

Rappaport, U., S. Sarig, and M. Marek. 1976. Results of tests of various aeration systems on the oxygen regime in the Genosar Experimental Ponds and growth of fish there in 1975. Bamidgeh 28:35-49.

Romaire, R.P., C.E. Boyd, and W.J. Collis. 1978. Predicting nightime dissolved oxygen decline in ponds used for Tialpia culture. Trans. Amer. Fish. Soc. 107:804-808.

Ruttner, F. 1953. Fundamentals of Limnology. U. of Toronto Press, Toronto. 242 pp.

Schaeperclaus, W. 1933. Textbook of Pond Culture: Rearing and Keeping of Carp, Trout and Allied Fishes. Fishery Leaflet 311. U.S. Fish and Wildlife Service. 261 pp.

Schroeder, G.L. 1975. Some effects of stocking fish in waste treatment ponds. Water Research 9:591-593.

Schuster, W.H. 1960. Synopsis of Data on Milkfish, Chanos chanos Forskil. FAO Fish. Biol. Synopsis No. 4. 2-3.

Sigler, W.F. 1958. The Ecology and Use of Carp in Utah. Utah Agric. Exp. Sta. Bull. 405, 63 pp.

Singh, V.P. 1980. Management of fish ponds with acid sulfate soils. Asia Aquaculture 3(3):4-7.

Stickney, R.R. 1979. Principles of Warmwater Aquaculture. John Wiley & Sons, New York. 375 pp.

Sverdrup, H.V., M.W. Johnson, and R.H. Fleming. 1942. The Oceans, Their Physics, Chemistry and General Biology. Prentice-Hall, Inc., Englewood, New Jersey.

Swingle, H.S. and E.V. Smith. 1947. Management of Farm Fish Ponds. Ala. Polytech. Inst. (Ala.) Agr. Exp. Sta. Bull. 254. 32 pp.

Thurston, R.V., R.C. Russo, C.M. Fetterolf, Jr., T.A. Edsall, and Y.M. Barber, Jr. (eds). 1979. A Review of the EPA Redbook: Quality Criteria for Water American Fisheries Society, Bathesda, MD. 313 pp.

Tucker, C.S. and C.E. Boyd. 1978. Consequences of periodic applications of copper sulfate and simazine for phytoplankton control in catfish ponds. Trans. Amer. Fish. Soc. 107: 316-320.

Weiss, R.F. 1970. The solubility of nitrogen, oxygen and argon in water and seawater. Deep-Sea Research 17: 721-735.

Weitkamp, D.E. and M. Katz. 1980. A review of dissolved gas supersaturation literature. Trans. Amer. Fish. Soc. 109(6):659-702.

Welch, P.S. 1948. Limnological Methods. McGraw-Hill Book Co., New York. 381 pp.

__________. 1952. Limnology. 2nd ed. McGraw-Hill Book Co., New York. 538 pp.

Weldon, L.W., R.D. Blackburn, and D.S. Harrison. 1969. Common Aquatic Weeds. U.S. Dept. of Agriculture Handbook 352. 42 pp.

Westlake, D.F. 1965. Some basic data for investigations of the productivity of aquatic macrophytes, pp. 231-248. In C.R. Goldman, (ed), Primary Productivity in Aquatic Environments. MEM. 1st. ITAL. IDROBIOL., 18 Suppl., University of California Press, Berkeley, CA.

Wetzel, R.G. 1975. Limnology. W.B. Saunders Co., Philadelphia. 743 pp.

Wirth, T.L. 1981. Experiences with hypolimnetic aeration in small reservoirs in Wisconsin, pp. 457-468. In F.L. Burns and I.J. Powling, (eds), Destratification of Lakes and Reservoirs to Improve Water Quality. Australian Government Publishing Service, Canberra.

Wunder, W. 1949. Fortschrittliche Karpfenteichwirtschaft. Erwin Nagele, Stuttgard.

POND PRODUCTION SYSTEMS: DISEASES, COMPETITORS, PESTS, PREDATORS, AND PUBLIC HEALTH CONSIDERATIONS

by

James A. Brock

INTRODUCTION

Production losses in warmwater fish pond culture can result from the occurrence of disease, competitors, pests, or predators in these systems. Reductions in yields are manifest by growth suppression and/or actual fish morbidity and mortality. It is not possible with the presently available data base to directly assess the economic impact of disease, competitors, pests, and predators on warmwater aquaculture production. This is primarily due to the lack of morbidity and mortality records in fish-raising industries. Klontz (1972) pointed out that in the U.S. livestock industry mortality due to disease rarely exceeds 0.5 - 1 percent per year, and he predicted that the annual mortality in fish culture settings exceeds 1 percent. The sporadic information available suggests that a mortality rate of 5 - 20 percent or greater is currently experienced in some warmwater pond culture systems. Without a doubt, survival rates and, therefore, production levels in warmwater pond aquaculture systems can be greatly improved. Understanding the diseases in these fish should result in improved survival and production yields.

This section briefly reviews the diseases (infectious and noninfectious), competitors, pests, and predators currently known to affect the following warmwater pond-cultured fish species: channel catfish (*Ictalurus punctatus*), common carp (*Cyprinus carpio*), grass carp (*Ctenopharyngodon idellus*), silver carp (*Hypophthalmichthys molitrix*), bighead carp (*Aristichys nobilis*), Indian carp (*Catla catla*), the tilapias (*Tilapia* spp. and *Sarotherodon* spp.), *Clarias* spp., mullet (*Mugil* spp.), and milkfish (*Chanos chanos*). For the diseases discussed the information provided is largely limited to the epidemiology and control. Specific information regarding diagnostic procedures, pathology, etc., can be found in the textbook *Fish Pathology*, edited by Ronald J. Roberts, or other references included in the bibliography. A discussion of the public health considerations in warmwater pond aquaculture is also included in this section.

INFECTIOUS DISEASES

Diseases of Viral Etiology

Viruses are obligate, intracellular organisms many of which are known to cause disease in host species. Wolf and Mann (1980) list 32 viruses that have been isolated from or are known to occur in fish. Six of these viruses are known to cause disease in such subtemperate and tropical freshwater pond fish species as common carp (*Cyprinus carpio*), grass carp (*Ctenopharyngodon idellus*), tilapia (*Sarotherodon* and *Tilapia* spp.), and channel catfish (*Ictalurus punctatus*). Viruses have not been reported from mullet (*Mugil* spp.), milkfish (*Chanos chanos*), silver carp (*Hypophthalmichthys molitrix*), bighead carp (*Aristichtys nobilis*), mud carp (*Cirrhinus molitorella*), catfish (*Clarias* sp.), or the Indian carp (*Catla catla*).

Control methods for viral diseases in fish are limited to quarantine and restriction of movement, test and slaughter, sanitation and disinfection, reduction of environmental stressors, and good nutrition. Vaccines are presently not available on a commercial scale, nor are there chemotherapeutic agents for treatment. Secondary bacterial infections may be dealt with using appropriate antibiotics.

Channel Catfish Virus Disease (CCVD) Channel catfish virus (CCV) is a herpes virus that causes a systemic disease in fry and fingerling channel catfish at water temperatures of 25-30°C. This disease can be catastrophic (mortality approaching 100 percent) in susceptible channel catfish populations when environmental conditions favor the spread of the virus. This disease is a serious threat to channel catfish culture (Plumb 1978).

The distribution of CCV is thought to be limited to the United States, although an isolated case of CCVD was reported in a batch of fry shipped to Central America from the United States (Plumb 1978). Reducing the temperature from 28 to 18°C has been found to reduce CCVD mortality from 95 to 24 percent (Plumb 1978). The best method of control is avoidance of infected stocks (Plumb 1978).

Carp Pox Carp pox is a relatively common benign cutaneous proliferative disease of cyprinids. It has been recognized as a disease of cultivated common carp (Cyprinus carpio) for more than 400 years (Liversidge and Munro 1978). The etiologic agent has been tentatively identified as a herpes virus. Carp pox only rarely causes mortality in carp but the cutaneous growths may reduce the value of the fish for aesthetic reasons. Carp pox is known to occur in Europe, but its geographical distribution is probably more extensive. Prevention is through destruction of infected stocks, and restriction of movement of potential carriers.

Lymphocystis Lymphocystis is a benign cutaneous proliferative disease of many fresh and marine fish species. Lymphocystis is caused by an iridovirus. The occurrence of this disease in fish populations is sporadic. Infection of fish by this virus results in chronic verrucose lesions on the skin and fins. It is rarely fatal, but infected fish may be unsuitable for sale for aesthetic reasons. The virus is thought to be transmitted horizontally. Lymphocystis has been reported to occur in Europe, North and South America (Wolf 1968), and Africa (Paperna 1974, cited in Balarin and Hatton 1979). Most species of fish are considered susceptible. Spontaneous recovery of infected fish commonly occurs. Prevention includes deriving stocks from fish populations free of lymphocystis and elimination of diseased individuals in cultivated populations.

Spring Viremia of Carp (SVC) and Swimbladder Inflammation (SBI) SVC and SBI are systemic diseases of common carp (Cyprinus carpio) and possibly other cyprinids caused by Rhabdovirus carpio. SVC is probably synonymous with infectious dropsy of carp (Ahne and Wolf 1977; Bucke and Finlay 1979). SVC is a highly contagious disease that causes extensive mortality in cultured carp populations. R. carpio is harbored in carrier fish and is horizontally transmitted to susceptible fish. Vertical transmission of R. carpio is considered probable (Ahne and Wolf 1977). Grass carp (C. idellus) are susceptible to experimental infections with R. carpio (Ahne and Wolf 1977). R. carpio has only been positively identified in Europe, the Soviet Union (Ahne and Wolf 1977), and Great Britain (Bucke and Finlay 1979). Fijan (1972, cited in Ahne and Wolf 1977) postulated that the virus is present wherever infectious dropsy of carp has been found. Bardach et al. (1972) mention infectious dropsy as a disease problem in Chinese carps. Prevention of SVC and SBI is through avoidance of infected carp stocks, quarantine, and restriction of movement of known infected fish. Control methods are limited to reduction of environmental stressors, good husbandry, and proper nutrition.

Grass Carp Rhabdovirus Ahne (1975) reported the occurrence of infectious dropsy in grass carp (C. idellus) caused by a rhabdovirus (grass carp rhabdovirus). This virus has been shown to be similar but not identical to Rhabdovirus carpio. Instead, grass carp rhabdovirus is considered to be the same as pike fry rhabdovirus (Wolf and Mann 1980). This virus has been reported in Holland and Germany, but the geographical distribution could be much greater. Horizontal and vertical transmission has been demonstrated (Liversidge and Munro 1978). Prevention is through avoidance of infected carrier fish. Control methods are limited to reduction of environmental stressors, good husbandry, and proper nutrition.

Carp Gill Necrosis Virus An iridovirus-like agent has been seen by electron microscopy in association with necrotic gill disease in common carp (Cyprinus carpio) (Wolf and Mann 1980). Little is known about this virus and its potential as a primary etiologic agent in gill disease of carp.

Diseases of Bacterial Etiology

There are eighteen recognized bacterial diseases of freshwater and marine fish species. Seven of these diseases are known to affect tropical pond aquaculture species. Lewis and Plumb (1979) state that bacteria are among the most important pathogens of cultured channel catfish. Carp raised in temperate climates are reported to suffer from Aeromonas sp. and Pseudomonas sp. infections. Haller (1974) and Scott (1977), cited in Balarin and Hatton (1979) reported presumed bacterial disease outbreaks in tilapia cultured in Kenya. Sairg (1971) points out that bacterial diseases do not seem to be much of a problem in pond culture in subtropic regions (Israel) and Africa. Reports of diagnosed cases of bacterial disease in pond-cultured milkfish and mullet are nonexistent in the literature. Tilapia are susceptible to experimental infections with Aeromonas hydrophila and Aeromonas salmonicida (Almeida et al. 1968).

Fin Erosion or Rot Fin erosion and ulceration in cultured fish is widely accepted as being a disease complex caused by overcrowding or exposure to other environmental stressors, poor nutrition, and bacterial infection. The species of bacteria isolated from ulcerative lesions of the fins is variable but is usually a gram-negative bacteria in the genus Aeromonas or Pseudomonas. Prevention and control methods include good husbandry practices, proper nutrition, and sanitation. The application of antibacterial drugs in the feed or water may be useful in the treatment of fin rot, but primary emphasis should be placed on correcting husbandry practices and reduction of stressors.

Bacterial Septicemia Bacterial septicemia is an acute-to-subacute systemic bacterial disease of fish caused by *Aeromonas* spp. (i.e., *A. hydrophila*) and *Pseudomonas* spp. All species of fish are regarded as susceptible (National Academy of Sciences 1977). Outbreaks of bacterial septicemia can result in high mortalities in fish populations and frequently - if not always - follow exposure of fish to stressors such as low oxygen, handling, or transfer. *Aeromonas* spp. and *Pseudomonas* spp. are ubiquitous in aquatic environments. Preventative measures include good husbandry, sanitation, and proper nutrition. Bacterins are presently not available. Treatment with systemic antibacterials such as terramycin or sulfamerazine can be effective.

Columnaris Disease Columnaris disease, usually a cutaneous but occasionally a systemic disease, is caused by Flexibacter columnaris. The gills and all cutaneous surfaces may be affected. Columnaris disease is widespread in young channel catfish (*Ictalurus punctatus*) and carp species. Most other freshwater fish species are regarded as susceptible (Richards and Roberts 1978). Columnaris disease is most common in catfish culture when water temperature is greater than 20°C (Lewis and Plumb 1979). Prevention of columnaris disease is through good husbandry and nutritional practices. Lewis and Plumb (1979) suggest potassium permanganate at 2-3 ppm in the pond water or terramycin at 83 gms per 100 lbs. of feed for ten to twelve days, or Furacin at 150 gm per 100 lbs. of feed for twelve days. Valuable fish (broodstock) may be injected with terramycin (25 mg/lb) or Erythromycin (4 mg/lb). Reichenbach-Klinke and Landolt (1973) suggest sulfamerazine (mixed into the feed 10-12 gm/liter:50 kg fish).

Mycobacteriosis Mycobacteriosis is a chronic systemic disease of fish caused by *Mycobacterium* spp. Outbreaks of mycobacteriosis occur infrequently in cultured cyprinids (Snieszko 1978). Mycobacteriosis has not been reported from pond-cultured mullet (*Mugil* sp.), milkfish (*Chanos chanos*), *Clarias* sp., *Tilapia* and *Sarotherodon* spp., or channel catfish (*Ictalurus punctatus*). Prevention is through avoidance of infection by quarantine and restriction of movement of infected fish and by the use of fish feeds which contain only sterilized fish flesh, rather than untreated fish products. Chemotherapeutics are not known to be effective for treatment of mycobacteriosis in fish. The species of mycobacterium which infect fish may infrequently cause skin infection in man. Mycobacteriosis has not been recognized as a problem in tropical pond aquaculture systems.

Edwardsiellosis Infections caused by *Edwardsiella tarda* are reported in Ictalurids, Cyprinids, and Anguillidae in the southern United States and Southeast Asia. *E. tarda* infections cause gas-filled lesions in the muscles of mature catfish (Lewis and Plumb 1979). Mortality in channel catfish cultured in ponds seldom exceeds 5 percent, but if fish are transferred to holding tanks losses may reach 50 percent (Lewis and Plumb 1979). *E. tarda* is widespread in organically polluted waters (Richards and Roberts 1978). Prevention is through reduction of environmental stressors, good husbandry, and proper nutrition. Treatment with antibacterials is reported to be effective. Terramycin, 83 gms per 100 lbs. of feed for ten to twelve days (Lewis and Plumb 1979), and Sulphonamide or Furacin (Richards and Roberts 1978) are reported to be effective.

Diseases of Fungal Etiology

The National Academy of Sciences (1977) lists five major mycotic diseases that cause serious epidemics in cultured and wild fish and shellfish. Three of these diseases are recognized as problems in tropical pond aquaculture species.

Branchiomycosis Branchiomycosis, or mycotic gill rot, is a necrotic gill disease of common carp (*Cyprinus carpio*), channel catfish (*I. punctatus*), and many other fish species caused by *Branchiomyces* spp. Sairg (1971) states that in areas of Europe where this disease commonly occurs it is one of the greatest threats to commercial fish farming. Branchiomycosis has been reported in fish from Europe, Israel, the Balkan states, U.S.S.R., and the southern United States. This disease is most prevalent in ponds high in decaying organic matter (Sairg 1971). Infection and morbidity rates of up to 50 percent may occur (Richards 1978). In Israel, branchiomycosis occurs infrequently but causes high mortalities (Sairg 1971). There is no known treatment for this disease. Control methods are limited to good sanitation and such management practices as increased water flow and prompt removal of dead fish.

Saprolegnosis and Achylosis Saprolegnosis and achylosis are acute fungal diseases of fish and fish eggs caused by fungi in the genera *Saprolegnia* and *Achlya*. Saprolegnosis is by far the most commonly encountered. All species of tropical pond-culture fish species are considered susceptible. These diseases generally arise as cutaneous infections (skin or gills) following insults such as traumatic wounds, bacterial infections, or nutritional deficiencies, that cause disruption to the mucous layer. Once established, the fungus can spread to adjacent healthy tissues and may eventually result in death to the host. In fish-egg incubators Saprolegnia usually develops on detritus or dead eggs and rapidly spreads to healthy eggs, killing them (Rogers 1978). Saprolegnia spp. and Achyla spp. are ubiquitous in natural waters. Preventative methods include good pond and hatchery practices and sanitation. Treatment of individually infected fish with malachite green (67 ppm dip for 10-30 sec.) has been reported as effective (Plumb 1979). The treatment of ponds for saprolegnosis with 0.1 ppm malachite green or 2 ppm potassium permanganate is reported to be occasionally useful (Plumb 1979).

Parasites

Approximately 100 genera of protozoan and metazoan parasites have been described as the etiologic agents of fish and shellfish diseases (National Academy of Sciences 1977). Fortunately, only a handful of these parasitic organisms are of serious consequence to pond-culture fish. Sairg (1971) states that ectoparasites are the largest group of disease organisms in warmwater fish ponds. The significant parasites in pond aquaculture include members from the Phyla Protozoa, Platyhelminthes, Aschelminthes, Arthropoda, and Annelidae.

Fish parasites can be grouped according to their usual location on the host. Ectoparasites are found on the external body surfaces, including the gills, and endoparasites locate in internal organs such as the liver, kidney, or intestines. According to Bauer (1961b), parasites affect fish populations by causing: (1) mortality (1-100 percent); (2) reduction in growth; (3) weight loss; or (4) suppression of reproductive activity or efficiency. In addition to these effects, which can have an impact on fish production, the poor carcass quality of fish infested with parasites can result in reduction of market value for aesthetic reasons (Rogers 1978).

Fish parasites can also be grouped according to life cycle. Those which have a direct life cycle do not require an intermediate host and can rapidly multiply and build up in high numbers. Other parasites have adapted to an indirect life cycle in which one or more intermediate hosts are required in an obligate fashion for the parasite to survive. Infection then depends on the presence of the appropriate intermediate hosts. An understanding of the parasite's life cycle is essential in developing rational prevention and treatment strategies for the control of parasitic disease problems.

Ectoparasites

Protozoa Costia sp. is reported to infect many species of fish and can cause high mortality, especially in younger age classes. This red-blood-cell-size protozoan flagellate attaches to the gills and external body surface of fish, causing irritation and host-cell death. Costia infections can occur at any time of the year, but they tend to be more prevalent in cooler weather in temperate regions. Costia sp. is cosmopolitan in distribution (Hoffman 1979) and has a direct life cycle. Resistant cysts are not found. Costia is a horizontally transmitted water-borne parasite. The organism is thought to be an obligate parasite of cold-blooded species. Prevention can be achieved through elimination of carrier fish and providing parasite-free incoming water. Costia infections can be effectively treated in ponds with formalin 15-59 ppm, potassium permanganate 2-3 ppm, malachite green 0.1 ppm, or malachite green 0.1 ppm/formalin 15 ppm (Hoffman 1979).

Ichthyophthirius multifilis (Ich) is reported to be the most detrimental parasite of pond-culture fresh- and brackish-water fish (Sairg 1971; Rogers 1978). Ichthyophthirius has been reported to destroy whole pond fish populations (Bauer 1961b; Sairg 1971; Rogers 1979a). All species of pond-cultured fish are thought to be susceptible. Ichthyophthirius multifilis is an obligate parasite of cold-blooded vertebrates. Ich is a large, ciliated protozoan with a prominent C-shaped nucleus. The adult parasite (Trophozoite) burrows into the skin and gills of fish. Once mature, it drops off the fish, settles on the substrate, and undergoes multiple divisions, producing 1,000 or more infective tomites. The tomites or "swarmers" are most susceptible to available chemotherapeutics for Ich treatment. Ichthophthirius multifilis is cosmopolitan in distribution (Hoffman 1972). Prevention is through elimination of exposure through destruction of carrier fish and through disinfection of potentially contaminated water or equipment. Vaccines for Ich are not presently available. In ponds, treatment with 0.1 ppm malachite green/15 ppm formalin for one to three treatments has been reported to be effective. Sairg (1971) reports that low oxygen (1-2 ppm) levels are inhibitory to I. multifilis.

Tricodina, Scyphidia, Glossatella, Epistylis, and Chilodonella These organisms are ciliated ectoparasitic protozoans that are frequently found to infect pond-cultured fish (Sairg 1971; Rogers 1979a). More than one of these parasite species may be found to infect the same fish species. These parasites attack both the gills and the external body surface. Heavy infections are not uncommonly associated with poor environmental conditions. Of these ciliates, Chilodonella and Tricodina are reported to be the most damaging (Sairg 1971). Heavy Chilodonella sp. infestation leads to high mortality in pond-cultured fish species in Israel (Sairg 1971) and common carp cultured in Russia (Bauer 1961b). Epistylis sp. is thought in many cases to be a secondary invader growing in fungus-like colonies on damaged areas of the fishes' bodies. Tricodina, Scyphidia, Glossatella, Epistylis, and Chilodonella are cosmopolitan in distribution. These parasites have a direct life cycle; cyst stages are not formed. Prevention of disease caused by Chilodonella and Tricodina can be achieved through the elimination of carrier fish or waterborne contamination. Good husbandry practices and proper nutrition are very important in reducing the impact of these parasites on fish populations. Effective treatments for ponds include copper sulfate 0.25-0.5 ppm (do not use if the water hardness is less than 20 ppm), potassium permanganate 2-3 ppm, formalin 15-25 ppm, or malachite green 0.1 ppm.

Platyhelminthes Monogenetic Trematodes Monogenetic trematodes are microscopic ectoparasitic flukes with a direct life cycle. Many genera are described as parasites of fish. Sairg (1971) states that monogenetic trematodes are the most numerous group of ectoparasites of cultured fish in Israel. Dactylogyrus vastator, D. anchoratus, and D. extensus are reported to be especially dangerous to susceptible fish species (Sairg 1971). Monogenetic trematodes are particularly pathogenic to fry and juvenile fish (Sairg 1971; Hoffman 1979), with mortalities reaching 80 to 100 percent in carp fry populations (Sairg 1971). While species of monogenes are usually relatively host-specific and may have a limited geographic distribution, as a group the monogenetic trematodes are cosmopolitan in distribution. Those monogenetic trematodes that infect fish are obligate parasites. Prevention is through elimination of carrier fish and use of parasite-free water. Treatments include the use of formalin 25-50 ppm (Hoffman 1979); potassium permanganate 2-3 ppm (Hoffman 1979); Masoten (Dipterex) (0,0-dimethyl 1-hydroxy 2-trichloromethyl phosphate) 0.25 ppm (Sairg 1971); D.D.V.P. (0,0-dimethyl 0-2-2 dichloroenyl phosphate) 0.25-0.40 ppm (Sairg 1971); or Bromex (1,2-dibromo-2,2-dichlorethyl-dimethyl-phosphate) 0.15 ppm (Sairg 1971).

Arthropoda Crustacea Several parasitic Crustacea, including the branchiuran Argulus and the copepods Lernaea, Achtheres, and Ergasilus, are considered as extremely dangerous and reported to have caused great loss to fish culturists (Hoffman 1979; Sairg 1971; Rogers 1979a). The crustacean parasites are large ectoparasites visible to the naked eye. As a group they are obligate parasites with a direct life cycle and are cosmopolitan in distribution. High mortality and growth reduction in pond fish species are attributed to parasitism by these organisms (Sairg 1971). Transmission is horizontal. Argulus lays its eggs on substrata in the pond. The copepods carry eggs until they hatch. Prevention includes elimination of carrier fish and avoidance of contact with parasite-contaminated water.

Treatment methods include:

1. Chemotherapeutics
   Lindane (1,2,3,4,5,6-Hexachlorocyclohexane 0.02 ppm) (Sairg 1971)
   Malthion 0.25 ppm (Sairg 1971)
   Dipterex 0.25 ppm (Sairg 1971)
   D.D.V.P. 0.25 ppm (Sairg 1971)
   Bromex 0.12 ppm (Sairg 1971)

2. Mechanical
   The placement, daily removal, and replacement of sticks in the ponds; these serve as "egg traps" on which Argulus lay their eggs. Most Argulus eggs can be destroyed this way (Balarin and Hatton 1979).

Annelidae Hirudinea (The Leeches) Leeches are reported to be a problem in cultured carp and channel catfish. Leeches are recognized as a problem in carp only in temperature regions. Piscicola geometra (a leech infecting common carp) is reported to have a wide distribution and is thought to serve as the vector for Trypanoplasma cyprini (hemoflagellate). T. cyprini infections are also restricted to cooler climates (Bauer 1961b).

Myxobdella lugubris affects catfish in the south-eastern United States and may result in fish mortalities (Hoffman 1979). Leeches have a direct life cycle with a cyst stage (cocoon) which is resistant to dessication (Bauer 1961b).

Bauer (1961b) recommends the following bath treatments for Piscicola infections in carp.

2.5 percent NaCl for 1 hour
0.2 percent Lysol for 5-15 seconds
0.2 percent solution of unslaked lime for 2-5 seconds
0.005 percent solution of $CuCl_2$ for 15 minutes.

Masoten 0.50 ppm has also been suggested for pond treatment of leech infestation in catfish (Hoffman 1979).

Endoparasites

Protozoa Myxosporidae Members of the subphylum Cnidospora, Class Myxosporidae, are parasitic to pond-cultured fish species. Sairg (1971) reports the occurrence of Myxobolus sp. infections in pond-cultured Mugil sp. in Israel. Myxobolus sp. and Henneguya sp. have been found to infect Cyprinidae, Cichlidae and Siluridae in Africa (Baker 1960 cited in Sairg 1971). Rogers (1979b) indicated that the interlamellar form of Henneguya sp. is thought to cause high mortality in catfish in the United States. Bauer (1961b) reported the occurrence of pernicious anemia and Hoferellasis in common carp in the Soviet Union as being caused by Myxobolus cyprini and Hoferellus cyprini respectively. Myxosporidian parasites can locate in many organ tissues but this varies from species to species. The mode of transmission is largely unknown. Prevention is through elimination of carrier fish and through pond disinfection (liming-calcium hydroxide 1,000-2,500 lbs. per acre) (Wellborn 1979). No effective chemical treatments have been reported.

Telosporea Coccidiosis is primarily an intestinal-tract disease reported to affect pond-cultured Cyprinids. Four species of the sporozoan parasite Coccidia infect the common carp in Europe (Bauer 1961b). Chinese carp are also reported to be susceptible to Coccidia infections (Bardach et al. 1972). Coccidia produce two disease syndromes in cultured carp: coccidiosis and coccidal enteritis (each disease is caused by a different Eimera sp.). Coccidal enteritis is reported to cause serious losses in one-year-old common carp (Bauer 1961b). Coccidia are obligate intracellular parasites. The reported distribution of this disease is Europe and Asia. Preventative procedures include elimination of carrier fish, disinfection of ponds with lime, good husbandry practices, and proper nutrition. Specific chemotherapy for coccidiosis in fish is not reported.

Zoomastigophora The protozoan hemoflagellate Trypanoplasma cyprini is the cause of trypanoplasmosis in common carp in Europe (Bauer 1961b). The disease is not reported as a problem in pond-cultured Chinese or Indian carp species. In carp, T. cyprini is believed to be transmitted by the leech Piscicola. Hemoflagellates (Trypanosoma and Trypanoplasma) are reported to affect channel catfish in the United States but are not a problem in pond culture settings (Rogers 1979b). Hemoflagellates are not reported to infect cultured tilapia, Mugil sp., or Chanos chanos. Apparently, there is disagreement concerning the effect this parasite has on infected carp and on its significance in causing disease (discussed in Bauer 1961b). Control methods are not reported.

Platyhelminthes Digenetic Trematodes Digenetic trematodes are endoparasites that require one or more intermediate hosts for completion of the life cycle. In this regard fish may serve as intermediate or final hosts for these parasites. Many species of digenetic trematodes are reported to infect fish. Sairg (1971) reports digenetic trematodes to be of little consequence except as a public health problem such as Heterophyes heterophyes infections in Mugilidae. Hoffman (1979) suggests that channel catfish infected with the ovarian fluke Acetodextra ameimi may have reduced egg production. Bauer (1961b) reports that the blood fluke Sanguinicola, which infects cultured common carp in the Soviet Union, causes mortality, particularly in young fish. Bauer (1961b) also indicates that in the southern areas of the Soviet Union cultured carp may be 100 percent infected with "Black-spot disease" caused by encysted trematodes (metacercaria). Prevention and control procedures for digenetic trematodes include breaking a link in the life cycle. This may be achieved by reducing or eliminating snail intermediate hosts or final hosts like the heron from the pond habitat. Chemotherapy for fluke infections in fish has not been developed.

Cestodes Cestodes (tapeworms) are reported to infect pond-cultured fish. Fish may serve as intermediate or final host for these parasites. Some tapeworms are rather benign, causing little damage to the host. Corallobothrium sp. infect catfish in the United States. This parasite is not reported to cause significant pathology or to be harmful to catfish (Hoffman 1979). On the other hand, other species of tapeworms are considered dangerous to fish. Bauer (1961b) maintained that Caryophylleaus finbriceps infections resulted in mass mortalities in older common carp. The Asian tapeworm (Bothriocephalus acheilognathi) is considered an extremely dangerous parasite to cultured fish species (Hoffman 1979). Preventative procedures for tapeworms include quarantine and restricted movement of infected stocks, and reduction or elimination of intermediate or definitive hosts. Control procedures include removal of infected fish from the pond followed by disinfection of the pond with calcium hydroxide 1,000 - 2,500 lbs. per acre

(Wellborn 1979). Chemotherapy is available for treatment of intestinal cestode infections. Di-n-butyl tin oxide and dibutyl tin dilurate 250 mg per kg of fish added to the feed or Yomesan (phenasal) at 0.5 percent of the weight of the feed for three days is reported as effective (Hoffman 1979).

Aschelminthes Nematodes Many species of Nematode have been reported to parasitize fish. Parasitic Nematoda life cycles are indirect, usually involving one or two intermediate hosts and a final host. Channel catfish cultured in the United States may harbor nematodes from four genera. Two of these, Spinitectus sp. and Contracaecum spiculigerum, are reported to be damaging to catfish (Hoffman 1979). Hoffman (1979) lists Philometra carassii as dangerous to cultured species of fish, causing great losses. Balarin and Hatton (1979) point out that these parasites may become a problem in situations where predatory birds that may serve as final hosts for nematodes are abundant. Prevention and treatment methods include breaking the life cycle (controlling intermediate or final hosts) (Balarin and Hatton 1979). Masoten has been reported as effective in controlling Camallanus sp. in aquarium fish (Hoffman 1979).

Acanthocephlans There are at least five genera of acanthocephlans that are known to infect fish. Fryer and Iles (1972 cited in Balarin and Hatton 1979) reported acanthocephlans as common in african cichlids. Acanthocephlans are reported to parasitize all species of pond-cultured fish in this discussion except Chanos chanos. Nevertheless, members of this phyla have not been reported as the cause of disease or problems under pond-cultured conditions. The acanthocephlans that infect fish have indirect life cycles with fish serving as intermediate or final hosts. Control methods have not been implemented for acanthocephlans in pond-cultured fish.

NON-INFECTIOUS DISEASES

Diseases of Nutritional Etiology

In general, nutritional diseases are not recognized as a problem in warmwater pond-fish culture unless stocking rates approach or exceed 4,000 kg/ha. This is believed to be due to the availability of natural foods in the pond environment. Launer et al. (1978) reported no significant difference in weight gain, feed conversion, or survival in channel catfish in ponds fed diets deficient in vitamins C and D_3 (Cholecalciferol) for five months compared to controls fed diets with these vitamins. Lovell and Lim (1978) reported that when channel catfish are stocked at densities of less than 4,000 kg/ha there is sufficient vitamin C in ponds to prevent "broken back syndrome" in fish fed diets without vitamin C.

Nutritional problems in fish populations may appear in the form of reduced fecundity, slowed growth, decreased appetite, increased susceptibility to infectious disease, frank morbidity with clinical signs and pathological lesions, mortality, or some combination of the above. It is well known that prolonged storage of feed may result in reduction of feed quality, particularly for vitamin C and the essential fatty acids (Lovell 1976; National Academy of Sciences 1977).

The National Academy of Sciences (1977) reported that a great deal of nutritional research data exists for channel catfish and the common carp, but that there is limited information for tilapias, milkfish, Chinese carps, and most other warmwater pond-cultured species. The following discussion on reported nutritional deficiency problems is necessarily centered on channel catfish and common carp.

Deficiency Diseases

Starvation Starvation refers to absolute nutrient deprivation resulting from inadequate intake or assimilation of feed. The energy needs of the organism are below maintenance levels. Typically, starved fish will have a large head and slender body and will be dark in coloration.

Protein and Amino Acids In warmwater fish species, specific protein and amino-acid deficiency diseases are generally not recognized as a problem in pond-culture settings. Dietary protein in these fishes supplies both essential amino acids and energy. Feeding experimental diets deficient in selected amino acids causes appetite depression and poor growth regardless of the type of amino acids omitted (Lovell 1976; National Academy of Sciences 1977; Cowey and Roberts 1978). Cridland (1960, cited in Balarin and Hatton 1979) reported growth suppression, skeletal deformities, and exophthalmia in S. esculentus fed maize meal, deficient in tryptophan (Gohl 1975, cited in Balarin and Hatton 1979). A tryptophan-deficiency-related scoliosis has been reported for salmonids (Ashley 1972). Lovell (1976) suggested the occurrence of tryptophan-deficiency-induced scoliosis was unlikely in warmwater fish fed practical diets. Lovell (1976) found growth suppression in pond-reared channel catfish resulting from feeding diets with high protein levels (42 percent) in the presence of low amounts of nonprotein energy.

Lipids Specific fatty-acid requirements have not been established for channel catfish (Lovell 1976) or for other warmwater species. Yet an essential fatty-acid-deficiency syndrome, which results in growth reductions and tissue build-up of 5,8,11-eicosatrienoic acid, has been reported in catfish, carp, and eel (National Academy of Sciences 1977). Lipoid liver degeneration, an essential-fatty-acid deficiency disease reported in salmonids (Cowey and Roberts 1978), is not known to occur in pond-cultured warm-water species. Tilapia and carp apparently can tolerate dietary lipid content above 25 percent (Lagler et al. 1962, cited in Balarin and Hatton 1979).

Carbohydrates The utilization of carbohydrate as an energy source by warmwater fish is unclear (National Academy of Sciences 1977). Research data suggest that dietary lipids and proteins are utilized preferentially to carbohydrates as energy sources in warmwater fish species (National Academy of Sciences 1977). Sekoke disease, described as a spontaneous diabetes in carp fed extremely high-starch diets, has been reported in Japan (Yokote 1970, cited in Lovell 1976). The disease can be prevented by eliminating the excess starch from the diet.

Vitamins Vitamins are essential in warmwater fish diets. If intake levels are inadequate, specific deficiency signs may become apparent (Table 1). Vitamin-deficiency disease is unlikely to occur in pond-culture settings unless stocking densities are considerable. "Broken back syndrome" is a well-known disease of channel catfish cultured under hyperintensive conditions. This disease has been reported in heavily stocked catfish ponds (6270 kg/ha) or in catfish ponds stocked with large numbers of tilapia (Lovell 1976). As previously mentioned in the Feeds and Feeding section, broken back syndrome results from inadequate intake of dietary vitamin C. The disease is characterized by deformed spinal column, opercula and gill filaments, caudal fin erosion, anemia, reduced growth rate, increased susceptibility to bacterial pathogens, and reduced tissue levels of vitamin C (Lovell 1976). Broken back syndrome arises if channel catfish are fed diets deficient in vitamin C for periods longer than eight weeks. The minimum dietary requirement for vitamin C in channel catfish is 30 mg/kg of diet (Lovell 1976).

Sato et al. (1978) found that the absence of vitamin C in the diet of carp cultivated under normal conditions for 84 weeks did not slow growth, increase mortality, or reduce liver concentrations of ascorbic acid. On the other hand, Mahajan and Agrawal (1979) reported poor growth, high mortality (42 percent), severe hemorrhages, fin necrosis, increased pigmentation, and spinal flexures in Indian carp (Cirrhina mrigala) fed diets lacking in vitamin C. These experiments were conducted in 12-liter plastic troughs and plastic pools that were cleaned each day to remove algae growth and detritus.

Minerals Fish absorb minerals from both feeds and the water (Lovell 1976; National Academy of Sciences 1977). Calcium and phosphorus are the minerals known to be required in the largest quantities by fish (Lovell 1976). Fish can obtain adequate levels of calcium from the water unless the water is very soft. Phosphorus is not absorbed as well from the water, and dietary intake of phosphate is required by warmwater fish (National Academy of Sciences 1977). Lovell (1976) found that experimental phosphorus deficiency in channel catfish caused decreased growth rate and increased mortality. In carp, inadequate phosphate intake results in slowed growth, scoliosis, lordosis, and skull and opercular deformities (National Academy of Sciences 1977). Scott (1977 cited in Balarin and Hatton 1979) described deformities in tilapia farmed in Kenya caused by calcium and phosphorus dietary imbalances. The minimum requirements of available phosphorus for carp are 0.8 percent and for channel catfish 0.45 percent of the diet (National Academy of Sciences 1977).

The dietary requirements of other minerals have not been established for warmwater fishes. Experimentally, high levels of potassium, iron, zinc, copper, iodine, and molybdenum have caused growth depression in fish. Diets low in iron and copper have been reported to cause anemia in fish (National Academy of Sciences 1977; Sakamoto 1978). Mineral deficiencies are not known to cause problems in pond-culture settings. Ishac and Dollar (1968) reported that tilapia kept in manganese-free water and fed manganese-deficient diets were sluggish, developed equilibrium problems, had poor appetites, lost weight, and began to die after three weeks. These authors calculated a minimum daily requirement of 1.7 mg manganese per kilogram of fish for normal growth.

Diseases with a Toxic Etiology

In tropical pond-aquaculture systems, toxicological problems are those resulting from exposure to inorganic, synthetic organic, or natural organic toxicants. Inorganic toxicants of significance to pond-aquaculture systems include ammonia, nitrite, and hydrogen sulfide. Synthetic organic toxicants are the pesticides and herbicides. Natural

Table 1

THE ESSENTIAL VITAMINS AND DEFICIENCY SIGNS IN WARMWATER FISHES[a]

Vitamin	Deficiency signs
Thiamine	Poor appetite, muscle atrophy, convulsions, instability and loss of equilibrium, edema, poor growth, congestion of fins and skin, fading of body color, lethargy.
Riboflavin	Corneal vascularization, cloudy lens, hemorrhagic eyes, photophobia, uncoordination, abnormal pigmentation of iris, striated constrictions of abdominal wall, dark coloration, poor appetite, anemia, poor growth, hemorrhage in skin and fins.
Pyridoxine	Nervous disorders, epileptiform fits, hyperirritability, ataxia, anemia, loss of appetite, edema of peritoneal cavity, colorless serous fluid, rapid onset of rigor mortis, rapid breathing, flexing of opercles, irridescent blue coloration, exophthalmos.
Pantothenic acid	Clubbed gills, necrosis, scarring and cellular atrophy of gills, gill exudate, prostration, loss of appetite, lethargy, poor growth, hemorrhage in skin, skin lesions and dermatitis.
Inositol	Distended stomach, increased gastric emptying time, skin lesions, poor growth.
Biotin	Loss of appetite, lesions in colon, altered coloration, muscle atrophy, spastic convulsions, fragmentation of erythrocytes, skin lesions, poor growth.
Folic acid	Lethargy, fragility of caudal fin, dark coloration, macrocytic anemia, poor growth.
Choline	Poor food conversion, hemorrhagic kidney and intestine, poor growth, accumulation of neutral fat in hepato-pancreas, enlarged liver.
Nicotinic acid	Loss of appetite, lesions in colon, jerky or difficult motion, weakness, edema of stomach and colon, muscle spasms while resting, sensitivity to sunlight, poor growth, hemorrhage in skin, tetany, lethargy, anemia.
Vitamin B_{12}	Poor appetite, low hemoglobin, fragmentation of erythrocytes, macrocytic anemia, reduced growth.
Ascorbic acid	Scoliosis, lordosis, impaired formation of collagen, abnormal cartilage, eye lesions, hemorrhagic skin, liver, kidney, intestine, and muscle, reduce growth.
Vitamin A	Ascites, edema, exophthalmos, hemorrhagic kidneys, poor growth.
Vitamin E (α-tocopherol)	Ascites, ceroid in liver, spleen, and kidney, epicarditis, exophthalmia, microcytic anemia, pericardian edema, fragility of red blood cells, poor growth.
Vitamin K	Anemia, prolonged coagulation time.

[a] From National Academy of Sciences 1977.

organic substances are the biotoxins. In this category only the phytotoxins have been recognized as toxicants in tropical aquaculture systems.

Low dissolved oxygen, the most significant cause of fish mortality in warmwater pond aquaculture, is discussed in the section on water quality.

Exposure of animals to sufficient concentrations of toxicants may result in death, illness, or physiopathologic alternation. In aquaculture systems these effects can cause decreased survival, reduced growth, increased susceptibility to disease, or impaired fecundity, resulting in suboptimal yields and loss of profits. At toxicant exposure levels that result in acute mortality the effect is readily apparent, and corrective procedures may be implemented. Sublethal effects of toxicants may be difficult to detect until some time after exposure when pond yields reflect poor performance of the stock.

Inorganic Toxicants

Ammonia Ammonia is present in most natural waters and results from biological degradation of nitrogeous organic matter (Environmental Protection Agency 1976). When ammonia dissolves in water a chemical equilibrium is established which contains unionized ammonia (NH_3), ionized ammonia (NH_4^+), and hydroxide ions (OH^-). The equilibrium shifts in favor of increased NH_3 with increasing pH (Environmental Protection Agency 1976). The un-ionized form of ammonia (NH_3) is toxic to fish. Toxicity varies with fish age and species.

Ammonia toxicity has been suggested as a limiting factor in channel catfish culture (ponds) in the United States (Avault 1978). Problems arise in catfish ponds at high stocking densities (5,000 catfish per acre). Recommended control methods include management practices, proper feeding and stocking levels, and rearing in ponds not deeper than 1.5 meter (Avault 1978).

The un-ionized ammonia 96 hr. LC50 values for channel catfish fry at 22°C, 26°C, and 30°C are 2.4 mg/L, 2.9 mg/L, and 3.8 mg/L (Colt and Tchobanoglous 1976). Growth suppression in channel catfish resulting from exposure to sublethal levels of NH_3 has been reported (Robinette 1973; Colt and Tchobanoglous 1978). The safe level of un-ionized ammonia for freshwater aquatic life is 0.02 mg/L at temperatures >5°C and pH 8.5 (Environmental Protection Agency 1976).

Common carp are considered to be relatively resistant to acute ammonia toxicity (Environmental Protection Agency 1976). On the other hand, differences in sensitivity among fish species to chronic exposure is considered small (European Inland Fisheries Advisory Commission, 1970 cited in Environmental Protection Agency 1976).

Acute ammonia toxicosis causes metabolic acidosis and hyperkalemia resulting in heart block (Buck et al. 1973). In chronic sublethal exposure, ammonia acts as a tissue irritant, causing gill epithelial hyperplasia and focal necrosis in the liver (Smith and Piper 1975). Burrows (1964, cited in Smith and Piper 1975) suggests that chronic sublethal exposure to un-ionized ammonia may predispose fish to infectious disease. Flis (1968, cited in Environmental Protection Agency 1976) reported that exposure of common carp to sublethal ammonia levels resulted in cellular necrosis in several organ systems.

Nitrite-Nitrate Nitrite toxicity (brown-blood disease) has been reported as a cause of significant losses in channel catfish pond culture in intensively fed ponds at the time of sharp decline in temperature in late fall or early winter (Lovell 1979). Nitrite is an intermediate compound in the biological oxidation of ammonia. Nitrate is relatively nontoxic but may be reduced to nitrite. Nitrite absorbed into the fish oxidizes hemoglobin to methemoglobin. Methemoglobin in the oxidized state is not available for oxygen transport, resulting in decreased oxygen-carrying capacity of the blood and death through tissue anoxia (Buck et al. 1973).

Both acute and chronic toxic effects of nitrite/nitrate are reported for domestic animals (Buck et al. 1973). Only the acute toxic effects have been reported for fish (Lovell 1979). The nitrite 96 hr. LC 50 level for channel catfish at 22 and 30°C are 42 and 43 ppm respectively (Colt and Tchobanoglous 1976). Nitrite nitrogen levels below 5 ppm (McCoy, 1972 cited in Environmental Protection Agency 1976) and nitrate nitrogen levels below 90 ppm (Knepp and Arbin 1973 cited in Environmental Protection Agency 1976) should be safe for most warmwater fish species.

Reduction of feeding rate is suggested by Lovell (1979) as a means to control nitrite levels in pond water. Increased water flow into the pond may also be effective in reducing nitrite levels in ponds.

Hydrogen Sulfide Hydrogen sulfide toxicity is considered to be a problem in intensive fish pond culture systems in the southern United States (Johnston 1975). The anaerobic decomposition of algae and uneaten feed is the major source of hydrogen sulfide in pond systems. In well-oxygenated water the effect of sulfide on fish is minimal because it is oxidized to nontoxic sulfate or sulfur. The degree of hazard to aquatic animals from sulfide depends on temperature, pH, and dissolved oxygen (Environmental Protection Agency 1976). At neutral or lower pH and low dissolved oxygen levels the hazard from sulfides is exacerbated. Sulfide levels that exceed 2.0 mg/L may constitute a long-term health hazard to pond-cultured fish (Environmental Protection Agency 1976.

Synthetic Organic Toxicants (Pesticides and Herbicides) Synthetic organic toxicants (pesticides and herbicides) may enter pond aquaculture systems through the purposeful or inadvertant activities of man. It is well documented that contamination with these substances may be highly detrimental to fish and members of the pond ecosystem. This is a recognized problem on a worldwide scale (Food and Agriculture Organization 1964). Koesoemadinata (1980) states that pesticides can be a major constraint in integrated farming by reducing productivity of fish or other aquatic animal crops reared simultaneously or as alternative crops.

The nature of some of these compounds, especially the chlorinated hydrocarbons, is such that they persist and are concentrated in the food we eat. Koesoemadinata (1980) points out that synthetic organic toxicants can affect fish production, the consumers of fish, and fisheries resources. The effect on fish may be direct or indirect through reduction of food organisms in the food web. The Food and Agriculture Organization has developed a policy statement regarding the use of pesticides as it relates to fish culture and fisheries resources (Food and Agriculture Organization 1964). Extreme caution is advised in the use of pesticides and herbicides where there is any likelihood of contamination of fish culture or fisheries resources areas. Table 2 shows suggested safe levels for some common pesticides.

Biotoxins (Phytotoxins)

Chrysomonadinae *Prymnesium parvum* is a phytoflagellate inhabiting brackish-water and produces powerful exotoxins capable of causing high mortality in pond-reared fish. Sairg (1971) reports the economic loss to fish farmers in Israel to have exceeded $100,000 in some years. Fish mortalities resulting from *Prymnesium parvum* have been reported in Israel, England, Europe, the United States, Bulgaria, and Japan (Sairg 1971). *P. parvum* does not grow at salinities below 10 percent (Shilo and Shilo 1962, cited in Sairg 1971).

The toxins produced by *P. parvum* affect many invertebrates and all cultured teleosts (Sairg 1971). Mortality in pond-cultured fish may reach 100 percent. The toxicity of *P. parvum* toxin to fish is highly dependent on complexing with a cation at pH 7 to 9. The icthythotoxin affects the gills by destroying the selective permeability of the epithelial cells. Death ensues due to osmotic shock and intravascular hemolysis.

Ulitzur and Shilo (1964, cited in Sairg 1971) developed a sensitive bioassay to detect the presence of levels of *P. parvum* toxin in pond water. This procedure which utilizes *Gambusia* sp. as a test fish, allows for the early detection and treatment of dangerous levels of toxins in the ponds. The application of this procedure has led to the prevention of fish mortality in 95 percent of infested ponds and has significantly reduced economic losses.

Sairg (1971) lists aqua-ammonia, ammonium sulfate, and copper sulfate as effective treatments of ponds to control *P. parvum*.

Cyanophycophyta (Blue-Green Algae)

Of the more than 50 genera of blue-green algae known, only 10 or so have been reported to cause toxicosis in fish, birds, or mammals (Collins 1978). Blue-green algae like *Microcystis aeruginosa* produce endotoxins that are released upon cell lysis. Others, like *Anabaena flos-aquae*, excrete a water-soluble exotoxin. Blue-green algae are reported to cause fish mortalities due to excretion of toxic substances and through oxygen depletion following mass algae die-offs (Sairg 1971). Not all isolates of a blue-green species will produce toxins, and toxin production can be activated or inhibited by a variety of environmental conditions (Collins 1978). Sairg (1971) reported that carp injected with *Microcystis aeruginosa* endotoxin developed central nervous system signs (loss of equilibrium and spiraling).

Sairg (1971) reports that a "muddy" flavor in fish flesh, which resulted in considerable economic loss due to lack of acceptability of product by consumers, was caused by the fish feeding on *Oscillatoria limosa* and *Anabaena spiroides*.

Sairg (1971) recommends the control of blue-green algae blooms with copper sulfate, but cautions that copper sulfate is not at all selective for blue-greens. In addition, application of copper sulfate may lead to acute algae die-offs, causing fish mortalities through oxygen depletion. Dissolving copper sulfate in warmwater at a 3 percent concentration and selectively spraying the corners and shores where floating blue-green algae accumulate is suggested. Total copper applied should not exceed 1.5 ppm (Sairg 1971).

COMPETITORS, PESTS AND PREDATORS

In pond aquaculture systems, competitors, pests, and predators are deleterious fauna that cause production or

Table 2

SUGGESTED SAFE LEVELS FOR SOME PESTICIDES FOR AQUATIC LIFE[a]

Chemical	Safe level[b]	Bioaccumulation index	Persistence index
Aldrin/Dieldrin	0.003 mg/L	High	High
Chlordane	0.01 mg/L (F) 0.004 mg/L (m)	High	High
DDT	0.001 mg/L	High	High
Demeton	0.01 mg/L	?	?
Endosulfan	0.003 mg/L (F) 0.001 mg/L (M)	?	Low
Endrin	0.004 mg/L	Low	Low
Guthion	0.01 mg/L	Low	Low
Heptachlor	0.001 mg/L	High	High
Lindane	0.01 mg/L (F) 0.004 mg/L (M)	Moderate	?
Malathion	0.1 mg/L	Low	Low
Methoxychlor	0.03 mg/L	Low	Low
Mirex	0.001 mg/L	High	High
Parathion	0.04 mg/L	Low	Low
Toxaphene	0.005 mg/L	High	High
Polychlorinated Biphenyls	0.001 mg/L	High	High

[a] Adapted from Environmental Protection Agency 1976

[b] (F) = freshwater, (M) = marine

financial losses. The impact of these organisms is particularly evident in extensive pond culture settings. Losses occur through reduced yields of desirable species or result from damage to pond surfaces or structures (Pillai 1970). Competitors reduce fish production through competition with cultured species for food and habitat, resulting in depressed growth and survival of the latter. Pests damage pond surfaces or structures in ponds. Predators decrease numbers of cultivated fishes, resulting in production losses. In certain circumstances predator biomass control can result in production increases (discussed in the section on stocking procedures).

Competitors

Snails are some of the most common competitors in fresh- and brackish-water ponds. Snails in the family Cerithidae are reported to be particularly troublesome in the Indo-Pacific region (Pillai 1970). Snails compete with herbivorous fish for available algae feeds, slow the growth of desirable algae species, and in some instances destroy habitat (Pillai 1970). Ling (1960, cited in Pillai 1970) states that snails compete for algae with milkfish in brackish-water fish ponds. Snails may also serve as intermediate hosts for several fish, bird, and mammal parasites.

If they are not abundant, snails may be beneficial as scavengers, but if they are allowed to multiply unchecked they may accumulate in large numbers. As many as 34 tons of snails have been recovered from fish ponds in East Java (Ling 1960, cited in Pillai 1970). Djajadiredja (1957, cited in Pillai 1970) found as many as 6,940 snails/m^2 in ponds in Djakarta.

The Polychaete worm Dendronereis pinnaticirris is reported to infest brackish-water ponds in the Philippines and seriously deplete algae biomass, thus reducing available algae for milkfish (Pillai 1970).

Another competitor is the larval Chironomid midge (Tendipes longilobus). These organisms are reported to occur in high numbers in brackish-water ponds in the Philippines and compete with milkfish fry and juveniles for blue-green algae (lab-lab) in the ponds (Ling 1977). Chronomid larvae are reported to consume 60-90 kg/algae/day (Bardach et al. 1972).

Bardach et al. (1972) mention that tadpoles and fairy shrimps (Streptocephalus texanus) are competitors with catfish fry for available feed in catfish ponds in the United States.

Undesirable species of fish in fresh- and brackish-water ponds are the most serious competitors to desirable species. Sarotherodon mossambica, Gambusia affinis, Mugil sp., Scatophagus sp. and Mollienesia latipinna are reported as competitors in milkfish ponds in the Philippines and Taiwan (Pillai 1970; Bardach et al. 1972).

Pests

In the context used here, pests are those co-inhabitants of fish ponds that mechanically damage fish pond surfaces or structures within ponds. Pillai (1970) mentions that oysters and barnacles occasionally pose a problem by fouling screens and sluices, thereby hindering free exchange of water in brackish-water ponds in the Indo-Pacific region.

Wood-boring organisms are reported to cause considerable damage to wooden structures in ponds (Pillai 1970). Molluscs in the family Teredinidae (genus Teredo - eleven species, genus Bankia - six species, and genus Nausitora - one species) and the family Philadidae (genus Martesia) attack and eventually destroy wooden objects in ponds (Pillai 1970). Wood-boring amphipods and isopods are also known to destroy wooden structures in ponds (Pillai 1970). In India the most important genera are Sphaeroma and Limnoria (Food and Agriculture Organization 1958, cited in Pillai 1970).

Through burrowing activities in the pond bottom, Bristleworm (Polychaetes) cause damage that results in water losses (Pillai 1970). Several species of burrowing crabs (Scylla serrata, Sesarma taeniolata, Uca sp. and the large crustacean Thalassina scorpinoides) are reported by Pillai (1970) to burrow into pond bottoms and dykes, causing leaks.

In the Philippines, the mucilaginous egg masses of the Polychaete worm Marphysa gravelyi are reported to cause appreciable losses to milkfish fry and fingerlings that accidently swim into and become stuck to these egg masses (Pillai 1970).

Predators

Piscivorous fish, amphibians, reptiles, birds, and mammals are reported to cause considerable losses of fish stocks in fresh and brackish-water pond culture systems (Bardach et al. 1972; Balarin and Hatton 1979; Pillai 1970).

Fishes Pillai (1970) lists numerous species of predatory fish that are a problem in brackish-water pond culture in the Indo-Pacific Region. Some of the major predatory species in this region include Elops hawaiiensis, Megalops cyprinoides, Epinephelus sp., and Sphyraena sp. Balarain and Hatton (1979) mention that predatory fishes cause losses to cultured tilapia in Africa.

Reptiles and Amphibians Piscivorous aquatic snakes cause substantial losses to juvenile fish in ponds. Sairg (1971, cited in Balarin and Hatton 1979) reported up to 300 snakes (Natrix natrix) caught in ten traps in two weeks in a 0.2 ha pond. The snake Cerberus rhynchops has been reported by Djajadiredja (1957, cited in Pillai 1970) to be a pond predator in Java. Balarin and Hatton (1979) reported that African toads of the genus Xenopus cause losses in tilapia nursery ponds.

Avian Predators Birds are reported to be some of the most destructive predators of pond-cultured fish (Balarin and Hatton 1979; and Pillai 1970). Chimits (1957, cited in Balarin and Hatton 1979) reported that a pelican can consume between one to three tons of fish in a year. Schaeperclaus (1933, cited in Pillai 1970) stated that herons may cause losses of up to 30-40 percent of fry and juvenile fishes cultured in ponds. Fryer and Iles (1972, cited in Balarin and Hatton 1979) report that a heron consumes as much as 100 kg of fish per year.

Mammalian Predators Otters (Lutra maculicollis and Aeonyx capensis) are considered extremely destructive and have been reported to reduce fish stocks as much as 80 percent (Marr et al. 1966, cited in Balarin and Hatton 1979).

PREVENTION AND CONTROL PROCEDURES FOR COMPETITORS, PESTS AND PREDATORS IN POND AQUACULTURE SYSTEMS

In pond aquaculture systems prevention and control procedures for competitors, pests, and predators can be classified as physical, chemical, or biological. The choice of a prevention and control procedure is frequently determined by the past experience of the pond manager. Once an effective method is known, efforts to find more efficacious procedures are often not attempted. Physical and chemical procedures are frequently used in combination.

Preventative Procedures

Physical and chemical methods are employed as preventative procedures in pond aquaculture systems. Physical methods to prevent the introduction of unwanted predatory or competitive species of fish in ponds include draining and drying of ponds prior to stocking and screening the inflow water.

Chemical methods are used to prevent the attack of wood-boring molluscs (Polychaetes) or crustaceans on wood structures in ponds. Pillai (1970) reports the treatment of wood with antiborer formulations such as creosote, other oil preservatives, copper compounds, and a thick coating of tar in the Indo-Pacific Region. As pointed out by Pillai (1970), these treatments are not indefinitely effective in preventing attack by the wood-boring pests.

Some species of predatory birds can be discouraged from entering ponds by extending fencing into the shallows of the ponds (Marr et al. 1966, cited in Balarin and Hatton 1979) or by stringing wire strands along the edges of or across ponds (Balarin and Hatton 1979). Scare devices such as scarecrows, bamboo rattles, empty-can rattles, bells, flashguns, sirens, klaxon horns, and gongs are reported by Pillai (1970) to be effective deterrents to avian predators. Marr et al. (1966, cited in Balarin and Hatton 1979) reported that fencing was effective in preventing entry of otters into fish ponds.

Physical Methods Ling (1960, cited in Pillai 1970) reported that the collection and removal of unwanted snails was effective in controlling these competitors in milkfish ponds in the Philippines. Pond drying in conjunction with the use of chemicals is mentioned by Pillai (1970) as beneficial in controlling snails.

Oysters, barnacles, and Polychaete worms that foul screens and sluices in ponds can be effectively controlled by periodic drying and physical removal by hand (Pillai 1970). Problems caused by wood-boring molluscs and crustaceans can be partially controlled by periodic removal and drying of wooden structures in ponds (Pillai 1970). Traps and other methods of physical removal are reported by Pillai (1970) to be effective in controlling burrowing crabs and other crustacean pests found in brackish-water ponds in the Indo-Pacific Region. Huet (1972, cited in Balarin and Hatton 1979) report that toads (genus *Xenopus*) in African ponds can be trapped or captured for removal. Sairg (1971) reported that predatory snakes can be controlled by trapping or can be killed by workers. Pillai (1970) stated that predatory birds and otters may be controlled by trapping or shooting.

Chemical Methods A number of chemicals are used to control competitors, pests, and predators in fresh- and brackish-water ponds. Both natural substances and synthetic compounds are used. Ling (1960, cited in Pillai 1970) reported that heavy application of green manure in ponds will result in destruction of snail competitors. Pillai (1970) reported that application of molasses into the pond has some beneficial effect in controlling snails. The mechanism of action of these substances on snails was not given.

Tea seed (*Camellia drupisera*) has wide use as a biocide in ponds in the Indo-Pacific region. The active principle is saponin. Pillai (1970) recommends it be used following pond drainage to control snails and unwanted fish. Reported application rates are 15-18 kg/ha (Tang 1967, cited in Pillai 1970); 180 kg/ha (Djajadiredja 1957, cited in Pillai 1970) and 200 kg/ha (Bardach et al. 1972). Saponin applied at 0.5 ppm has been recommended to control competitive and predatory species of fish in Philippines milkfish ponds prior to stocking (Bardach et al. 1972).

Tobacco wastes (tobacco dusts) are also widely advocated as a biocide to control snails, polychaete worms, and competitor and predator fish in Indo-Pacific ponds (Bardach et al. 1972; Pillai 1970). The active principle in tobacco byproducts is nicotine. In milkfish ponds in the Philippines tobacco dust is applied at 12-15 kg/ha (Tang 1967, cited in Pillai 1970). Pillai (1970) reports that polychaete worms are also controlled by 2 ppm nicotine in the water.

Rotenone, derived from the derris plant, is widely used as a natural piscicide. Reported application rates vary; 0.5 ppm (Hall 1949, cited in Pillai 1970), 4 gm/m^3 (Djajadiredja 1957, cited in Pillai 1970) and 20 ppm (Alikunki 1957, cited in Pillai 1970).

Quicklime applied to the pond bottom after draining and drying is recommended as an effective biocide to control snails, unwanted fish, and disease organisms. Application rates of 100 kg/ha are reported as effective (Bardach et al. 1972; Tang 1967, cited in Pillai 1970).

Synthetic organic and inorganic toxicants are used to control unwanted competitors, pests and predators in dirt ponds. Pillai (1970) recommends the use of Bayluscide at 3 ppm in partially drained pond water to control unwanted snails and polychaete worms.

Phenol is added to pond water after most of the water is drained to control polychaete worms, even those which are deeply burrowed into the mud (Pillai 1970). Chemical techniques used to control crab pests in the Philippines include spraying a 10 percent solution of technical BHC (containing 6.5 percent gamma isomer) or applying a few cc of kerosene into each crab burrow.

Endrin applied 0.1 ppm (Hickling 1962, cited in Pillai 1970) or 340 gms in 45 liters of water in drained ponds (Hickling 1962, cited in Pillai 1970); D.D.T. applied at 0.03 gm/liter (Rabanal and Hosillos 1957, cited in Pillai 1970); and 2,4-D applied at 0.13 gm/liter (Rabanal and Hosillos 1957, cited in Pillai 1970) have been advocated to control unwanted fish in ponds.

Tadtox is mentioned by Bardach et al. (1972) as being used to control unwanted tadpoles in ponds in the United States.

Pillai (1970) mentions the use of poisoned baits to control reptilian and avian predators, but the types of poisons are not given.

PUBLIC HEALTH CONSIDERATIONS IN WARMWATER POND AQUACULTURE

Public health diseases in pond aquaculture are zoonoses (diseases of animals which are transmitted to man) or diseases of man whose prevalence is increased through aquaculture activities. These diseases can be classified as (1) those that affect the consumer and (2) those that affect pond workers or people living in close proximity to ponds.

Consumer-Related Diseases

There are five bacterial diseases of man (Salmonellosis, Shigellosis, *Vibrio* gastroenterities, Botulism, and *Clostridium perfringens* enterotoxemia) that have been associated with the consumption of contaminated fish products (World Health Organization 1968; Food and Drug Administration 1977a; Hobbs 1979; Bryan et al. 1979; and Sakazaki 1979). Specific cases of human bacterial disease that resulted from consumption of pond-raised tilapia, catfish, carps, mullet, or milkfish are apparently not present in the literature. This is surprising in view of the common practice in many areas of the world of fertilizing with human and animal wastes. Apparently this practice has not resulted in significant outbreaks of human bacterial infection or intoxications, or at least reports regarding such outbreaks are lacking in the published literature.

A survey of fresh and frozen commercially caught and pond-reared channel catfish in the southern United States indicated that 93-94.5 percent of the fresh and frozen catfish met the proposed quality standards of less than or equal to ten organisms per gram (Food and Drug Administration 1977a). The report further stated that catfish have never been definitely incriminated in human outbreaks of food-borne illness caused by *Salmonella* (Food and Drug Administration 1977a). For additional information regarding human bacterial infections or intoxications the reader is referred to Bryan et al. 1979; Hobbs 1979; and Sakazaki 1979.

Fish, particularly freshwater fish, can be intermediate hosts for a number of metazoan parasites that can infect man. The more important parasites transmissible to man through the consumption of raw or improperly processed fish flesh include:

1. The broadfish tapeworm (*Diphyllobothrium latum*), which is spottily distributed in parts of Europe, Asia, Australia, and North and South America (Healy and Juranek 1979).

2. The flukes *Clonorchis sinensis*, *Opisthorchis* sp., members of Heterophyidae, and *Metagonimum yokagawai*. These fluke parasites are common in Asian countries and are nonspecific in fish species they infect (Healy and Juranek 1979; Bauer 1961a).

Fish are the second intermediate hosts for each of these parasites. These parasites encyst in muscle or other organ tissues of the fish. Human infection results from ingestion of improperly cooked or processed contaminated fish tissues. Prevention and control of human disease from these parasites include proper sewage disposal, snail control, adequate cooking, and adequate freezing of fish products (Food and Agriculture Organization 1973).

The use of raw human excrement as a pond fertilizer is thought to result in a high prevalence of intestinal parasitism among Chinese people who eat pond fish (Hickling 1962). Wykoff and Winn (1965, cited in Healy and Juranek 1979) estimate that over 3.5 million people are infected with Opisthorchis viverrini in Thailand, China, Laos, and Vietnam. These infections resulted from the consumption of improperly cooked or processed fish flesh. Whether these fish were pond-cultured or from streams was not mentioned.

Detailed information regarding these parasitic diseases of man can be found in Bauer (1961a) and Healy and Juranek (1979).

Fish are not known to transmit virus or fungal organisms pathogenic to man (Janssen 1970).

A potential problem is human toxicosis resulting from the consumption of fish flesh contaminated with drugs and drug metabolites, toxic elements, or contaminants of natural or industrial origin (Food and Drug Administration 1977b). Specific cases of human toxicosis following consumption of aquacultured channel catfish, Chinese catfish, milkfish, mullet, tilapia, and carps were not found by this author. Nevertheless, caution is advised regarding the addition of any element or compound which could be selectively absorbed and concentrated in fish flesh and which may constitute a public health hazard to the consumer.

Nonconsumer-Related Diseases

Leptospirosis Leptospirosis, a sometimes fatal, systemic bacterial disease of mammals, is known to be transmitted through contact with urine or water contaminated with viable Leptospira (Kenzy and Ringen 1971). This zoonotic disease has a world wide distribution. Leptospirosis is considered an occupational hazard of agricultural workers, abattoir and fish handlers, sewer workers, veterinarians, and other groups that are in frequent contact with wild or domestic mammals.

While pathogenic Leptospira are known to survive in freshwater for in excess of one month, these bacteria are destroyed within twenty hours in 20 percent brackish-water (Chang et al. 1948, cited in Kenzy and Ringen 1971).

Neither the isolation of pathogenic Leptospira from pond water used in fish culture nor human cases of Leptospirosis positively shown to have resulted from exposure to fishpond water have been reported. Nevertheless, in many human cases of Leptospirosis the route of exposure is never established, and Leptospirosis as a known water-borne disease should be considered as a potential occupational disease for aquaculture pond workers. No evidence exists to support a relationship between human Leptospirosis and the consumption of pond-cultured or wild caught fish.

Schistosomiasis (Bilharzia) Schistosomiasis is a severe, debilitating, parasitic disease of man and animals. Human infection by these blood flukes is acquired through skin penetration by infective cercaria while a person is immersed in infested water. The distribution of human Schistosomiasis is correlated with specific species of snails which are intermediate hosts for the parasite. More than 100 million people are affected with this disease in tropical and subtropical regions of the world (National Academy of Sciences 1977).

If water harbors the appropriate snail intermediate host, then contamination with human or animal wastes containing schistosome eggs constitutes a public health hazard. Prevention and control procedures include: (1) proper disposal of sewage and animal wastes, (2) snail control, and (3) prevention of contact with cercaria-infested waters (protective clothing or nonentry). Snail control methods employed are as follows: (1) Removal of grass growing in the pond using ducks or grass carp to reduce habitat suitable for snails (Proginin and Lipshitz 1957); (2) Stocking species of fish that feed on molluscs to reduce snail populations. Appropriate fish species include Black carp (Mylopharyngodon piceus), African lung fish (Protopterus sp.), and the Cichlids Serranochromis macrocephala, Haplochromis mellandi, H. bimaculatus, and Astatereochromis alluadi (Hickling 1962); (3) Periodic draining and drying out of the pond in conjunction with the application of liming; (4) Prevention of pond contamination by human and animal feces and urine containing schistosome ova (sanitation).

Malaria Fish ponds as standing or slow-moving bodies of water may assist in the spread of malaria by providing a breeding ground for anopheline mosquitoes (Hickling 1962). Pielou (1946, cited in Hickling 1962) concluded from studies carried out in Northern Rhodesia that well-managed fish ponds do not significantly increase the danger from malaria. In large ponds minimum management to achieve this goal included keeping the banks clear of grasses and unshaded, while small ponds are kept unshaded, stocked with larvivorus fish, and drained immediately if left unattended (Hickling 1962). It was mentioned that drainage ditches and other small bodies of water near the pond (or resulting from pond leakage) would have to be managed also.

Other control measures involve proper construction of ponds, with a minimum depth of two feet. Also, extensive efforts should be made to educate persons in underdeveloped countries as to the risks involving snails and malaria from standing water (Hickling 1962).

A Special Case: Manuring

Animal waste as a natural fertilizer in fish ponds to enhance fish productivity is routinely used in many areas of the world. In recent years the dynamics of manure on pond ecosystems and fish production have been the focus of many scientific studies. The problem of public health diseases resulting from the practice of applying manure to fish ponds has received some attention (Bhattacharya and Taylor 1975; Food and Drug Administration 1977b). The occurrence of infectious disease agents, drugs, and chemicals in animal wastes is well established, but it is uncertain that the practice of applying manure to ponds actually results in human disease. Reports to show this are simply lacking. Either clinical disease does not result, or the data base is inadequate to make this assessment. Clearly, studies need to be undertaken to determine if a public health hazard actually exists.

RECOMMENDATIONS FOR FUTURE PROGRAMS

Disease-control programs are practiced in animal-husbandry systems, because these programs result in enhanced productivity through increases in growth and survival of farmed animals. Warmwater pond aquaculture as an animal-husbandry system should benefit by increased productivity with the implementation of progressive herd health programs. The following operations are needed to increase the level of disease control, and therefore production, in warm-water pond aquaculture systems.

Surveillance

A quantitative data base on the status of diseases in warmwater pond aquaculture systems should be established. This program could be in the form of morbidity and mortality reporting on local, regional, and larger scales. The data base would need to be continuously updated and analyzed to determine trends. Standardized animal identification methods, disease nomenclature, and disease classification would be required prerequisites.

Services and Research

To advance the level of knowledge of the diseases affecting pond aquaculture species, increases in suitably trained manpower and facilities to provide disease diagnosis and control, services, and research are needed.

Regulations

Regulatory programs to minimize the spread of infectious diseases of warmwater pond aquaculture species need to be designed and implemented.

LITERATURE CITED

Ahne, W. 1975. A rhabdovirus isolated from grass carp (Ctenopharyngodon idella Val.). Arch. Virol. 48:181-185.

________, and K. Wolf. 1977. Spring Viremia of Carp. Fish Disease Leaflet 51. United States Department of the Interior, Washington, D.C. 11 pp.

Alikunki, K.H. 1957. Fish culture in India. Fun. Bull. Indian. Coun. Agric. Res. 20:144.

Almeida, L.J., E.J. Da Silva, and Y.M. Freitas. 1968. Microorganisms from some tropical fish diseases. J. Fish. Res. Bd. Can. 25:197-201.

Anon. 1977. Nutrient Requirements of Warmwater Fishes. Subcommittee on Warmwater Fish Nutrition, National Academy of Sciences. 77 pp.

Ashley, L.M. 1972. Nutritional pathology, pp. 439-531. In J.E. Halver (ed), Fish Nutrition. Academic Press, New York.

Avault, J.W. 1978. Ammonia: A limiting factor in fish production. The Commercial Fish Farmer and Aquaculture News 4:46-47.

Baker, J.R. 1960. Trypanosomes and dactylosomes from the blood of freshwater fish in East Africa. Parasitology 50:515-526.

Balarin, J.D., and J.P. Hatton. 1979. Tilapia: A Guide to Their Biology and Culture in Africa. University of Stirling, Scotland. 172 pp.

Bardach, J.E., J.H. Ryther, and W.O. McLarney. 1972. Aquaculture: The Farming and Husbandry of Freshwater and Marine Organisms. Wiley-Interscience, New York. 868 pp.

Bauer, O.N. 1961a. Fishes as carriers of human helminthoses, pp. 320-363. In V.A. Dogiel, G.K. Petrushevski, and Yu. I. Polyanski (eds), Parasitology of Fishes. Oliver and Boyd, Ltd., London.

__________. 1961b. Parasitic diseases of cultured fishes and methods of their prevention and treatment, pp. 265-298. In V.A. Dogiel, G.K. Petrushevski and Yu. I. Polyanski (eds), Parasitology of Fishes. Oliver and Boyd, Ltd., London.

Bhattacharya, A.N., and J.C. Taylor. 1975. Recycling animal waste as a feedstuff: a review. Journal of Animal Science 41(5):1438-1454.

Bryan, F.L., M.J. Fanelli, and H. Riemann. 1979. Salmonella infections, pp. 74-121. In H. Riemann and F.L. Bryan (eds), Food-Borne Infections and Intoxications, 2nd ed. Academic Press, San Francisco.

Buck, W.B., G.D. Osweiler, and G.A. Van Gelder. 1973. Clinical and Diagnostic Veterinary Toxicology. Kendall/Hunt Publishing Co., Dubuque, Iowa. 287 pp.

Bucke, D., and J. Finlay. 1979. Identification of spring viremia in carp (Cyprinus carpio L.) in Great Britain. The Vet. Record 104:69-71.

Burrows, R.E. 1964. Effects of accumulated excretory products on hatchery-reared salmonids. U.S. Fish Wildlife Serv. Res. Rep. 66:1.

Chang, S.L., et al. 1948. Studies on Leptospira icterohaemorrhagiae. IV. Survival in water and sewage: destruction in water by halogen compounds, synthetic detergents, and heat. Jour. Inf. Dis. 82:256-266.

Chimits, P. 1957. The Tilapia and their culture, a second review and bibliography. FAO Fish. Bull. 10(1):1-24.

Collins, M. 1978. Algal toxins. Microbiological Reviews 42(4):725-746.

Colt, J., and G. Tchobanoglous. 1976. Evaluation of the short-term toxicity of nitrogenous compounds to channel catfish, Ictalurus punctatus. Aquaculture 8:209-224.

Colt, J. and G. Tchobanoglous. 1978. Chronic exposure of channel catfish, Ictalurus punctatus, to ammonia. Effects on growth and survival. Aquaculture 14:353-372.

Cowey, C.B., and R.J. Roberts. 1978. Nutritional pathology of teleosts, pp. 216-226. In R.J. Roberts (ed), Fish Pathology. Bailliere Tindall, London.

Cridland, C.C. 1960. Laboratory experiments on the growth of tilapia species. 1. The value of various foods. Hydrobiologia 15:135-160.

Djajadiredja, R. 1957. A Preliminary Report on the Introduction of the Philippine-Type Nursery in Indonesia. Tech. Paper. Indo-Pacific Fisheries Council. IPFC/C57/-Tech./17. 23 pp. (mimeo).

Environmental Protection Agency. 1976. Quality Criteria for Water. U.S. Government Printing Office, Washington, D.C. 256 pp.

European Inland Fisheries Advisory Commission. 1970. Water Quality Criteria for European Freshwater Fish. Report on Ammonia and Inland Fisheries. EIFAC Technical Paper No. 11, 12 pp; Wat. Res. 7:1011 (1973).

Fijan, N. 1972. Infectious dropsy in carp -- a disease complex, pp. 39-51. In L.E. Mawdesley-Thomas (ed), Diseases of Fish. Symp. Zool. Soc. Land. No. 30. Academic Press, London.

Flis, J. 1968. Anatomicohistopathological changes induced in carp (Cyprinus carpio L.) by ammonia water. Acta Hydrobiol. Part I. Effects of toxic concentrations, 10:205. Part II. Effects of subtoxic concentrations, 10:225.

Food and Agriculture Organization. 1958. Report to the Government of India on the Protection of Wood Against Marine Borers. Based on the Work of G. Becker, Rep. FAO, (795). 111 pp.

______________. 1964. Fish and pesticides: A general statement of FAO policy. FAO Fisheries Technical Papers. 45:1-5.

______________. 1973. Fish and Shellfish Hygiene--Report of a World Health Organization Expert Committee Convened in Cooperation with FAO. Geneva. 62 pp.

Food and Drug Administration, Department of Health, Education, and Welfare. 1977a. Survey of Salmonella and Parasites in Catfish. FDA Quarterly Act. Report, Fourth Quarter. Washington, D.C. 2pp.

_____________. 1977b. Recycled Animal Waste. Federal Register. Washington, D.C.

Fryer, G., and T.D. Iles. 1972. The Cichlid Fishes of the Great Lakes of Africa: Their Biology and Evolution. Oliver and Boyd, Edinburgh. 641 pp.

Gohl, B. 1975. Tropical Feeds. FAO, Rome. 661 pp.

Hall, C.B. 1949. Ponds and Fish Culture. Faber and Faber, London. 244 pp.

Haller, R.D. 1974. Rehabilitation of a Limestone Quarry. Report of an Environmental Experiment. Publication by Bamburi Portland Cement Co., Ltd. Mombasa, Kenya. 32 pp.

Healy, G.R., and D. Juranek. 1979. Parasitic infections, pp. 343-382. In H. Riemann and F.L. Bryan (eds), Food-Borne Infections and Intoxications, 2nd ed. Academic Press, San Francisco.

Hickling, C.F. 1962. Fish Culture. Faber and Faber, London. 286 pp.

Hobbs, B. 1979. Clostridium perfringens gastroenteritis, pp. 131-167. In H. Reimann and F.L. Bryan (eds). Food-Borne Infections and Intoxications, 2nd ed. Academic Press, San Francisco.

Hoffman, G.L. 1972. Intercontinental and transcontinental dissemination and transfaunation of fish parasites with emphasis on whirling disease (Myxosoma cerebralis), pp. 69-81. In S.F. Snieszko (ed), A Symposium on Diseases of Fishes and Shellfishes. American Fisheries Society Special Publication, No. 5, Washington, D.C.

___________. 1979. Helminthic parasites, pp. 40-58. In Principal Diseases of Farm-Raised Catfish. Southern Cooperative Series No. 225. Auburn University, Auburn, Alabama.

Huet, M. 1972. Textbook of Fish Culture: Breeding and Cultivation of Fish. Translated by H. Kohn, Fishing News Ltd., Farnham, Survey, England. 436 pp.

Ishac, M.M., and Dollar, A.M. 1968. Studies on manganese uptake in Tilapia mossambica and Salmo gairdneri in response to manganese. Hydrobiologia 31:572-584.

Janssen, W.A. 1970. Fish as potential vectors of human bacterial diseases, pp. 284-290. In S.F. Snieszko (ed), A Symposium on Diseases of Fishes and Shellfishes. American Fisheries Society Special Publication No. 5, Washington, D.C.

Johnston, S.K. 1975. Mortality in Texas Farm-Reared Catfish Caused by Natural Toxins. FDDL-F6 Texas Agricultural Extension Service. Texas A&M University. 3 pp.

Kenzy, S.G., and L.M. Ringen. 1971. The spirochetes, pp. 486-502. In I.A. Merchant and R.A. Packer (eds), Veterinary Bacteriology and Virology, 7th ed. The Iowa State University Press, Ames, Iowa.

Klontz, G.W. 1972. Veterinary medical aspects of maintaining the health of aquatic food animals. JAVMA 161(11):1489-1491.

Knepp, G.L., and G.F. Arbin. 1973. Ammonia toxicity levels and nitrate tolerance of channel catfish. The Progressive Fish-Cult. 35:22.

Koesoemadinata, S. 1980. Pesticides as a major constraint to integrated agriculture - aquaculture farming systems, pp. 45-51. In R.S.V. Pullin and Z.H. Shehadeh (eds), Integrated Agriculture Aquaculture Farming Systems. International Center for Living Aquatic Resources Management, Proceedings 4, Manila, Philippines.

Lagler, K.F., J.E. Bardach, and R.R. Miller. 1962. Icthyology. Wiley and Sons, New York. 545 pp.

Launer, C.A., O.W. Tiemeier, and C.W. Deyoe. 1978. Effects of dietary addition of vitamins C and D on growth and calcium and phosphorus content of pond-cultured channel catfish. The Progressive Fish-Cult. 40(1):16-20.

Lewis, D.H., and J.A. Plumb. 1979. Bacterial diseases, pp. 19-24. In Principal Diseases of Farm Raised Catfish. Southern Cooperative Series No. 225. Auburn University, Auburn, Alabama.

Ling, S.W. 1960. Control of Competitors and Predators and Disease and Parasites. Lectures presented at the Third International Inland Fisheries Training Center, Bogor, Indonesia, FAO Rome 1, pag. var.

_________. 1977. Aquaculture in Southeast Asia: A Historical Overview. University of Washington Press. 108 pp.

Liversidge, J., and A.L.S. Munro. 1978. The virology of teleosts, pp. 114-143. In R.J. Roberts (ed), Fish Pathology. Bailliere Tindall, London.

Lovell, R.T. 1976. Nutritional Diseases in Channel Catfish. FAO Technical Conference on Aquaculture. Food and Agriculture Organization of the United Nations, Rome Italy. 10 pp.

___________, and C. Lim. 1978. Vitamin C in pond diets for channel catfish. Trans. Am. Fish. Soc. 107(2):321-325.

Lovell, T. 1979. Brown blood disease in pond raised catfish. The Commercial Fish Farmer and Aquaculture News. 5(3):3.

Mahajan, C.L., and N.K. Agrawal. 1979. Nutritional requirements of ascorbic acid by Indian major carp, Cirrhina mrigala, during early growth. Aquaculture 19:37-48.

Marr, A., M.A.E. Mortimer, and I. van der Lingen. 1966. Fish Culture in Central East Africa. FAO Publ. 53608-66/E. 158 pp.

McCoy, E.F. 1972. Role of Bacteria in the Nitrogen Cycles of Lakes. Water Pollution Control Research Series (EP 2.10:16010EHR 03/72), U.S. Environmental Protection Agency, U.S. Government Printing Office, Washington, D.C.

National Academy of Sciences. 1977. Aquatic Animal Health. Subcommittee on Aquatic Animal Health. National Research Council. Washington, D.C. 46 pp.

Paperna, I. 1974. Lymphocystis in fish from East African lakes. J. Wildl. Dis. 9(4):331-335.

Pielou, D. 1946. Anopheline Mosquitoes Breeding at Chilanga. Appendix V to the Annual Report for 1946 of the Game and Tsetse Control Dept. of Northern Rhodesia.

Pillai, T.G. 1970. Pests and predators in coastal aquaculture systems of the Indo-Pacific region, pp. 456-470. In T.V.R. Pillay (ed), Coastal Aquaculture in the Indo-Pacific Region. FAO Fishing News (Books) Ltd., London, England.

Plumb, J.A. 1978. Epizootiology of channel catfish virus disease. Marine Fisheries Review 40(3):26-29.

_________. 1979. Fungal diseases, pp. 25-27. In Principal Diseases of Farm-Raised Catfish. Southern Cooperative Series No. 225. Auburn University, Auburn, Alabama.

Pruginin, J., and N. Lipshitz. 1957. The control of noxious weeds in fish ponds. Bamidgeh 9(3).

Rabanal, R.R., and L.V. Hosillos. 1957. Control of Less Desirable Fish Competing with or Harmful to Desirable Indigenous Species in Inland Waters in the Philippines. Indo-Pacific, Fish. Coun. Tech. Pap. IPFC/C57/Tech./20. 23 pp. (mimeo).

Reichenbach-Klinke, H., and M. Landolt. 1973. Fish Pathology. T.F.H. Publications, Inc. Neptune, New Jersey. 512 pp.

Richards, R.H. 1978. The mycology of teleosts, pp. 205-215. In R.J. Roberts (ed), Fish Pathology. Bailliere Tindall, London.

___________, and R.J. Roberts. 1978. The bacteriology of teleosts, pp. 183-204. In R.J. Roberts (ed), Fish Pathology. Bailliere Tindall, London.

Roberts, R.J. (ed). 1978. Fish Pathology. Bailliere Tindall, London. 318 pp.

Robinette, R.H. 1973. The Effect of Selected Sublethal Levels of Ammonia on the Growth of Channel Catfish (Ictalurus punctatus). Ph.D. Thesis. Southern Illinois University.

Rogers, W.A. 1978. Parasitic diseases of freshwater fishes. Marine Fisheries Review 40(3):56-57.

__________. 1979a. Crustacean parasites, pp. 38-39. In Principal Diseases of Farm-Raised Catfish. Southern Cooperative Series No. 225. Auburn University, Auburn, Alabama.

__________. 1979b. Protozoan parasites, pp. 28-37. In Principal Diseases of Farm-Raised Catfish. Southern Cooperative Series No. 225. Auburn University, Auburn, Alabama.

Sairg, S. 1971. Diseases of Fishes. Book 3: The Prevention and Treatment of Diseases of Warmwater Fishes Under Subtropical Conditions, with Special Emphasis on Intensive Fish Farming. T.F.H. Publications, Inc., Ltd. Neptune, New Jersey. 127 pp.

Sakamoto, S. 1978. Iron deficiency symptoms of carp. Bull. Jpn. Soc. Sci. Fish. 44(10):1157-1160. (Fish Res. Lab. Kyushu Univ. Tsuyazaki, Fukuoka, 811-33, Japan).

Sakazaki, R. 1979. Vibrio infections, pp. 174-206. In H. Riemann and F.L. Bryan, (eds), Food-Borne Infections and Intoxications, 2nd ed. Academic Press, San Francisco.

Sato, M., R. Yoshinaka, Y. Yamamoto, and S. Ikeda. 1978. Nonessentiality of ascorbic acid in the diet of carp. Bull. Jpn. Soc. Sci. Fish. 44(10):1151-1156. (Dep. Fish. Fac. Agric. Kyoto Univer. Kyoto, Japan).

Schaeperclaus, W. 1933. Textbook on Pond Culture. (English translation by F. Hund). Fish. Leaflet, Washington, (311), 260 pp.

Scott, P.W. 1977. Preliminary Studies on Diseases in Intensively Farmed Tilapia in Kenya. M.S. Thesis, Stirling University, Scotland. 159 pp.

Shilo, M., and M. Shilo. 1962. The mechanism of lysis of Prymnesium parvum by weak electrolytes. J. Gen. Microbiol. 29:645-658.

Smith, C.E., and R.G. Piper. 1975. Lesions associated with chronic exposure to ammonia, pp. 497-514. In W.E. Ribelin and G. Migaki (eds), The Pathology of Fishes. The University of Wisconsin Press, Madison.

Snieszko, S.F. 1978. Mycobacteriosis (Tuberculosis) of Fishes. Fish Disease Leaflet 55. United States Department of the Interior. 9 pp.

Tang, Y. 1967. Improvement of milkfish culture in the Philippines. Curr. Aff. Bull. Indo-Pacific Fish. Coun. 49:14-22.

Ulitzur, S., and M. Shilo. 1964. A sensitive assay system for determination of the Ichthyotoxin of Prymnesium parvum. J. Gen. Microbiol. 36:161-169.

Wellborn, T.L. 1979. Control and therapy, pp. 61-89. In Principal Diseases in Farm-Raised Catfish. Southern Cooperative Series No. 225. Auburn University, Auburn, Alabama.

Wolf, K. 1968. Lymphocystis Disease of Fish. Fish Disease Leaflet B. United State Department of the Interior, Washington, D.C. 4 pp.

__________, and J.A. Mann. 1980. Poikiliothermic vertebrate cell lines and viruses: a current listing for fishes. In Vitro 16(2):168-179.

World Health Organization. 1968. Microbiological Aspects of Food Hygiene--Report of a WHO Expert Committee with the Participation of FAO. Technical report Series No. 399. Geneva. 64 pp.

Wykoff, D.E., and M.M. Winn. 1965. Opisthorchis viverrini in Thailand - the life cycle and comparison with O. felineus. J. Parasitol. 51:207.

Yokote, M. 1970. Sekoke disease, spontaneous diabetes in carp, Cyprinus carpio, found in fish farms. 5. Bull. Freshwat. Fish. Res. Lab. Tokyo. 26(2):161.

POND CULTURE PRACTICES

by

John Colt

Pond culture practices have generally evolved in a slow empirical manner. Improved pond operations and management will serve as a basis for increased pond productivity and the improved nutritional status in the developing countries. Direct transfer of technology from the temperate regions (i.e. United States, Europe, Israel) is not feasible because of changes in the physical and biological phenomena, socio-political environments, and the level of management skills and objectives. Increased productivity will be based on an improved understanding of the physical, chemical and biological processes in the pond. Key unknowns in pond culture practices and research needs are considered briefly in the following section. The effect of growth-limiting variables on fish production and pond engineering are considered in subsequent sections.

OVERVIEW

The material presented in this section is intended to serve as an overview of pond practices and to identify critical areas where more research is needed.

Hatchery Operations

One of the major problems in pond aquaculture is the lack of adequate fry for stocking. Greater availability of fry is a single factor that would result in a rapid increase in pond production in many developing countries.

Increasing the fry supply will depend on a better understanding of the reproductive biology, better techniques for the rearing of larvae and fry, and better production hatcheries. The advantages and disadvantages of regional, village, or individual hatcheries will need to be determined on an individual country basis. The adaptation of systems used for trout, salmon, or catfish may not be possible because of differences in the environmental requirements, behavior, and feeding requirements. While most of the potential culture species have been spawned under hatchery conditions, emphasis must be placed on the production aspects under varying size levels. In some areas, small-scale hatcheries operated by farmers (Osborn 1977b) may be economic but in many areas, regional or national hatcheries will produce fry and distribute them to the farmers. Improved transportation methods using pure oxygen (Johnson 1979; U.S. Fish and Wildlife Service 1978) can result in a significant increase in fry survival at a moderate cost. Legal, social and political constraints to hatchery production will need to be considered in areas where local fry collection and distribution systems exist.

Stocking Practices in Pond Fish Culture

Rational stocking procedures will depend on a knowledge of the food preference of the various species and the amount (or production) of the different types of fish food in the pond. Better information on the food preferences of fish as a function of size and location is needed. The utilization of natural foods in the ponds is central to the objectives of increased pond productivity. Supplemental foods may be added to the pond, but the major source of food will be from natural sources. Better information on the estimation and prediction of standing crop and production of the various food items is needed. Emphasis should be placed on the development of rational stocking procedures for polyculture.

Feeds and Feeding Practices in Warmwater Fish Ponds

In most developing countries only supplemental feeding with locally available products will be economically feasible. The fish will obtain a significant amount of their food from natural pond production. The key problem in formulation of supplemental feeds in warmwater ponds will be the ingestion and digestion of the food. The digestion process in the tilapia and filter-feeding fish may be inherently different from trout and catfish. These differences may especially be critical when dealing with vegetable proteins, algae, and detritus. While the conduct of digestibility experiments with filter-feeding fish is very different, this information will be critical to the design of cost effective supplemental diets.

The formulation of supplemental diets for pond systems is complicated because undigested or partially digested food may be consumed by another species or form the basis for another food chain. The formulation of supplemental diets is an ecological problem rather than a purely nutritional problem. Thus, a knowledge of the various food webs is needed to assess the impact of supplemental feed in the warmwater pond.

Food preferences, feeding habits, and feeding behavior will also be important in pond feeding. Better information on the feeding processes as a function of size and species is needed.

Local processing of local feed items to increase acceptability and digestion may be desirable. Processing techniques for local input need to be within the management and economic resources of the developing countries. Transportation of local input to a central site and then back to the local pond may not be desirable or feasible. The use of animal or fish wastes will be common in the developing countries. The passage of food items through the animal or fish may be viewed as a form of processing for local items.

Digestion and Assimilation

The majority of fish nutritional research has been conducted with fish on the top of the food chain such as rainbow trout. The mechanisms of digestion and assimilation of the filter-feeding fish or herbivorous fish are not well known. This is important because the basic processes in the digestion and assimilation for these fish may be different. *Tilapia nilotica* digests blue-green algae by lysing the cells with low-pH gastric secretions (Moriarty 1973) rather than using digestive enzymes. While the Tilapia species may digest algae, the efficiency is lower than for animal tissue (Mironova 1974; Pandian and Raghuraman 1972). The optimum growth of *Tilapia aurea* is at a protein level of 36 percent (Davis and Stickney 1978). Information on the digestibility of foods can be used to formulate cheaper feeds using local materials.

Fertilization Practices in Warmwater Fish Ponds

Fertilization with inorganic fertilizers, animal manures, or plant material is commonly used to increase production. Type, method of application, and interval between applications vary from country to country. Over-fertilization can commonly result in algal blooms and oxygen depletion problems. Improved fertilization practices will depend on a better knowledge of local water chemistry, determination of optimum nutrient levels, and development of a better understanding of the oxygen balance in the ponds.

Over-application of inorganic fertilizers can result in the precipitation or loss of much of the fertilizer. Many of these reactions can be predicted from a knowledge of the local water chemistry. Water test kits have been found to be accurate enough for management decisions for pond aquaculture in the U.S. The validity and use of these kits should be evaluated in the developing countries. The use of these kits by extension or government workers would be very helpful.

Due to changes in the water chemistry, water flow rate, and bottom sediments, the application frequency and rate will change significantly from region to region or even from pond to pond. The measurement of critical nutrient concentration during fertilization experiments may allow management of fertilization on a concentration basis rather than a gross application basis.

The application of manure and plant matter to a pond may increase the importance of bacteria and protozoa as a source of food for the fish. A better understanding of the food web in the warmwater pond will be necessary for a fundamental understanding of the fertilization process. Increased fertilization leading to increased natural food production (algae, zooplankton, bacteria, protozoan, etc.) can also result in oxygen depletion and massive mortality.

Water Quality Management Practices

For small pond operators, water quality management in the developing countries may take the form of long-term management plans rather than plans for daily or weekly management. The knowledge that the dissolved oxygen level is going to be 0.5 mg/L in the morning is of little use to a farmer who cannot increase the flow to the pond or aerate.

In large village or commercial ponds, both aeration or pumping may be feasible. Development of low or intermediate technology pumps and aerators should be a high priority. Special emphasis should be directed toward the use of available power sources (wind, water, or animals).

Improvements in the predictive ability of pond management models will aid in the development of better stocking, fertilization, and feeding practices. Three areas of interest are prediction of dissolved oxygen, management of alkalinity and pH, and management of the organic buildup on pond bottoms.

Development of predictive dissolved oxygen models will be critical in highly manured and fertilized ponds. Emphasis must be placed on evaluation of the oxygen consuming and producing components (i.e., algae, bacteria, sediment). The importance of the various components may be very different in different countries or in different types of ponds. Models developed in the U.S. should not be used without experimental verification.

Alkalinity and pH prediction will be important in areas of acid-sulfate soils or in prediction of pH due to algal activity. Extreme pH values can stress the fish, but probably more importantly can greatly increase the toxicity of ammonia, nitrite, and hydrogen sulfide. At low pH, the solubility of iron, aluminum, and heavy metals is increased. The prediction of pH depends on a complete understanding of the local water and soil chemistry and buffer systems. The buffer systems in the areas of acid-sulfate soils may significantly differ from those found in the temperate regions. Better information on water chemistry will also help in the reclamation of acid-sulfate soil.

Draining and drying of the pond is a significant feature of water quality management in pond aquaculture. This allows oxidation and mineralization of the accumulated organic matter and reduction of the oxygen demand of the pond sediment. Development of better criteria for draining would be useful for pond management.

Diseases, Competitors, Pests, Predators, and Public Health Considerations

Prevention of disease problems in the developing countries will be a management problem rather than a treatment problem. Primary emphasis should be placed on better culture practices and reducing stress.

Public health considerations will greatly influence the use of aquaculture to increase food production in the developing countries. This includes diseases of animals that are transmitted to man and diseases of man whose prevalence is increased through aquaculture activities. Major effort will be needed to monitor the impact of aquaculture on public health. The impact of pond fertilization with human wastes needs careful study.

Harvesting

Improvements in harvesting will have little effect on the productivity of ponds. In subsistence ponds, labor-intensive harvesting using either modern nets or locally produced nets is adequate. In many areas, the use of better harvesting techniques may not be economic because of the lack of transportation or processing facilities. In large production ponds mechanical harvesting, grading, loading, and hauling techniques may be useful (Greenland 1974). The use of traps for some species of fish should be investigated (Greenland and Gill 1974a, 1974b).

Pond Engineering

The technology required for pond engineering is well-defined and available at the present time. The significant problem facing the developing countries is the application of technology within the technical, social, political, and legal structure of the country. Innovative engineering will be required to provide simple but effective solutions. Pond engineering is considered further in a subsequent section.

EFFECT OF GROWTH-LIMITING VARIABLES ON FISH PRODUCTION

The growth and production of fish in a pond culture system will depend on complex interactions of physical, chemical, and biological parameters. The major fish species that have potential for pond culture in developing countries are listed in Table 1. Local species may also have potential or have a high market value. In many areas, the main emphasis of aquaculture will be the production of high-priced products for the export market (Kutty 1980). Freshwater and marine shrimp, salmon, trout, and eels would be important species in this group.

The species listed in Table 1 feed low on the food chain or are omnivores. These fish are also very hardy and can survive a wide range of environmental conditions. The scientific data base on these species ranges from excellent to none. Also, much of the published literature on the silver and bighead carp is in Russian, Chinese, or East European languages. The Food and Agriculture Organization (Rosa 1965) has prepared reviews of some of the species, but these reports are not routinely updated nor easily available. Excellent reviews are available for the Tilapia (ICLARM 1980; Balarin 1979; Colt et al. 1979), common carp (Colt et al. 1979), and mullet (De Silva 1980). For other species such as the Chinese carp, very little is known or available.

To develop pond aquaculture on a rational basis, the effects of physical, chemical, and biological parameters must be available to the researchers prior to the start of research. Emphasis must be placed on the synthesis of the data, rather than fancy computer techniques. A preliminary outline of the topics of interest is presented in Table 2. This

Table 1

POTENTIAL SPECIES FOR POND CULTURE

Common Name	Scientific Name
Common Carp	Cyprinus carpio
Grass Carp	Ctenopharyngodon idellus
Silver Carp	Hypothalmichthys molitrix
Bighead Carp	Aristichthys nobilis
Mud Carp	Cirrhina molitorella
Snail Carp	Mylopharyngodon piceus
Walking Catfish	Clarias lazera
	Clarias fuscus
	Clarias betrachus
Thai Catfish	Pangasius sutchi
	Pangasius larnaudi
Java Tilapia	Sarotherodon mossambica
Nile Tilapia	Sarotherodon nilotica
	Sarotherodon aurea
Zill's Tilapia	Tilapia zilli
Grey Mullet	Mugil cephalus
Milk Fish	Chanos chanos
Tambaqui	Colossoma macropomum
Pirapitinga	Colossoma bidens

list is based on experience in the temperate zone, so other parameters may be added as data becomes available. Documentation of the effects of key variables on the growth and production of the important species is necessary. This can best be done at a central site. This center could also be responsible for custom literature searches and supplying researchers in the field with needed articles and reports. The key purpose of preparing literature reviews and data synthesis is the identification of key gaps in the data.

There are two critical areas that limit fish production in ponds:

1. the effects of ammonia, nitrite, nitrate, dissolved oxygen, hydrogen sulfide, and pH on the growth and mortality of fish

2. control of the reproduction and early larval development of fishes.

A better understanding in these two areas will have a significant effect on increasing pond production. The importance of these areas will be discussed in more detail.

Ammonia, Nitrite, Nitrate, Dissolved Oxygen, Hydrogen Sulfide, and pH

Better information on the effects of these parameters on growth and mortality will be necessary for the development of both research growth models and the management plans. These parameters are critical in ponds receiving animal manures and waste products. The over-fertilization of ponds can result in low dissolved oxygen, and high ammonia and hydrogen sulfide at the same time (Boyd et al. 1979; Hollerman and Boyd 1980).

Ammonia Dissolved ammonia gas is a weak base; the equilibrium expression for this reaction can be written as:

$$NH_3 + H_20 = NH_4^+ + 0H^-$$

The un-ionized ammonia (NH_3) is the toxic form. The concentration of un-ionized ammonia depends on pH, temperature, and salinity. The source of ammonia in ponds is ammonia excretion of fish and the addition of fertilizers and manures. High temperatures and pH favor the un-ionized form. In water with low alkalinity, the removal of inorganic carbon algae can produce pH in excess of 10.0. As the ammonia level of the ambient water increases, the ammonia excretion of most aquatic animals decreases and the ammonia level in the blood and tissue increases. The lethal levels of un-ionized ammonia range from 0.4 mg/L NH_3-N for trout to 3.0 mg/L NH_3-N for channel catfish (Colt and Armstrong 1981). Un-ionized ammonia reduces the growth of channel catfish in a linear manner over the range of .048 to 1.00 mg/L NH_3-N (Colt and Tchobanoglous 1978). At least for channel catfish, there does not appear to be a "no effect" level for un-ionized ammonia. Except for work on the lethal tolerance of Tilapia (Redner and Stickney 1979) and the common carp (Flis 1968a, 1968b), little is known about the effects of ammonia on the species listed in Table 1. Information on the effects of ammonia on mortality and growth is needed. The effect of daily increases in pH and un-ionized levels must be determined. Fluctuating un-ionized ammonia concentrations are more toxic than exposure to a constant value equal to the mean of fluctuating concentrations (Thurston et al. 1981a). The effects of fluctuating ammonia concentration on the growth of fish are unknown. It is unknown if the peak un-ionized ammonia level, the mean un-ionized ammonia level, or some combination of the two would be the best predictor of growth reduction. Ammonia toxicity is increased at low dissolved oxygen (Thurston et al. 1981b). Therefore, the joint toxicity of these two parameters will be more important than the effects of the individual parameters. Ammonia toxicity may be a key parameter for the determination of fertilizer and

Table 2

PARAMETERS OF INTEREST

Physical	Chemical	Biological	Metabolic Wastes
Temperature	Dissolved Gases		
Light Intensity	Oxygen	Feeding Habits	
	Carbon Dioxide		Ammonia
Photoperiod	Hydrogen Sulfide	Digestion and Assimilation	Nitrite
Sound		Growth	Nitrate
Water Depth/Pond Size	Dissolved Gas Supersaturation	Off-Flavor	Fecal Wastes
	Salinity	Ecology	Bacteria
	Heavy Metals	Reproduction	Solids
	pH	Disease	Phenomones
	Biocides	Predation	

manure application rates because both high un-ionized ammonia concentrations and low dissolved oxygen levels may be produced.

Nitrite Under aerobic conditions, nitrite is commonly produced by nitrification of ammonia. Nitrite is the ionized form of nitrous acid, a weak acid. This reaction can be written as

$$HNO_2 = H^+ + NO_2^-$$

The toxicity of nitrite may be due to the nitrous acid concentration (Colt et al. 1981; Russo et al. 1981). The nitrous acid level depends on temperature, pH, and salinity. Low pH and temperature favor the formation of nitrous acid. The major effect of nitrite (or nitrous acid) is the oxidation of the iron in the hemoglobin molecule from Fe^{+2} to Fe^{+++} (Colt and Armstrong 1981). Since the oxidized form of hemoglobin is unable to act as an oxygen carrier, if sufficient ferrihemoglobin is formed, hypoxia and cyanosis may result. The toxicity of nitrite (or nitrous acid) will be greatly increased at low dissolved oxygen levels and high temperatures. Except for rainbow trout and channel catfish, little is known about the effect of nitrite on fish (Colt et al. 1981; Colt and Armstrong 1981). The growth of channel catfish was reduced at 1.60 mg/L NO_2^--N and above (Colt et al. 1981). Information on the effects of constant and fluctuating levels of nitrous acids on both growth and mortality is needed. This information may prove critical in areas of acid sulfate soil where very low pH may be present. Because of the interaction with dissolved oxygen, the joint toxicity of these parameters will need to be investigated.

Nitrate Nitrate levels may build up in ponds from the nitrification of ammonia. Nitrates can be considered completely dissociated and have very little toxicity to aquatic animals. Under anaerobic conditions, the reduction of nitrate to nitrite may occur. Also, under some conditions, algae may produce high levels of nitrite (Kinne 1977).

Dissolved Oxygen Dissolved oxygen is a key parameter that may limit production or result in high mortality in ponds. The lethal dissolved oxygen for fish ranges from 2 to 3 mg/L for trout and salmon and 0.5 to 1.0 mg/L for warmwater fish. The growth of several species of fish is reduced when the dissolved oxygen is less than 5 mg/L (Brett and Blackburn 1981). Dissolved oxygen levels in ponds typically are supersaturated in the afternoon and decrease during the night (Hollerman and Boyd 1980). The effects of these fluctuations on mortality and growth are critical to development of pond management models. As discussed in the previous section, dissolved oxygen can greatly increase the toxicity of ammonia and nitrite. Therefore, the joint toxicity of dissolved oxygen + ammonia and dissolved oxygen + nitrite are the key parameters to be studied.

Hydrogen Sulfide Hydrogen sulfide (H_2S) may be produced on the pond bottoms by anaerobic bacteria. Mass mortality of fish may result if the pond water is mixed. The concentration of H_2S depends on the pH, temperature, and salinity. Low pHs favor H_2S. Lethal H_2S levels range from 0.8 mg/L for channel catfish (Bonn and Follis 1967) to 0.008 mg/L for trout (Reynolds and Haines 1980). H_2S levels above 2.0 mg/L have been measured in ponds (Bonn and Follis 1967). The toxic effect of H_2S may be due to interference with enzyme action or to mitochondrial changes (Smith and Oseid 1974).

The accumulation of organic matter on the pond bottom or the failure to remove vegetation before filling the pond may result in anaerobic conditions and the production of H_2S. Data on the effects of H_2S on the growth and mortality of those species that may be raised in areas of low pH or in manured ponds is needed.

pH For short periods of time, fish may be able to tolerate pH values ranging from 3.5 to 10.5 (Haines 1981, Colt et al. 1979). Over the range of 6 to 9, pH may have little effect on fish. At a chronic low pH, reproduction and normal larval development are prevented (Haines 1981). Low pH increases the toxicity of nitrite and hydrogen sulfide, while high pHs increase the toxicity of ammonia. Therefore, the major effect of pH may be to influence the concentration of the weak acids and bases and the solubility of iron and aluminum compounds (Singh 1980). The joint toxicity of these parameters with a constant or diurnally fluctuating pH concentration needs to be defined.

Control of Reproduction

The lack of adequate fry is a major problem in pond culture in many parts of the world and therefore is a key topic in increasing pond production (Kutty 1980). Emphasis must be placed on both controlled reproduction, the early growth of the fry, and production hatchery techniques. The control of overbreeding by the tilapia species should be investigated by the use of androgens (Anderson and Smitherman 1978; Guerrero 1975), sterile hybrids (Pruginin et al. 1975), or local predators. Technological control of overbreeding of tilapia is of little use if the techniques can not be used by the developing countries. The implementation of technology into the developing countries is a critical issue that needs to be addressed in research planning.

POND ENGINEERING

The design of ponds for aquaculture depends on information from a large number of fields including meteorology, hydrology, hydraulics, soil mechanics, civil engineering, and biology. The technology required for the engineering, construction, and management of ponds is well defined and available at the present time. The significant problem facing the developing countries is the application of technology within the technical, social, political, and legal structure of the country. Therefore, pond engineering practices will vary with country and the size (or scale) of operation. In subsistence ponds, improved pond engineering may have little impact. In the large-scale commercial pond systems, the level of engineering will be similar to Israel, Taiwan, or the southern United States. Improvements in pond engineering in the small-scale commercial or village ponds will have the greatest potential to increase pond production, but will require the most innovative engineering. Emphasis must be placed on intermediate technology that is both simple and effective.

Pond engineering will be discussed in terms of pond design, pond construction, pond maintenance, and pond operation. Areas of research will be discussed in each section. Emphasis will be placed on areas where research can improve engineering practices and increase production.

Design of Ponds and Other Facilities

Pond design includes not only the design of the pond itself, but the water collection system, water conveyance system, water control structure, and the hatchery facilities.

Water Source The most common source of water will be surface streams or tidal water. The salinity, temperature or silt load may vary significantly due to high winds, rain, or flooding. This water may also contain adults, larvae, or eggs of undesirable predators.

In Central America, South America, or Africa (Lovshin 1980; Grover et al. 1980; Huet 1972), a small dam is constructed and part of the stream flow is diverted into the pond. The water flows to the pond by gravity. During high-flow conditions, the major water flow passes over the dam. The pond should be circled by a ditch to intercept surface runoff and prevent excessive flow to the pond.

In Indonesia brackish-water ponds are typically sited so that the ponds are filled during high tides and emptied during low tides. The source of water is from tidal streams or the ocean. Therefore, the siting and operation depend on the tidal cycle. The water level within the main levee is controlled by a series of secondary gates.

For large pond systems, it may be necessary to measure streamflow or estimate water yield from precipitation data. The diversion dams can be designed with local contractors using simple methods (Soil Conservation Service 1971).

The use of gravity-flow systems offers a simple, cheap, and reliable method for ponds constructed in a suitable site. This requirement for brackish-water farms may require that the ponds be located in areas that may be subject to frequent flooding or storm surge damage. The use of pumps may allow greater flexibility in the siting of brackish-water ponds. The use of low head/high volume pumps for aquaculture should be investigated. Axial flow (Jamande 1977), hydraulic pumps (Jamande 1977), or low-technology pumps (Tamiyavanich 1977; Watt 1976) may be used. The use of windmills may be economic in area with adequate wind energy (Tamiyavanich 1977; Fraenkel 1975). The economics of electrical or gasoline pumps must be carefully formulated to reflect the cost of bringing the power to the pond, availability of spare parts, and availability of people to repair these pumps. A proper analysis may find that low-technology pumps with low efficiencies may be more economic than modern pumps.

Because of the variation in water quality of surface water, this source may be unacceptable without pretreatment for hatchery systems. Drilled wells or subsand wells (Scholes 1980) may be required. Because of the impact of the loss of the water supply, gravity reservoirs or standby generators may be necessary for hatcheries.

Distribution System Unlined ditches are the most common method for water conveyance. In large projects, the design should be based on standard hydraulic practices (Aisenbrey et al. 1978; Kraatz and Mahajan 1975a 1975b). Because of low cost and ease of installation, the use of PVC pipe should be investigated. The design of water conveyance systems is simple and within the technical ability of the developing countries.

Water Pretreatment In coastal fish farms, the introduction of predators or trash fish into the pond is a serious problem. These fish may be introduced as adults, larvae, or eggs and may directly feed on the cultured fish, compete for food, or have low market value. Multiple bamboo screens are installed on the main gate (Yamashita and Sutardjo 1977). Because of the importance of preventing the induction of predators, the use of more durable plastic or plastic-coated mesh may be economic. In ponds where the water is introduced into the pond from a ditch or pipe, the saran box filter or saran sock filter could be used (U.S. Fish and Wildlife Service 1973).

In hatcheries, the removal of silt, suspended matter, and parasites is required (Osborn 1977a; Cook 1977). Modern pressure filters of the mixed media type should be investigated for large-scale regional hatcheries. The development of better hatchery systems is a high priority.

Water Control Structures The water level in ponds is controlled by overflow devices of the "monk" (Huet 1972) or the turn-down drain pipe type (Stickney 1979). The turn-down type is superior to the monk both in ease of construction and use. The pond bank under the inlet pipe should be riprapped or concreted to prevent erosion.

In brackish-water farm ponds, the main and secondary gates are major structures up to 1.5 m wide (Tang 1979). The water level is controlled by stop logs. The main gate is constructed from concrete or wood depending on local practices and the bearing strength of the soil. The design and operation of this unit is straightforward.

Levees The design of levees for ponds depends on the desired depth of the water, the probability of overtopping due to flooding, and the hydraulic and physical characteristics of the levee material.

In areas with a long history of fish culture, the design of ponds has evolved in an empirical manner and may represent an optimum for the local management skills and practices. In other areas where pond design has been adapted from other areas, there may be more potential for increased production and efficiency. In any case, the local design should be used as a starting point.

Pond depths typically range from 30 to 200 cm (Tang 1979; Hickling 1971). The optimum pond depth may be determined by complex interaction between the requirements of the fish and principal food organisms. For tidal milkfish culture some of the factors influencing pond depth are listed in Table 3. Because of the complexities of the effect of pond depth on product, the design depth in most regions is based on experience. Rational selection of depth will be based on growth models for both the food and fish, the energy balance (Szumiec 1979), and dissolved oxygen (Romaire and Boyd 1979; Boyd et al. 1978). Separate ponds for the food and fish and artificial support for the benthic algae (Crance and Leary 1979) could be used to increase production, but require a higher level of management.

The design of levees depends strongly on the local characteristics of the soil. Therefore, the design may vary from country to country and within a country. If available, the levees should be constructed from an impermeable clay or

Table 3

FACTORS INFLUENCING THE OPTIMUM POND DEPTH IN MILKFISH PONDS

Factor	Problem	Reference
Depth	Optimum depth for benthic algae is 10-15 cm	Korringa 1976
	Older fish will not enter shallow water, ponds must be > 30 cm	Korringa 1976
	Self shading by algae may limit the useful depth of the ponds	Szumiec 1979
	Increased depth may produce undesirable anaerobic bottom conditions	Korringa 1976
Temperature	Shelter trenches required for fish	Huet 1972
Salinity	High salinity has serious effect on benthic algae, but not the fish	Tang 1979; Korringa 1976
Productivity	Heavy organic fertilization may increase the production of non-photosynthetic organisms	Moav et al. 1977

with a clay core (Tang 1979; Hechanova 1977). The bottom of the cut-off trench or clay core must be extended through the top soil. The use of plastic films (such as Hypalon or butyl rubber) or bentonite clay to reduce seepage should be investigated. Improved testing procedures must be developed to determine the suitability of soil for levee construction, especially with regard to acid-sulfate soils. In exposed coastal areas, research on the use of breakwaters or natural buffer zones is needed to prevent erosion. The use of pumps in coastal areas may allow construction of ponds away from the near-tidal areas where erosion and acid-sulfate soil are common.

Pond Size and Orientation Pond size typically ranges from 0.1 to 2.0 ha. Pond size depends on the availability of land and on harvesting and operational flexibility. Nursery ponds for milkfish or Chinese carp (Korringa 1976; Hickling 1971) may be only 10-50 m^2. Production ponds for milkfish range from 4 to 6 ha. Ponds of rectangular shape are more convenient to harvest (Tang 1979). Their long sides should be aligned perpendicular to the prevailing winds to reduce wind-induced bank erosion. Larger ponds are cheaper to build, but small ponds are cheaper to maintain and more convenient to manage.

Siting of Ponds In coastal areas (Tang 1979) the siting of ponds will depend on 1) ground elevation and tidal range, 2) soil type, and 3) type and density of vegetation. The use of mangrove swamps may be restricted due to environmental concerns and erosion problems (Gatus and Martinez 1977; Thompson and Tai 1977). Also, uncontrolled pond construction may increase flooding and sedimentation problems (Guanzon and Basa 1977). Pesticide pollution may be a major constraint in some areas (Rudayat and Oetomo 1977). In many areas, flooding and loss of fish is a serious problem. This flooding may occur due to heavy rains or more commonly due to large tropical storms that result in high tides and winds. A methodology for risk/benefit analysis is needed as a management tool.

The procedure for the siting of coastal ponds is well developed (Tang 1979; Hechanova 1977). The use of pumps of aeration may allow siting of ponds in areas that are presently unsuitable.

In tropical Asia there are 13 million hectares of acid-sulfate soils that could be developed for aquaculture (Singh 1980). The reclamation of these areas requires periodic drying and flushing to oxidize the pyrite and leach the acidity and fertilization with lime and manure (Singh 1980). This process may require three to four years. Improved techniques for both the assessment of the acid-producing potential of a soil and reclamation of acid sulfate soils may allow development of large areas of unused land.

In noncoastal areas (Maar et al. 1966), the siting of ponds will depend on 1) a source of surface water, 2) soil type, and 3) topography. It is desirable that the ponds can be drained by gravity. In larger pond systems, the use of pumps may be economic and useful.

Pond Construction

The construction of ponds and associated structures is a well-developed art in most developing countries. Construction methods may range from manual methods (Denila 1977) to modern capital-intensive methods (Food and Agriculture Organization 1975). Several excellent construction manuals for manual and modern methods are available (Soil Conservation Service 1971; Maar et al. 1966; Chakroff 1976). In some cases, mechanical methods are unsuitable for use in coastal areas because of the low bearing strength of the soil (Lijauco 1977).

The choice of construction method will depend strongly on local conditions including the availability of labor and capital, political constraints, time requirements, and the method of economic analysis. Mechanically intensive methods tend to be cheaper, of better quality, and faster than manual methods (Tang 1979). The manual method will generate local employment and save foreign exchange. The impact of increasing local employment has not been adequately evaluated in relationship to the overall economics of the construction methods.

Pond Maintenance

Important aspects of pond maintenance include levee stabilization, predator control, and weed control. Pond maintenance problems are more serious in the coastal zone. Pond maintenance will require daily or weekly inspection of levees and the water system.

Maintenance of ponds will require adjustment of water flow, checking inlet and outlet structures, and checking for leaks. In milkfish culture, leakage and stabilization problems may prevent the stocking of fish until three to four years after construction. Levee stabilization may be improved by planting of grasses or trees on the levee (Comacho 1977; Korringa 1976). Species include mangroves, mulberry, or Bermuda grass. Bermuda grass shows a promise in the stabilization of acid sulfate soil, a serious problem in some coastal areas. In most areas, pond control of both predator fish and terrestrial predators such as otter, herons, frogs, and snakes can prevent significant loss of fish. Screens will have to be checked to prevent entry of predator fish. Draining and drying of ponds will be required to kill predators prior to stocking. Because some tropical fish can burrow into the mud and survive, ponds may need to be dried and filled several times. Grass and trees on the levees will have to be trimmed to avoid hiding places for snakes and other predators. Birds and otters can be controlled by guns. Weed control in the ponds may depend on proper construction, manual removal, or stocking with herbivorous fish.

Pond Operation and Management

Improved pond operations and management plans will form the basis for improved production in ponds. For subsistence ponds, pond management plans need to be formulated in terms that the operators can understand and use. While these plans will not be optimum, they should be optimum in terms of the management level available.

In the larger ponds, the management plan can be more sophisticated to reflect both the increase in management skills and management tools. Water test kits (Boyd 1980; Boyd 1977) are adequate for management decisions, although their use in tropical areas should be evaluated. The monitoring as well as the use of aeration systems and pumping will allow significant improvements in production. Development of predictive water quality management models (Boyd and Lichtkoppler 1979; Boyd 1979; Boyd et al. 1978; Romaire and Boyd 1979) should be developed for the tropics.

Most pond management practices such as liming, fertilization, manuring, or draining and drying have developed from empirical growth trials using fish production as the figure of merit. The formulation of management plans on a more fundamental basis (i.e., maintenance of total hardness > 20 mg/L rather than addition of 1000 kg/ha$\cdot$y) may offer significant cost savings to the larger operations. Development of rational management criteria will also aid in the selection of local materials (i.e., hog manure versus cow manure).

Integrated fish culture and animal production will allow increased efficiency and reduced fertilization costs (Pullin and Shehadeh 1980). In the subsistence ponds, the animals will be housed over the ponds or upslope from the pond. In larger ponds, the manure will be collected and distributed to the ponds. This may allow higher levels of production and flexibility.

LITERATURE CITED

Aisenbrey, Jr., A.J., R.B. Haynes, H.L. Warren, D. L. Winsett, and R.B. Young. 1978. Design of Small Canal Structures. Bureau of Reclamation, U.S. Department of the Interior, Denver, Colorado. 435 pp.

Anderson, C.E., and R.O. Smitherman. 1978. Production of normal male and androgen sex-reversed Tilapia aurea and T. nilotica fed a commercial catfish diet in ponds, pp. 34-42. In R.O. Smitherman, W.L. Shelton, and J.H. Grover (eds), Symposium on Culture of Exotic Fishes. Fish Culture Section, American Fisheries Society, Auburn, AL.

Balarin, J.D. 1979. Tilapia - A Guide to Their Biology and Culture in Africa. University of Stirling, Scotland. 174 pp.

Bonn, E.W., and B.J. Follis. 1967. Effects of hydrogen sulfide on channel catfish, Ictalurus punctatus. Trans. Am. Fish. Soc. 96:31-36.

Boyd, C.E. 1977. Evaluation of a water analysis kit. J. Environ. Qual. 6(4):381-384.

_________. 1979. Water Quality in Warmwater Fish Ponds. Agricultural Extension Station, Auburn University, Auburn, Alabama. 359 pp.

_________. 1980. Reliability of water analysis kits. Trans. Am. Fish. Soc. 109(2):239-243.

_________, and F. Lichtkoppler. 1979. Water Quality Management in Pond Fish Culture. International Center for Aquaculture, Research and Development Series No. 22, Auburn University, Auburn, Alabama. 30 pp.

_________, R.P. Romaire, and E. Johnston. 1978. Predicting early morning dissolved oxygen concentrations in channel catfish ponds. Trans. Am. Fish. Soc. 107(3):484-492.

_________, R.P. Romaire and E. Johnston. 1979. Water quality in channel catfish production ponds. J. Environ. Qual. 8(3):423-429.

Brett, J.R., and J.M. Blackburn. 1981. Oxygen requirements for growth of young coho (Oncorhynchus kisutch) and Sockeye (O. nerka) salmon at 15°C. Can. J. Fish. Aquatic Sci. 38:399-404.

Chakroff, M. 1976. Freshwater Fish Pond Culture Management. Vita Publications Manual Series Number 36E, Washington, D.C. 191 pp.

Colt, J. and G. Tchobanoglous. 1978. Chronic exposure of channel catfish, Ictalurus punctatus, to ammonia: effects on growth and survival. Aquaculture 15:353-372.

_________, A. Mitchell, G. Tchobanoglous, and A. Knight. 1979. The Environmental Requirements of Fish, Appendix B, the Use and Potential of Aquatic Species for Wastewater Treatment, Publication No. 65, California State Water Resources Control Board, Sacramento, CA 240 pp.

_______, and D. Armstrong. 1981. Nitrogen toxicity to crustaceans, fish, and mollusca, pp. 34-37. In L.J. Allen and E.C. Kinney (eds), Proceedings of the Bio-Engineering Symposium for Fish Culture. Fish Culture Section, American Fisheries Society, Bethesda, Maryland.

________, R. Ludwig, G. Tchobanoglous, and J.J. Cech, Jr. 1981. The effects of nitrite on the short-term growth and survival of channel catfish, Ictalurus punctatus. Aquaculture 24:111-122.

Comacho, A.S. 1977. Implications of acid sulfate soils in tropical fish culture, pp. 97-102. In Aquaculture Engineering, South China Sea Fisheries Development and Coordinating Programme, SCS/GEN/77/14, Manila, Philippines.

Cook, H.C. 1977. Small-scale shrimp hatchery project, pp. 115-135. In Aquaculture Engineering, South China Sea Fisheries and Development and Coordinating Programme, SCS/GEN/-77/14, Manila, Philippines.

Crance, J.H., and D.F. Leary. 1979. The Philippine Inland Fisheries Project and Aquaculture Production Project - Completion Report. International Center for Aquaculture Research and Development Series No. 25, Auburn University, Auburn, Alabama.

Davis, A.T., and R.R. Stickney. 1978. Growth response of Tilapia aurea to dietary protein quality and quantity. Trans. Am. Fish. Soc. 107:479-483.

Denila, L. 1977. Improved methods of manual construction of brackish water fishponds in the Philippines, pp. 238-259. In Aquaculture Engineering, South China Sea Fisheries Development and Coordinating Programme, SCS/GEN/77/14, Manila, Philipines.

DeSilva, S.S. 1980. Biology of juvenile grey mullet: a short review. Aquaculture 19(1):21-36.

Flis, J. 1968a. Anatomicohistopathological changes induced in carp (Cyprinus carpio L.) by ammonia water. Part I. Effects of toxic concentrations. Acta Hydrobiol. (Krakow) 10:205-224.

_______. 1968b. Anatomicohistopathological changes induced in carp (Cyrpinus carpio L.) by ammonia water. Part II. Effects of subtoxic concentrations. Acta Hydrobiol. (Krakow) 10:225-238.

Food and Agriculture Organization of the United Nations. 1975. Aquaculture Planning in Africa. Aquaculture Development and Coordination Programme, ADCP/REP/75/1. United Nations Development Programme. Rome. 114 pp.

Fraenkel, P.L. 1975. Food from Windmills. Intermediate Technology Publication, London, England. 56 pp.

Gatus, A.R., and E.S. Martinez. 1977. Engineering considerations in the release of mangrove swamps development into fishponds, pp. 85-95. In Aquaculture Engineering, South China Sea Fisheries Development and Coordinating Programme, SCS/GEN/77/14, Manila, Philippines.

Greenland, D.C. 1974. Recent developments in harvesting, grading loading and hauling pond raised catfish. Trans. Am. Society Ag. Engin. 17(1):59-62.

_______________, and R.L. Gill. 1974a. Trapping pond-raised channel catfish. Prog. Fish-Cult. 36(2):78-79.

_______________, and R.L. Gill. 1974b. A diversion screen for grading pond-raised channel catfish. Prog. Fish-Cult. 36:78-79.

Grover, J.H., D.R. Street, and P.D. Starr. 1980. Review of Aquaculture Development Activities in Central and West Africa. International Center for Aquaculture, Research and Development Series No. 28, Auburn University, Auburn, Alabama. 31 pp.

Guanzon, J.C., and S.S. Basa. 1977. Fishpond development and its relation to flooding in the Philippines with particular reference to the Pampanga River Delta and Candaba Swamp Area, Luzon, pp. 59-74. In Aquaculture Engineering, South China Sea Fisheries Development and Coordinating Programme, SCS/GEN/77/14, Manila, Philippines.

Guerrero, R.D. 1975. Use of androgens for the production of all-male Tilapia aurea (Steindacher). Trans. Am. Fish. Soc. 104(2):342-348.

Haines, T.A. 1981. Acidic precipitation and its consequences for aquatic ecosystems: a review. Trans. Am. Fish. Soc. 110:669-707.

Hechanova, R.G. 1977. Practical applications of the basic principles of hydraulics and soil mechanics in aquaculture engineering, pp. 433-452. In Aquaculture Engineering, South China Sea Fisheries and Development and Coordinating Programme, SCS/GEN/77/14, Manila, Philippines.

Hickling, C.F. 1971. Fish Culture, 2nd ed. Faber and Faber, London. 317 pp.

Hollerman, W. D., and C.E. Boyd. 1980. Nightly aeration to increase product of channel catfish. Trans. Am. Fish. Soc. 109(4):446-452.

Huet, M. 1972. Textbook of Fish Culture. Fishing News (Books), Farnham, England.

ICLARM. 1980. ICLARM Conference on the Biology and Culture of Tilapias. Bellagio, Italy, 1-6 September 1980.

Jamande, T.J., Jr. 1977. Pumps for brackishwater aquaculture, pp. 393-422. In Aquaculture Engineering, South China Sea Fisheries Development and Coordinating Programme, SCS/GEN/77/14, Manila, Philippines.

Johnson, S.K. 1979. Transport of Live Fish. Texas Agricultural Extensive Service, Fish Disease Diagnostic Laboratory, FDDL-F14, College Station, Texas. 13 pp.

Kinne, O. (ed). 1977. Marine Ecology. Vol. III, Part 1. John Wiley, New York.

Korringa, P. 1976. Farming the milkfish (Chanos chanos) in Indonesia, pp. 65-90. In Farming Marine Fishes and Shrimp. Elsevier, Amsterdam.

Kraatz, D.B., and I.K. Mahajan. 1975a. Small Hydraulic Structures. Irrigation and Drainage Papers, 26/1, Food and Agricultural Organization of the United Nations, Rome, 293 pp.

___________, and I.K. Mahajan. 1975b. Small Hydraulic Structures. Irrigation and Drainage Papers, 26/2, Food and Agricultural Organization of the United Nations, Rome, 293 pp.

Kutty, M.N. 1980. Aquaculture in South East Asia: Some points of emphasis. Aquaculture 20(3):159-168.

Lijauco, N.M. 1977. Plans and programmes of the SEAFDEC Legones Station with emphasis on varied layout and design of ponds, pp. 207-232. In Aquaculture Engineering, South China Sea Fisheries and Development and Coordinating Programme, SCS/GEN/77/14, Manila, Philippines.

Lovshin, L. 1980. Progress Report on Fisheries Development in Northeast Brazil. International Center for Aquaculture, Research and Development series No. 26, Auburn University, Auburn, Alabama. 15 pp.

Maar, A., M.A.E. Mortimer, and I. Van Der Lingen. 1966. Fish Culture in Central East Africa. Food and Agriculture Organization for the United Nations, Rome. 160 pp.

Mironova, N.V. 1974. The energy balance of Tilapia mossambica. J. Ichthyology 14(3):431-438.

Moriarty, D.J.W. 1973. The physiology of digestion of blue-green algae in the cichlid fish, Tilapia nilotica. J. Zool. Lond. 171:25-39.

Moav, R., G. Wohlfarth, G.L. Schroeder, G. Hulata, and H. Barash. 1977. Intensive polyculture of fish in freshwater ponds. I. Substitution of expensive feeds by liquid cow manure. Aquaculture 10(1):25-43.

Osborn, P.E. 1977a. Design of freshwater fish hatchery for small fish farmers, pp. 103-114. In Aquaculture Engineering, South China Sea Fisheries Development and Coordinating Programme, SCS/GEN/77/14, Manila, Philippines.

___________. 1977b. A survey of aeration devices, pp. 428-432. In Aquaculture Engineering, South China Sea Fisheries Development and Coordinating Programme, SCS/GEN/77/14, Manila, Philippines.

Pandian, T.J., and R. Raghuraman. 1972. Effects of feeding rate on conversion efficiency and chemical composition of the fish Tilapia mossambica. Mar. Biol. 12:129-136.

Pruginin, Y., S. Rothbard, G. Wohlfarth, A. Halevy, R. Moav, and G. Hulata. 1975. All-male broods of Tilapia nilotica x T. aurea hybrids. Aquaculture 6(1):11-21.

Pullin, R.S.V., and Z.H. Shehadeh (eds). 1980. Integrated Agriculture - Aquaculture Farming Systems. ICLARM Conference Proceedings 4, Manila, Philippines, 258 pp.

Redner, B.D., and R.R. Stickney. 1979. Acclimation to ammonia by Tilapia aurea. Trans. Am. Fish. Soc. 108:383-388.

Reynolds, F., and T.A. Haines. 1980. Effects of chronic exposure to hydrogen sulfide on newly hatched brown trout Salmo trutta L. Envir. Poll. 22a:11-17.

Romaire, R.P., and C.E. Boyd. 1979. Effects of solar radiation on the dynamics of dissolved oxygen in channel catfish ponds. Trans. Am. Fish. Soc. 108(5):473-478.

Rosa, H. 1965. Preparation of Synopses on the Biology of Species of Living Aquatic Organisms. FAO, FIB/SI Rev. 1, Rome, 82 pp.

Rudayat, N., and K.S. Oetomo. 1977. The design and construction of freshwater fishponds in Indonesia, pp. 177-192. In Aquaculture Engineering, South China Sea Fisheries Development and Coordination Programme, SCS/GEN/77/14, Manila, Philippines.

Russo, R.C., R.V. Thurston, and K. Emerson. 1981. Acute toxicity of nitrite to rainbow trout (Salmo gairdneri): effects of pH, nitrite species, and anion species. Can. J. Fish. Aquatic Sci. 38:387-393.

Scholes, P. 1980. The sea-water well system at the fisheries laboratory, Lowestoft, and the methods in use for keeping marine fish. J. Mar. Biol. Assn. U.K. 60(1):215-225.

Singh, V.P. 1980. Management of fishponds with acid sulfide soils. Asian Aquaculture 3(4):46.

Smith, L.L, and D.M. Oseid. 1974. Effect of hydrogen sulfide on development and survival of eight freshwater fish species, pp. 417-430. In J.H.S. Baxter (ed), The Early Life History of Fish. Springer-Verlag, New York.

Soil Conservation Service. 1971. Ponds for Water Supply and Recreation, Agricultural Handbook No. 387, Soil Conservation Service, U.S. Department of Agriculture, Washington, D.C. 55 pp.

Stickney, R.R. 1979. Principles of Warmwater Aquaculture. Wiley-Interscience, New York. 375 pp.

Szumiec, M.A. 1979. Hydrometeorology in pond fish culture, pp. 117-120. In T.V.R. Pillay and W.A. Dill (eds), Advances in Aquaculture. Fishing News Books, Farnham, England.

Tamiyavanich, S. 1977. Water pumps and their use for aquaculture in Thailand, pp. 423-425. In Aquaculture Engineering, South China Sea Fisheries Development and Coordinating Programme, SCS/GEN/77/14, Manila, Philippines.

Tang, Y.A. 1979. Planning, design and construction of a coast milkfish farm, pp. 104-117. In T.V.R. Pillay and W.A. Dill (eds), Advances in Aquaculture. Fishing News Books, Farnham, England.

Thompson, G.B., and Y.C. Tai. 1977. Environmental impact of coast development arising from reclamation and associate activities, with particular reference to fisheries and aquaculture, pp. 55-58. In Aquaculture Engineering, South China Sea Fisheries and Development and Coordinating Programme, SCS/GEN/-77/14, Manila, Philippines.

Thurston, R.V., C. Chakoumakos, and R.C. Russo. 1981a. Effect of fluctuating exposures on the acute toxicity of ammonia to rainbow trout (Salmo gairdneri and cutthroat trout (S. clarki). Water Res. 15:911-917.

______________, G.R. Phillips, and R.C. Russo. 1981b. Increased toxicity of ammonia to rainbow trout (Salmo gairdneri) resulting from reduced concentrations of dissolved oxygen. Can J. Fish. Aquat. Sci. 38:983-988.

U.S. Fish and Wildlife Service. 1973. Second Report to the Fish Farmers. Resource Publication 113, Bureau of Sport Fisheries and Wildlife, Washington, D.C. 123 pp.

__________________. 1978. Fish Transportation. Section G, Manual of Fish Culture. Washington, D.C. 88 pp.

Watt, S. 1976. Twenty-one chain and washer pumps. Intermediate Technology Publications, London, England. 49 pp.

Yamashita, M., and Sutardjo. 1977. Engineering aspects of brackish-water pond culture in Indonesia, pp. 261-280. In Aquaculture Engineering, South China Sea Fisheries and Development and Coordinating Programme, SCS/GEN/77/14, Manila, Philippines.

PART 3

MODELLING OF POND CULTURE SYSTEMS

MODELLING THE HYDROMECHANICAL AND WATER QUALITY RESPONSES OF AQUACULTURE PONDS

A Literature Review

by

Nikola Marjanovic and Gerald T. Orlob

INTRODUCTION

Development of a quantitative methodology for the assessment of aquaculture pond performance and design suggests the need for a suitable mathematical model. Such a model should have the capability for simulation of the important characteristics of pond dynamics that influence overall protein production, i.e., fish biomass. Hence, the model might be expected to include hydrodynamic behavior, water quality-ecological interactions, chemical-biochemical kinetics and certain environmental factors that determine ecologic response at the trophic level of prime concern.

Such models, dealing at least in part with these notions and principles, are known to exist, although few are apparent that are readily transferable to the case of pond aquaculture. As a point of beginning, it was determined that a careful review of the state of the art of modelling, specifically directed to ponds, and small lakes or reservoirs that might resemble ponds in scale at least, should be conducted.

Scope of Review

An earlier review of mathematical modelling of surface water impoundments (Orlob 1977) conducted for the Office of Water Research and Technology (OWRT) Department of the Interior provided a substantial foundation for the present undertaking. It included more than 400 references, dealing with the broader subject area of impoundments, and a digest of 90 specific models of impoundments that were considered to be at a level of development and documentation adequate for transfer to aquatic systems. Among these were included the models most widely used by the major water agencies in the United States (e.g., Corps of Engineers, Environmental Protection Agency, Bureau of Reclamation, etc.) to evaluate environmental impacts of water development and pollution control projects. In addition, models of more specialized scope, such as those of the International Biological Program were referenced and documented.

Notwithstanding the scope of this earlier effort, it did not focus sufficiently on the smaller water bodies, such as aquaculture ponds, and it did not treat questions of primary concern in the present research. Specifically , it did not address questions directed to enhancement of the productivity of ponds designed to generate a source of protein.

It was decided to examine in depth the extant published literature dealing with a selected list of topics which would encompass all aspects of pond technology relevant to construction of a mathematical model, or models, or an aquaculture system. The following topics were identified:

-- Hydromechanics of ponds, small lakes and reservoirs, including wind-induced circulation, thermal effects, stratification and destratification, turbulence and mixing, and artificial circulation systems.

-- Water quality, including temperature, dissolved oxygen, BOD, chemical constituents, abiotic nutrients, toxics, sediment, etc.

-- Ecological interactions, including biota at all trophic levels, bacteria through nekton, kinetics of growth, respiration, nutrient limitation, toxicity, photosynthesis, predation, and resilience.

-- Model development, including models specifically concerned with hydromechanics, water quality, or ecology of ponds, or combinations of these, noting degree of transferability and documentation.

The review was designed to cover at least the most recent ten years of publication in selected scientific media, although through cross-referencing and tracing of cited references, the chronological scope was inevitably extended. Also, certain limited circulation documents, notably project reports, which were found in the earlier review (Orlob 1977) to be prime sources of "working" models, were carefully reviewed when accessible.

The major journals covered in the present review included but were not limited to the following:

1) Water Resources Research

2) Journal of Water Pollution Control Federation
3) Journal of Hydrology
4) American Soc. of Civil Engineers
   --Jour. of Hydraulic Division
   --Jour. of Irrigation and Drainage Division
   --Jour. of Environmental Engineering Division
5) Transaction of the Amer. Fishery Society
6) Hydrobiologia
7) Jour. of Limnology and Oceanography
8) Journal of Ecology
9) Journal of Environmental Quality
10) Aquaculture
11) Journal of the World Mariculture Society

Organization of Literature Review

The following literature review is organized into four major sections: Summary, Annotated Literature Review, Discussion of Findings, and Literature Cited. The Summary is intended to provide a concise overview of the Annotated Literature Review for those wishing a brief introduction to the subject. The Annotated Literature Review is presented in the form of an annotated narrative. It is organized into four major subsections dealing with 1) General Ecology and Water, 2) The Hydraulics and Hydromechanics of Ponds, Small Lakes, and Reservoirs, 3) Water Quality of Small Lakes, Ponds, and Reservoirs and Effects on Fish, and 4) Models of Aquatic Systems. The section entitled Discussion of Findings is intended to provide a further summary of the insights gained from the conduct of this review. The articles, papers, reports, books, etc., cited in the Literature Review section are arranged alphabetically by author. Multiple citations by the same author are arranged chronologically.

SUMMARY

A review of literature was performed to determine the state of research and development related to mathematical modelling of aquaculture ponds with specific reference to hydromechanical and water quality responses. The results of this review are summarized below in four categories dealing with 1) General Ecology and Water Quality; 2) The Hydraulics and Hydromechanics of Ponds, Small Lakes, and Reservoirs; 3) Water Quality of Ponds, Small Lakes, and Reservoirs, and 4) Models of Aquatic Systems.

General Ecology and Water Quality

General literature sources provide useful background for model development, but comparatively little that specifically deals with the quantitative aspects of aquaculture. The classical treatise of Hutchinson (1975) and the textbook by Wetzel (1975) are among the most useful. Textbooks and monographs which provide specific examples of models - usually of lakes or reservoirs - that are analogous to pond systems are useful in guiding future model development. For example, an excellent review of mixing and circulation processes in natural water bodies is found in Fischer et al. (1979), which also includes a detailed description of a unique one-dimensional reservoir model, DYRESM (Imberger et al. 1978), that may be useful in pond modelling.

Temperature effects on impoundments, including major heat exchange processes represented in quantitative form, are thoroughly reviewed in a report by the TVA Engineering Laboratory (1972). Physicochemistry of stratified lakes and reservoirs has been modelled with some success, and extension of the classical BOD-DO relationships to pond systems appears practical. However, such models do not integrate water quality and aquatic ecology, i.e., they are not comprehensive. Nutrient balances and limitations have been subjects of much research and in a few instances have been utilized with some success in ecological model building (Chen and Orlob 1975; DiToro et al. 1975; Park et al. 1974). However, there are few references which deal with water quality - ecological interactions at a level of detail suitable for structuring of an aquaculture pond model.

Hydraulics and Hydromechanics of Ponds, Small Lakes, and Reservoirs

Published literature relevant to the hydromechanics of ponds or pond systems is virtually nonexistent. The developer of aquaculture pond models will have to draw on experience with modelling of larger water bodies, even though this may not be particularly fruitful. The earliest lake and reservoir models, for example, were usually one-dimensional, applied to highly stratified systems. They considered only the most rudimentary hydromechanical principles. Subsequent development in hydrodynamic modelling appears to have been focused on very large lakes, such as the Great Lakes, where circulations are induced by a combination of wind stress at the surface and convective mixing induced by seasonal and diurnal thermal cycles. Mixing processes in such systems were generally treated empirically from the modelling viewpoint, coefficients of "effective diffusion" being derived from observation on actual lakes and reservoirs.

The state of the art of modelling wind-driven circulation in large lakes is probably best represented by the models of Simons (1973) and Simons et al. (1977). Two-dimensional models of stratified flows, common in both natural and manmade impoundments, are exemplified by the work of King et al. (1975) and Edinger and Buchak (1975). While these experiences are useful in the general sense of hydrodynamic modelling, they provide little that is directly transferable to design of pond systems. The most promising research identified in this review, in terms of applicability to ponds, may well be that concerned with small-scale wind-mixing effects in the epilimnion (Stefan et al. 1976). Otherwise, the hydromechanics literature is nonspecific as regards the particular kind of problems likely to be encountered in design of a pond model.

Water Quality of Ponds, Small Lakes, and Reservoirs

The thermal behavior of lakes and reservoirs has been a subject of considerable research and some excellent models exist for prediction of temperature changes and calculation of heat energy transfer (Water Resources Engineers 1967; Huber et al. 1972; TVA 1972). Pond temperature regimes have been well characterized for both temperature and tropical climates (Young 1975), and effects on fish have been investigated under a wide variety of conditions.

Effects of low dissolved oxygen on fish in ponds have been analyzed in a series of well-executed experiments on catfish ponds (Tucker and Boyd 1977; Tucker et al. 1979; Romaire and Boyd 1979; Hollerman and Boyd 1980). A model was developed to estimate DO fluctuations in ponds over a diurnal cycle due to photosynthesis and respiration by plankton, fish, and benthic communities. Experiments were performed on methods of raising DO.

Chemical characteristics of ponds in relation to physical and biological properties are not well documented in the extant literature. On the other hand, considerable information is available on toxic effects of heavy metals, chlorine, ozone, etc. on fish (Cairns et al. 1975; Katz 1977), including behavioral responses to pollutants (Larrick et al. 1978; Morgan 1979). Usually these experiments have been under closely controlled conditions in the laboratory; few have been performed in actual ponds.

Modelling of Aquatic Systems

The stimulus for development of mathematical models of aquatic systems came largely from pollution control efforts in the United States during the 1960s. Simple gross-nutrient budget models were first devised to evaluate eutrophication control strategies for lakes (Vollenweider 1969; Snodgrass and O'Melia 1975; Larsen and Mercier 1975). These were complemented subsequently by more descriptive models that attempted to include important water quality - ecological interactions. The development of these descriptive models began with one-dimensional representations of stratified impoundments, either lakes or reservoirs, with thermal structure as the primary concern (WRE 1967; Huber et al. 1972; Baca and Arnett 1976). Once a capability to model temperature changes in simple impoundments was achieved, other water quality and ecological phenomena were added (Markovsky and Harleman 1971). A versatile water quality - ecological model known as LAKECO was developed for the Environmental Protection Agency (Chen and Orlob 1975). This model, in one-dimensional form, was capable of simulating twenty or more water quality and biological state variables to characterize their temporal and spatial distributions in reservoirs and lakes under realistic environmental and operating conditions. It was, however, a one-dimensional model.

The limitations of the one-dimensional representation for ecological models were examined by Ford and Thornton (1979), who noted that the characteristic time scales of hydrodynamic, chemical, and biological phenomena dictate upper and lower bounds for which the one-dimensional approximation is acceptable. Later the concepts of LAKECO were incorporated in WQRRS, a comprehensive river-reservoir model of the Corps of Engineers. It was extended to three dimensions by Chen et al. (1975) for simulation of large lakes. Similar models were developed by DiToro et al. (1975) and Thomann et al. (1973, 1974, 1975).

An important modelling effort with stronger emphasis on biological aspects was initiated in the International Biological Program by a group at Rensselaer Polytechnic Institute (Park et al. 1974). It resulted in a series of multicompartment models, CLEAN, CLEANER, MS. CLEANER, etc. (Youngberg 1977; Desormeau 1978; Leung 1978), which have been applied to a wide variety of situations in the United States and Europe. The careful biological structuring of these models places them in a leading position, along with recent versions of LAKECO, as the most promising foundations for development of an aquaculture pond model.

Concluding Comments

A review of the literature reveals some large gaps in the science and technology needed for development of a mathematical model that can be used to aid in the design of aquaculture ponds. The greatest deficiencies are in the lack of comprehensiveness of available models. This is found in failure to deal with the complete ecosystem (not just specific species), neglect of hydromechanical factors, and lack of attention to the interaction of water quality constituents and biota of the pond system. The relationship between the aerobic, facultative, and anaerobic zones in an aquaculture pond has not been adequately studied, at least from a modelling viewpoint.

It appears that mathematical modelling of aquaculture pond systems is a potentially fruitful area of research that could yield useful tools for enhancing the overall productivity of such systems. Such models, if they can be developed, would surely improve design of ponds, facilitate structuring of field demonstration and data programs and generally contribute to better understanding of the behavior of aquaculture systems.

ANNOTATED LITERATURE REVIEW

The findings of the literature review are presented in the form of an annotated narrative. The review is highlighted by specific references to publications that are considered important background for development of a modelling capability. Although over 230 individual references were reviewed, an effort was made to cite the most relevant works (180 references) although it is recognized that some may have been overlooked inadvertantly.

The review is organized in four subsections. The first deals with publications of general interest in the fields of ecology and water quality, usually books, monographs, or treatises. The second section deals with the hydraulics and hydromechanics of ponds, small lakes, and reservoirs. The third section covers water quality of ponds, small lakes, and reservoirs, and the fourth deals with ecosystems. Emphasis in all sections is on models and modelling techniques and the feasibility of developing an aquaculture pond model or models.

General Ecology and Water Quality

All publications dealing with aquatic ecosystems in general or larger water bodies, such as lakes and big reservoirs, but which can be useful in analyzing and in constructing models of small lakes, ponds, and reservoirs, are classified in this group.

General Ecology, Limnology, and Water Quality There are a few important books dealing with general limnology and ecology of fresh waters which contain information useful in model development. Among the more notable are the treatise on limnology by Hutchinson (1975) and the textbook by Wetzel (1975). There is some tendency in these and other general works to treat the study of inland waters with specific regard to geographic conditions or as unique case studies. However, such references provide useful comprehensive and up-to-date accounts of the physical, chemical and biological processes operating in inland waters (or particular parts of them), very often with emphasis on their ecologic effects (Bayly and Williams 1973; Reid and Wood 1976; Aleyev 1977).

Also, there is a formidable body of information on the effects of man on ecosystems, especially on the environmental impact of thermal electric power facilities. Usually, one encounters in these reviews of thermal ecology discussions at various levels - individual organisms, groups or populations, and ecosystems. These reviews can be grouped in several major categories: thermal tolerance, temperature and fish behavior, environmental impact of electric-power facilities, impingement, entrainment, and electric power facilities (SCOPE Report 2 1972; Esch and McFarlane 1976).

Information on ecological principles to be followed in economic development is also available, with special emphasis on ecosystems that are currently subject to heavy development pressure (Dasmann et al. 1973).

In another large category of general publications, authors deal with a specific measure of quality of aquatic environment and the effects of pollutants on that quality. Information on the broad topics of bacteriological, chemical, and radioactive pollutants are available, as well as on the problems caused by pollution and specific methods of pollution control. In this category are classified publications dealing with biological indicators of environmental quality (the practical aspects of interpreting the biological manifestations of deteriorated environmental conditions), and with the quantitative techniques for the assessment of water body quality, especially of lakes (MacKenthum 1973; Thomas et al. 1973; Reckhow 1979; Cooper et al. 1976).

An important grouping of publications in this area are articles concerned with limnology and physico-chemical characteristics of specific water bodies in particular regions or countries of the world. Data are usually analyzed to provide descriptions of temperature, thermal stratification and location of thermocline, alkalinity, pH, turbidity, conductivity, dissolved oxygen, chlorides, nutrient conditions, algal species and numbers, distribution of flora and fauna, etc. Important general sources are Mishra and Yadav (1978); Eccles (1974); Brezonik and Fox (1974); Forsyth and McCall (1974); Rai (1974); Driver and Peden (1977); and Olsen and Sommerfeld (1977).

Hydrodynamics In this group are classified publications dealing with mixing and circulation of water and modelling of natural hydrosystems. Especially notable is the monograph by Fischer et al. (1979), which deals comprehensively with natural mixing phenomena from a hydromechanical viewpoint. A reservoir model, DYRESM, developed by Imberger et al. (1978) is described in some detail. The approach used in this model, a conservation of total energy including wind shearing effects, may be useful for modelling of shallow ponds. The mechanisms of stratification in lakes are carefully analyzed by a number of researchers. Valuable sources include Jassby and Powell (1975); Johnson and Merritt (1979); Bedford and Babajimopoulos (1977); Powell and Jassby (1974); and Moretti and McLoughlin (1977). The hydromechanical principles considered in these publications are generally useful for modelling of water bodies at all scales.

Temperature and Its Effects Temperature is probably the single most important physical characteristic of a water body in determining hydromechanical as well as water quality and ecological behavior. This parameter is the most often measured, and analyses of data are available in the literature for many different lakes and reservoirs.

The mechanics of heat exchange through the air-water interface is comprehensively treated by the TVA Engineering Laboratory (1972). Sometimes, general relationships between mean surface temperature and mean air temperature are utilized as predictive tools (Webb 1974).

In many natural water bodies, an annual cycle of stratification occurs due to differences in temperature and changes in the heat budget. Many factors influence the shape of the vertical temperature profile including heat exchange by diffusion and processes within the water mass, heat exchange due to inflow and outflow of water in reservoirs, the temperature profile at the time of ice formation, etc. (Rahman 1978). Some publications deal generally with prediction of temperature on the basis of time-variable meteorological conditions and lake morphology (Stefan and Ford 1975; Rahman and Marcotte 1974). Thermal stability and its effect on phytoplankton are also analyzed (Biswas 1977). A model which includes these effects has been developed by Chen and Orlob (1975) and tested on Lake Washington near Seattle.

Physico-Chemistry of Lakes and Reservoirs BOD and DO are the water quality parameters of lakes and reservoirs most often analyzed. These parameters have been modelled for lakes and values compared with observed values (Banks 1976; Markovsky and Harleman 1971; Chen and Orlob 1975). Field data indicate generally that oxygen-demanding substances, both biotic and abiotic, tend to be concentrated in the metalimnia of stratified lakes and reservoirs. The result is that high rates of oxygen depletion occur at these levels and below, and that dissolved oxygen is often fully depleted (Gordon and Skelton 1977). Concentration of dissolved oxygen close to the bottom of such water bodies is often very low, even zero. To alleviate adverse effects downstream, methods have been developed for increasing dissolved oxygen concentrations in the turbine releases from hydroelectric power plants. Systematic procedures have been used for evaluating promising solutions to low DO concentrations. One of the more popular methods is oxygen injection through small-pore diffusers located upstream from the turbine intakes (Ruane et al. 1977). Laboratory tests were conducted on commercially available fine-pore diffusers to select those diffusers worthy of field testing. Laboratory tests included determination of bubble sizes and oxygen transfer efficiency. Field tests for oxygen transfer efficiency were performed for two different diffusers.

Total injection of air at the bottom of thermally stratified lakes and reservoirs might cause destratification, which can be quantitatively predicted by means of mathematical and physical models (Kranenburg 1979). The agreement between theory and model experiments is found to be satisfactory. Increasing the number of injection points was an effective means of speeding up the destratification process, but increasing the air flow rate was not. Another system of hypolimnetic aeration called sidestream pumping, which uses liquid oxygen and a conventional water pump, was also tested. It was shown that hypolimnetic oxygen concentration increased from less than 0.5 mg/l to over 8.0 mg/l during two months of operation, while thermal stratification was maintained. These improvements created a suitable habitat for coldwater fish (Fast et al. 1975a and 1975b; Fast et al. 1977). Another experimental program in turbine aspiration was conducted to develop techniques for improving the dissolved oxygen concentration of hydroelectric releases. The program, developed by Raney (1977), involved water tunnel modelling of aspiration systems, in addition to prototype installations on existing hydroelectric facilities.

As noted, many methods for increasing dissolved oxygen concentrations of turbine releases have been suggested. In order to make the choice of one among them, an economic analysis was conducted. Fast et al. (1976) compared some of the methods on the economic basis, and results are available.

Effects of aeration on water quality were examined also, particularly the effect of hypolimnetic aeration on nitrogen and phosphorus concentration (Garrel et al. 1977).

There have been a few attempts to relate some traditional trophic state indices, such as phosphorus concentration, chlorophyll-a concentration, and transparency, to hypolimnetic dissolved oxygen, which is of direct relevance to existing water quality standards, particularly for fisheries management (Walker 1979).

Data on limnological responses of impoundments to acid mine drainage are also available. Koryak et al. (1979) showed that in spite of only moderate vertical thermal gradients in the reservoirs, these inflows penetrate the impoundment as well-defined temperature-density currents. The depth of penetration and resulting mixing patterns depend on design and operation of the dam. The internal hydrodynamics of the reservoir, in turn, influences the chemistry and biology of both the impoundment and the outflow.

Nutrients There is such a wide variety of data on nutrients that it is difficult to classify them. Problems which are most often analyzed are related to potentially limiting nutrients, such as phosphorus, nitrogen, carbon, silicon, and trace metals. These problems include mechanisms of nutrient limitation of primary biological productivity, sources of nutrients and relative abundance, and regeneration and reuse of nutrients in ecosystems (Middlebrooks et al. 1976; Fuhs 1974). Some researchers involved in calculation of nutrient budgets have used the U.S. Geological Survey's flow-duration curve method developed for determining the long-term average sediment loads transported by streams. Kothandaraman and Evans (1979) found that controlling phosphorus at the point sources is a major step in avoiding accelerated aging of the lake. Among the possible sources of nutrients are river and lake sediments. Decomposition of benthic deposits may have adverse effects on the quality of natural waters by exerting an oxygen demand and by releasing organics and nutrients into the overlying water. The oxygen uptake rate and the nutrient release rate from lake and river sediments were measured in long-term experiments, and results are available (Fillos and Swanson 1975). On the basis of data, it can be concluded that the release of nutrients from natural sediments is controlled by physical-chemical reactions.

The sorption capacity for phosphorus of lake sediments depends on the oxygen concentration in the overlying water. Fillos and Biswas (1976) showed that in some lakes, when the phosphate concentration in the overlying water is high, above 1-2 mg/l, the sediments will sorb phosphorus from the overlying water under both aerobic and anaerobic conditions. Below this concentration, the sediments will release phosphorus under anaerobic conditions while they will sorb phosphorus under aerobic conditions.

There are data on vertical diffusion and nutrient transport for some tropical lakes. Robarts and Ward (1978) noted that during a period of 4.5 months when the water was stratified, nutrients were transported upward at a significant rate. The oxygen deficit rate can be correlated positively with phosphorus loading. Phosphorus budget studies have been conducted for Lake Michigan (Sager and Wiersma 1975; Sridharan and Lee 1974). Results indicated the major sources of phosphorus, but it was concluded that the relative importance of the major sources varied considerable on a seasonal basis. It was possible to find an empirical method for estimating the retention of phosphorus in lakes on the basis of the relationship between phosphorus retention and several other lake and watershed parameters. Kirchner and Dillon (1975) reported that the predicted and measured values were in close agreement (r=0.94). A direct method of predicting summer levels of total phosphorus was suggested by Jones and Bachmann (1976). A strong correlation was found between average July-August chlorophyll-a concentrations and total phosphorus concentrations (Chapra and Tarapchak 1976).

A model based on conservation of mass was developed to simulate total phosphorus budgets for the Great Lakes (Chapra 1977). Phosphorus loadings were generated from variables indicative of human development, such as population and land use. Loadings are input to a budget model that can be solved for total phosphorus concentration as a function of time. Projections indicated that if point-source effluents were reduced to 1 mg/l of total phosphorus, all the lakes will show some improvement in trophic condition.

The potential trophic benefits to lakes of a 1 mg/l of total phosphorus effluent standard for municipal sewage treatment plants were examined by using two phosphorus mass balance models (Gakstatter et al. 1978). The analysis included 255 lakes and reservoirs receiving municipal sewage treatment plant effluents and located in the eastern half of the United States. It was shown that 18-22 percent of the water bodies would benefit from a 1 mg/l effluent standard. If the requirement was zero phosphorus, 78 percent of the water bodies would benefit.

Some studies of seasonal and spatial distribution of nutrients have been conducted by Stewart and Markello (1974) and Gruendling and Malanchuk (1974). Results of measurements indicated that variations may reflect the geochemistry of the area, local differences in lake hydrology and mixing characteristics, and the supply of nutrients from soil, man, and precipitation. Specific values for nutrient loading derived from these calculations do indicate the relative importance of human contributions and provide an aid in management considerations.

Ecological Interaction All publications dealing with different ecosystems and interaction between limnological characteristics of lakes and reservoirs and flora and fauna are classified in this group.

In eutrophic lakes, the supply of light and the extent of turbulent mixing near the water surface often control the actual occurrence and disappearance of algal blooms (Stefan et al. 1976). While nutrients provide necessary material for a high growth potential, light and mixing provide energetic controls. It is possible to find an expression for an equilibrium phytoplankton concentration under given physical conditions. Also, effects of temperature on growth constants of some particular alga are examined, as well as response of lake algae to addition of nutrients, such as phosphate, nitrate, and silica, and to iron ore processing wastes. Based on changes in the rate of carbon fixation, phosphate stimulated approximately 80 percent of the cultures to which it was added, and nitrate and silica were non-stimulating. Iron ore processing wastes were found to be neither toxic nor stimulating (Plumb and Lee 1975).

Changes in temperature, nutrients, etc. can cause ecological changes, so monitoring of the environment may be very useful. Van Belle and Fischer (1977) noted that the data are assumed to consist of a species-frequency list collected at several sites, in order to enable statistical analysis. The major statistical problem with analyzing such lists is the establishment of an appropriate frequency distribution under some suitable hypothesis. After assuming an appropriate frequency distribution, statistical tests can be applied to determine whether biological communities or changes in biological communities are connected with ecological changes some distance apart from these communities. Wiederholm (1980) provides information on the rationale behind the use of biological variables in environmental monitoring of lakes, as well as the principles of variable selection and the limitations of data usability. Difficulties might arise in the use of saprobic or indicator organism systems in developing water quality criteria when organisms are not known at the species level. Generic identifications are generally useless, because particular species may vary widely in ecological tolerance (Resh and Unzicker 1975). Some authors suggest cyanophage analysis as a biological pollution indicator. The detection and control of enteric viruses in wastewater is a primary health concern. Present wastewater standards do not involve practical assays for, or require virus removal from, discharged effluents. Cyanophages are nonpathogenic and present throughout the year in wastewater. A proposed cyanophage detection procedure is a practical and inexpensive animal virus indicator and could serve as a replacement for the coliform test. A description of the procedure is available (Smedberg and Cannon 1976).

Hydraulics and Hydromechanics of Ponds, Small Lakes and Reservoirs

The literature reviewed under this heading deals primarily with the movement of water in ponds, small lakes, and reservoirs and the effects of hydraulic structures on the aquatic environment.

Hydrodynamics and the Effects of Geometry and Hydrology Circulation induced in a water body results primarily from wind stresses and horizontal temperature gradients. A three-dimensional model was developed by Uzzel and Ozisik (1977) for the prediction of steady-state circulation induced in the far field regions of shallow lakes. For the case of lakes with rectangular geometry and uniform depth, explicit analytical solutions for the velocity distribution are available. The result of the analysis shows that a negative temperature gradient along the lake in a given direction gives rise to a flow in the same direction at the upper layer of the lake and to a reverse flow at the lower layer. Circulations induced by horizontal temperature gradients may be as important as those generated by wind stresses.

The shear stress due to wind action on a shallow lake or lagoon creates a velocity distribution in the vertical direction. Some information is available on the relationships between vertical turbulent diffusion, surface reaeration and wind velocity, depending on the wind shear stress coefficient (Banks 1975). It is noted that publications on this topic are comparatively few in number.

The effects of both the inflow and outflow on the dynamics of a reservoir are described and a new Lagrangian one-dimensional model is proposed by Imberger et al. (1978) for deep hypolimnetic mixing. The model contains only four universal constants. Data from one complete annual cycle of the salinity and temperature distribution from a Western Australian reservoir are used to calibrate the model. A complete discussion of this model, DYRESM, appears in Fischer et al. (1979).

Effect of Hydraulic Structures This group of publications concerns defining what types of structures should be selected for study in view of the potential effect on the environment, and identifying how hydraulic structures are interrelated with their environment. Information on categories of hydraulic structures and their specific environmental effects are available. An example developed by King (1978) examines the effects of impoundments on downstream water quality.

The major hydraulic structures affecting the environment are hydroelectric power plants. They change the natural conditions of these water bodies through both short-term and long-term impacts. Therefore, changes in water quality, rates of erosion fluctuations in water temperature and water flow, and eutrophication must be studied in planning future hydroelectric installations. Descriptions of reported physical, chemical and hydrodynamic changes associated with dams are available, as well as relationships of these changes to fish and benthic communities. El-Shamy (1977) briefly considered several aspects of the impacts of the hydroelectric projects and changes in the lower Susquehanna River. The sites of four large dams were evaluated in light of disruption of the upstream runs of the previously existing migratory fishes. Changes in the structure of the benthic community were also analyzed.

Dams and impoundments may affect the migration of fish. This problem is recognized, and data on migrations of juvenile chinook salmon and steelhead from tributaries of the Snake River as far downstream as The Dalles Dam on the Columbia River are available (Raymond 1979). New dams constructed on the Snake River adversely affected survival and delayed migrations of juveniles. Significant losses of juveniles in 1972 and 1973 were directly responsible for record-low returns of adults to the Snake River in 1974 and 1975. Major causes of mortality were passage through turbines at dams, predation, delays in migration through reservoirs in low-flow years and prolonged exposure to lethal concentrations of dissolved gases caused by spilling at dams during high-flow years. It was found that mortality of chinook salmon and steelhead resulting from new dams differed with respect to area and cause. In order to avoid mortality, different fish-handling facilities were analyzed, such as angled screens and louvers (Taft and Mussalli 1978). They may be used for diverting fish at power plants. Louvers have been shown to be more than 90 percent effective in diverting a variety of fish species in both laboratory and prototype studies. Studies with angled traveling screens, conducted in 1978, showed that they are 100 percent effective in guiding fish to bypasses. The inherent difficulties of screens, i.e., the necessity of lowering the approach velocity, poor velocity distribution across the screen, and danger to fish, may be overcome by the installation of a perforated-pipe inlet with an added internal perforated sleeve, as concluded by Richards and Hroncich (1976). The benefits of such a device, i.e., relative ease of maintenance, uniform approach velocity, uniform inflow, and protection for fish, are greater than when ordinary physical screening is used.

Generally, the information available on hydromechanical behavior of small water bodies is sparse. Few models exist which are addressed specifically to questions of concern in hydrodynamic modelling of aquaculture ponds.

Water Quality of Ponds, Small Lakes, and Reservoirs and Its Effects on Fish

The impact of water quality characteristics and parameters such as temperature and dissolved oxygen on pond aquaculture are examined in this subsection.

Temperature In a study of pond temperatures in temperate and tropical climates, seasonal changes played major roles (Young 1975). Four different periods were recognized: periods of low, rising, high, and declining temperature. In the course of a year, the range between the lowest and highest average temperatures was greater in the temperate pond. The spans between the average minimum and the average maximum weekly water temperatures in the warmest months of the year in the tropical pond were greater than those found at any time of the year in the temperate pond. The average weekly air and water temperatures showed the same pattern of seasonal fluctuations. In the tropical pond, the average weekly air temperatures were always less than the average minimum weekly water temperatures, where in the temperate pond they were below, within, or above the spans between the average minimum and average maximum weekly aquatic temperatures, according to the time of the year. In both ponds, diurnal fluctuations were absent during the cooler months; the amplitudes of the fluctuations in the warmer months varied according to the time of year and were greater during the warmest months in the tropical pond. In both ponds, lowest temperatures were recorded sometime between 02:00 and 10:00 and highest between 12:00 and 20:00 hours. The influence of temperature on the life-cycle of fish was acknowledged and discussed.

In an assessment of the effects of stratification on the depth distribution of gizzard shad, white crappie, freshwater drum, and black bullhead (Gebhart and Summerfelt 1976), it was noted that they were markedly affected by conditions of stratification. During the period of greatest stratification, the distribution of all species except the bullhead was limited largely to oxygenated water above the hypolimnion. Fish depth distribution increased significantly after the fall overturn. Also, it was observed that the diet of channel catfish varied considerably with water temperature (McNeely and Pearson 1977).

Dissolved Oxygen Effects of low dissolved oxygen concentration on channel catfish were analyzed by Tucker et al. (1979) and results are available. Test fish were stocked in 0.02 and 0.04 hectare ponds without aeration at three rates (4,962, 10,007, and 20,385 fish per hectare) and fed daily. Each treatment was replicated six times. Maximum feeding rates of 34, 56 and 78 kg/ha, respectively, were reached by midsummer. At the lowest level of treatment, no dissolved oxygen (DO) problems occurred and survival was 99 percent. At the medium treatment level, the lowest recorded DO concentration frequently was below 2.0 mg/l and some fish suffocated during an oxygen depletion in one pond, but survival still averaged 93 percent. For the high level treatment, DO at dawn was usually below 2.0 mg/l in August and September. Fish mortalities resulted from DO depletion in three ponds at the high level of treatment, but survival averaged 83 percent. Nitrate-nitrogen and un-ionized ammonia never reached concentrations recognized to be lethal to channel catfish in any of the treatments. However, concentrations of un-ionized ammonia were possibly high enough to have adversely affected growth. Even though the average weight of individual fish decreased, harvest weight of fish increased from low to high treatment. An economic analysis was also conducted and results are available.

Consumption of oxygen by planktonic communities of pond waters may be predicted from chemical oxygen demand (COD) and temperature (Boyd et al. 1978). Correlation was found to be good (r^2 = 0.85) in ponds where planktonic organisms are the major source of turbidity. Secchi disc visibility may also be used to estimate COD (r^2 = 0.81) or consumption of O_2 by planktonic communities (r^2 = 0.82). To calculate nighttime declination in dissolved oxygen for channel catfish ponds, a computer simulation model was developed. This model incorporated data on plankton respiration from the study of the same authors and data from the literature on respiration by fish, respiration by organisms in the mud and O_2 diffusion. The validity of the model was tested for two catfish ponds. Measured and calculated DO concentrations usually agreed within + 10 percent. Secchi disc visibility and COD values may be used in the computer simulation to estimate O_2 consumption by plankton, depending on conditions in a particular pond. A series of tables was prepared which give the minimum acceptable Secchi disc visibility and the maximum permissible COD which may be tolerated in a particular pond without danger of DO depletion during the night. These tables will be useful to fish pond managers. A simple graphic technique for estimating DO concentrations is also suggested.

The daytime increase in dissolved oxygen concentration of pond water may be estimated from solar radiation, chlorophyll-a concentration, and percentage O_2 saturation at dawn, or from solar radiation, Secchi disc visibility, and percentage O_2 saturation at dawn (Romaire and Boyd 1979). A computer simulation model for predicting the effects of cloudy weather (low solar radiation) on DO depletion was developed for channel catfish ponds. The model combines daytime DO increase estimated on the basis of solar radiation data and Secchi disc visibility data and the nighttime decline in DO due to respiration by the plankton, fish and benthic communities. Results from the study demonstrated that the combination of dense plankton blooms and low levels of light intensity were closely related to low concentrations of DO.

Negative effects of low dissolved oxygen concentration can be avoided by aeration, as concluded by Hollerman and Boyd (1980). Channel catfish were stocked in 0.02 and 0.04 hectare ponds at 20,385 fish per hectare and fed daily. Six ponds were aerated nightly (two to six hours) and six ponds were not. In aerated ponds, no dissolved oxygen problems occurred and survival was 92 percent. In all unaerated ponds, DO depletion was observed and fish mortalities resulted; survival was 40 percent. Nitrite-nitrogen concentrations were significantly higher in the aerated ponds, but never at levels reported as lethal to channel catfish. Concentrations of un-ionized ammonia were high enough in both aerated and unaerated ponds to adversely affect growth. Results of an economic study are also available.

To examine the effectiveness of different techniques of emergency aeration, dissolved oxygen concentrations were depressed in 0.57 hectare fish ponds through algicide treatment (Boyd and Tucker 1979). The most effective device for emergency aeration was a paddlewheel aerator powered by a tractor. Different types of pumps were tested and results are presented. Their efficiencies were shown to be lower than that of the paddlewheel aerator.

Another procedure for raising dissolved oxygen concentration is to reduce phytoplankton density (Tucker and Boyd 1977). Treatment of channel catfish production ponds with biweekly applications of 0.84 kg/hectare copper sulfate was ineffective, but three periodic applications of simazine totaling 1.3 mg/l drastically reduced phytoplankton density. However, extended periods of low dissolved oxygen concentrations following simazine applications resulted in decreased fish yields and poor conversion ratios as compared with control ponds.

The catfish processing industry as a whole has been examined for its pollution potential. A single modern plant was studied by Mulkey and Sargent (1974) to determine the waste characteristics of in-plant and total effluent streams. It was found that volume and concentration of the wastewater may be reduced by more effective solids removal and recycling of process waters. Process changes and water flow modifications were also proposed.

Physico-chemical Parameter Data on seasonal and diurnal fluctuations of the physico-chemical parameters in temporary ponds are available. Physico-chemical parameters studied were pond volume, air, water and substratum temperature, pH, conductivity, and dissolved oxygen concentrations (Khalaf and MacDonald 1975). A study of unfertilized ponds in pastures and woodlands was also conducted. It was shown that unfertilized ponds in pastures are more eutrophic than unfertilized ponds in wooded locations. It was suggested that fertilizer application rates currently used in bass-sunfish ponds in some regions of the USA may be greatly reduced in many pasture ponds (Boyd 1977).

The effect of draining was also analyzed (Boyd 1976). Percentages of small particles and organic matter in pond muds increased along transects from shallow to deep water in four fish ponds which had never been drained. Gradients in texture and organic matter were similar to those reported by nearby ponds which had been drained periodically.

Toxics and Heavy Metals Combined effects of chemicals and various temperatures on aquatic organisms were studied by Cairns et al. (1975). The frequent association of power plant and other heated wastewater discharges with potentially toxic chemicals justifies an evaluation of these combined stresses on aquatic organisms. Substances such as ammonia, cyanide, trace metals, pesticides and herbicides, phenolics, and chlorine exert their toxic effects through a wide variety of mechanisms which, in turn, are influenced by temperature. Also of considerable interest are the effects of interaction between temperature and substances such as antibiotics on microorganisms. Information on these effects, available in the literature, may assist management personnel in setting standards and providing the background for designing new studies.

A critical evaluation of material and methods to be used for the study of behavioral responses of fish to polluted water was made by Larrick et al. (1978). In this connection, the feasibility of using aberrations in fish movement patterns in response to intoxication (measured by infrared light beam interruption) to monitor and detect pollution was investigated. Acute and sublethal toxic effects were deleted for copper, cadmium, cyanide, mercury, phenol, and ammonia. The system proved satisfactory for detecting phenols and ammonia, much better than previous monitoring systems based on fish respiratory responses to intoxication, but was less effective for the other toxicants investigated (Morgan 1979).

Effects of different toxic materials on aquatic organisms were examined by Katz (1977). In his study, 23 nitrogenous compounds were selected for the determination of chlorine dissipation in relation to toxicity prior to and after the addition of fish. These toxicities, which were much less than that of free chlorine, varied directly with the logarithm of fish mortality and the rate of chlorine dissipation in the presence of fish. Thus, it was concluded that nitrogenous compounds alter the toxicity of total residual chlorine.

The use of chlorine by the power industry for slime control in freshwater systems generally differs from the use of chlorine for wastewater disinfection, since it is used intermittently rather than continuously. A typical power plant will chlorinate three times a day with 0.5 to 0.75 mg/l of chlorine with a 15-to-30 minute deviation at each chlorination period. Although extensive research programs have been conducted evaluating the toxicity of continuous exposure to chlorine, limited literature is available to evaluate the toxicity of intermittent exposures to chlorine. A laboratory study designed to identify the relationship between frequency of exposure, concentration, duration of exposure, and the expression of lethality in fish, showed that total time of exposure was the most important parameter in predicting lethality (Dickson et al. 1977; Mattice and Zittel 1976).

Intermittent chlorination has different effects on different fish species. Data from laboratory studies of the behavior of different species of fish, exposed to either free chlorine or monochloramine, are available (Heath 1977). It was shown that temperature has relatively little effect on the toxicity of intermittent chlorine to the species tested. Also, it was shown that free chlorine was significantly more toxic than monochloramine, and it was suggested that water quality criteria for the protection of fish should, in the future, take this differential toxicity into consideration.

Some species may acclimate to chlorine. Acclimatization of fathead minnows and lake trout to residual chlorine and bromine chloride was studied by DeGraeve and Ward (1977). Chronic effects of nickel and hydrogen cyanide on the fathead minnows were examined, also. Nickel (as the chloride salt) at concentrations ranging from 1.6 to 0.08 mg/l in a continuous-flow chronic bioassay had no effect on survival or growth of first-generation fathead minnows (Pickering 1974). However, the number of eggs per spawning and the hatchability of these eggs were affected. Methods for estimating toxicity of nickel for other species of fish are suggested. Egg production rate was most sensitive when hydrogen cyanide was applied. The highest level of toxicant with no adverse effect was between 12.9 and 19.6 mg/l of hydrogen cyanide (Lind et al. 1977).

Fish eggs and larvae may suffer from dissolved ozone. To find concentrations of dissolved residual ozone lethal to fish eggs and larvae during brief exposures, Asbury and Coler (1980) performed continuous-flow toxicity tests with eggs of yellow perch and fathead minnows, eggs of white suckers, and larvae of bluegill sunfish. The 50 percent and 99 percent lethal concentrations with confidence limits were calculated. Eggs of the species tested were more tolerant than larvae. It was concluded that because of the sensitivity of the larvae, residual ozone concentrations in natural waters should remain well below 50 mg/l (or micrograms per liter).

Examinations of the uptake rates by some fish species of methylmercury, biosynthesized in sediments, were conducted by Shin and Krenkel (1976) to find the effects of various environmental conditions, such as temperature, degree of organic pollution, chloride ion concentration and degree of mercury pollution. An attempt was made to quantify the effects on the overall methylation activities of microorganisms of varying environmental parameters such as temperature, BOD, Cl^-, degree of mercury pollution and pH. A rate equation and reaction order were introduced. It was found that for the conditions examined, the methylation reaction was first order with respect to the molar concentration of mercury. Rate constants of the methylation reactions of mercury and the half-lives of mercury contained in the sediments under given conditions were determined.

The locomotor response of goldfish to a steep gradient of copper ions was examined in Westlake et al. (1974).

<u>Ecological Interactions</u> Some general publications dealing with small lakes, reservoirs, and ponds exist for different regions of the world. These publications discuss some aspects of inland fisheries, such as biological monitoring with respect to detecting changes in the aquatic biota, its use as a basis for the classification of waters, and also as a guide to potential fish production (Beadle 1974; Salanki and Ponyi 1975; Alabaster 1977; Lowe-McConnel 1975).

Primary production of fish ponds was analyzed by Jana (1979). Estimates of primary production in relation to bacterioplankton growth kinetics were made for three fish ponds with mono- and polyculture in the same region of the USSR. Because the basic process of primary production is dependent on conditions such as temperature, light, and nutrients rather than on fish species, almost similar production rates were obtained in all three ponds. A maximal gross production, coupled with bacterioplankton peak was recorded in July with the rise of temperature of the water, while minimal production was recorded in September.

It has been noted by Boyd et al. (1978) that during windy weather in March and early April, dense populations of blue-green algae frequently develop in fish ponds in Alabama. Massive die-offs of this algae occur during prolonged periods of calm, clear, warm weather between mid-April and mid-May. Dissolved oxygen concentrations decline following die-offs, and fish kills may result.

Data on different ecosystems, such as pond ecosystems in India, prairie lakes in Nebraska, a salmon lake on Kamchatka Peninsula, a pond in Canada and a pond in the United Kingdom are available (Munawar 1974, 1975; Sreennasan 1974; Schwartzkopf and Hergenzader 1978; Sorokin and Paveljeva 1978; Kwei 1977; Idso and Foster 1974). Parameters usually involved are alkalinity hardness, electrical conductivity, pH, Secchi disc turbidity, temperature, and chlorophyll-a concentrations. Descriptions are provided to indicate how these parameters affect primary production of the ecosystem, vertical distribution of fish species, phytoplankton, standing crops, etc. Effects of rainfall, evaporation, tides, solar radiation, etc., are also examined.

An important effect on ecosystems is the activity of man. One of the major disturbances of aquatic ecosystems is caused by merely damming a river. The relative compositions of the major fish families in Lake Kainji in Nigeria between 1974 and 1975 were analyzed by Blake (1977) and Lewis (1974) and compared with pre- and post-impoundment of Niger River data. The dominant commercially valuable fish was changed and data on other related changes are also available. A similar process is recorded after damming the Volta River in Africa by Petr (1974).

Another major activity of man affecting ecosystems is the discharge of thermal effluents. Data on this effect on Lake Wabamun, Alberta, Canada, are reported by Hickman (1974).

Also, artificial destratification of lakes may affect ecosystems (Wilhm and McLintock 1978). Changes in species composition and diversity of benthic macro-invertebrates during summer and fall were compared in an area of an artificially destratified lake and in an area which was not destratified. Numbers of species, diversity, and density were significantly correlated with the concentration of dissolved oxygen, while none of the biotic variables were correlated with temperature.

Comment A substantial body of literature on water quality in lakes and reservoirs exists, but little deals directly with the problems of very small water bodies such as ponds. Basic water quality-ecological interactions in such systems are fairly well known and appear to be transferable to aquaculture ponds for the purpose of model development.

Models of Aquatic Systems

For ease of analysis, the review of the models of aquatic systems is organized into subsections dealing with various types of one-, two-, and three-dimensional models. Nutrient cycle and aquaculture models are considered in the final two subsections.

One Dimensional Temperature Models Simulation of temperature variations in deep, thermally stratified reservoirs or lakes has been successfully accomplished by application of the one-dimensional advection-diffusion equation and the heat energy conservation equation. Two mathematical models of very similar structure, one developed by Water Resources Engineers, Inc. (1967) and the other by the Parsons Laboratory at MIT (Huber et al. 1972), are most widely used today. Both have utilized the heat energy budget calculation procedures developed by the Engineering Laboratory of TVA (1972). Solution techniques for the models are different--one is explicit, the other implicit--but the results are comparable, generally in good agreement with prototype observations. Both models are well documented.

Swensson (1978) suggested a numerical procedure for the prediction of the thermal structure of the ocean and lakes. The mathematical formulation of this problem is based on the heat energy equation in its one-dimensional form. A turbulence model provides a means of calculating the turbulent transport coefficient for heat. Two momentum equations for the mean flow are also solved, as the turbulence model requires information about velocity gradients. Through the process of verification, it is shown that the thermal structure of lakes is predicted very accurately. This is demonstrated by a comparison of measured and predicted seasonal thermal structure of Lake Velen in Sweden.

One Dimensional Water Quality Models Modelling of other water quality characteristics of reservoirs, a logical extension of thermal modelling, was carried out by both the WRE and MIT groups in the early 70's. A water quality-ecological model of stratified impoundments was proposed originally by Chen in 1968 and led to the development under OWRR and EPA sponsorship of the model LAKECO (Chen and Orlob 1975). This model, which includes some 22 different state variables, both biotic and abiotic, has been applied successfully by the developers and government agencies in simulations of water quality changes in Lake Washington and a variety of artificial impoundments.

The MIT group extended their temperature model to include simulation of DO and BOD and demonstrated its application on Fontana Reservoir in the TVA system (Markofsky and Harleman 1971). An improvement in the solution technique for one-dimensional water quality models for impoundments was the introduction by Baca and Arnett (1976) of the finite element method. The resulting model, which has most of the capabilities of LAKECO, avoids problems of numerical mixing, instability, and adapting to steep gradients.

A recent development which is uniquely different than previous modelling approaches is the Lagrangian slab model of Imberger et al. (1978). Based on budgeting turbulent kinetic energy in horizontal slabs of varying thickness, the model is capable of simulating simultaneously both temperature and salinity distributions in small to medium-sized reservoirs. The model utilizes only four "universal" constants that can be evaluated from field performance.

Two-Dimensional Circulation Models Excellent reviews of modelling wind-driven circulation in lakes are provided by Simons (1973) and Cheng et al. (1976). These show that a well-developed capability exists for modelling two-dimensional, vertically mixed impoundments. It is represented in the simplest form by so-called single-layer, storm-surge models such as have been applied to the Great Lakes (Platzman 1963) and to shallow coastal seas (Hansen 1962; Reid and Bodine 1968). Models of this type are capable of providing approximate descriptions of circulations induced by winds and pressure differences associated with major storm events.

Improved descriptions of circulation in shallow systems are available through application of models for vertically mixed estuaries, like those of Leendertse (1967), Brandes and Masch (1971), Waldrop and Farmer (1973), and Codell (1973). Based on the phenomenological equations of motion in two-dimensions, such models are capable of providing good representation of current velocities and water surface elevations under highly unsteady conditions related to wind, tide, or hydrologic fluxes. The models are orthogonal, using central finite differencing, usually of an alternating direction or leapfrog type. Both implicit and explicit schemes are employed. Models of this type are well documented and widely applied.

An attractive alternative for two-dimensional, vertically mixed systems is a finite element model developed by King et al. (1973). The model utilizes elements of triangular, rectangular, or curved configuration and of variable size to represent irregular boundaries and topographic features of the prototype. Liggett (1969), Gallagher (1975), and Cheng et al. (1976) have demonstrated the potential of the method for shallow lake systems.

Multilayer Impoundment Models Density stratification induced by temperature, salinity, or suspended solids requires more rigorous mathematical representation of the hydromechanics of impoundments. The multilayer lake models of Simons (1973), referred to by Cheng et al. (1976) as Type I, are most representative. They have been applied with reasonable success to several of the Great Lakes and to Lake Vanern in Sweden (Simons et al. 1977). In the latter application it was recognized that model performance was most sensitive to empirically determined coefficients, e.g., eddy diffusivities, and that reduction in model detail and conceptual elegance may be justified as a trade-off against costs of computation.

Two-dimensional stratified flow models designed to simulate the hydrodynamic behavior of long, narrow impoundments are represented by the finite element model, RMA2, developed by King et al. (1975) and by a finite difference model under development by Edinger and Buchak (1975). Such models have been tested primarily against laboratory experiments and, in the case of RMA2, adapted for simulation of estuarial systems. The lack of prototype data has prevented direct verification of model capability.

Two- and Three Dimensional Water Quality-Ecologic Models A broad range of publications deal with modelling of aquatic ecosystems, both in theory and with applications. It is possible to find descriptions of the application of mathematical modelling techniques to practical problems of water quality management of lakes, rivers, and estuaries. Pertinent questions concerning the environmental impact and control of effluents are explored through presentation of examples where models have been used. Some studies contain results or ideas not currently being applied by practitioners, but which have potential--and which may be part of the next generation of models (Canale 1976). Publications dealing with how biological concepts can be formulated in mathematical terms and how the resulting formulations are used in research of ecosystems are available also (Gold 1977). Problems considered in these works are: the basic concepts of system states, variables and parameters, system decomposition, graphic representation of relations between variables, input-output relations, scale dimensions and similarity as related to biological systems, the concept of probability, the translation of a conceptual model, and the relationship of the model to the real world system.

The vertical one-dimensional assumption of the earlier lake and reservoirs models was examined with respect to its validity for mathematical ecological models (Ford and Thornton 1979). An analysis of time and length scales characterizing the hydrodynamics, chemistry, and biology shows that the two scales are necessarily coupled and that their interaction dictates both upper and lower bounds for the lake size that can be described by a one-dimensional model. According to Niemeyer (1978), an efficient procedure for solving an arbitrary number of coupled vertically integrated transport equations can be achieved by using some features of both finite difference and finite element methods.

Modelling of the transport and dispersion of conservative substances in shallow lakes is represented in a model by Lam and Simons (1976) that has been demonstrated on Lake Erie. Nonconservative substances, including phytoplankton and nutrients, have been treated in a model of Green Bay by Patterson et al. (1975) and a phytoplankton productivity model of Western Lake Erie by DiToro et al. (1975). In each of these examples the models are driven by a flow field derived either from field observation or a companion circulation model. The Lam and Simons model treated the lake system as either vertically mixed (one layer) or stratified (two layers), while the other models assumed vertical homogeneity.

The phenomena of eutrophication have been modelled by the gross-nutrient budget approach advanced first by Vollenweider (1969) and later extended by Snodgrass and O'Melia (1975), Larsen and Mercier (1975), and Bella (1973). More complete representations of the two-and three-dimensional and temporal variations of nutrient-biota interactions in lake systems are exemplified by the models of Thomann et al. (1975) and Chen et al. (1975) of Lake Ontario.

Three different models were envisioned by Thomann, ranging from a simple three-layer (epilimnion, hypolimnion, benthos) model to one with 7 layers and 67 segments and up to 15 variables. Only LAKE 1, the three layer model, was verified against prototype measurements.

The Chen Lake Ontario model considered some 15 different classes of biotic and abiotic substances, and considerably expanded biological compartmentalization from that of his earlier model, LAKECO. The lake was represented by 41 surface elements and 7 layers, a total of 109 discrete physical units. The model was tested to demonstrate functional capability but was not verified.

CLEANER, a comprehensive lake ecological model with 34 state variables, is a product of the International Biological Program by researchers at Renssalaer Polytechnic Institute (Park et al. 1974). It conceptualizes the "lake" as a one meter-square column which may be divided into as many as ten cells to provide vertical resolution. The model is capable of simulating, for both the epilimnion and hypolimnion, three types of each of the following groups: phytoplankton, zooplankton and fish. Also, it is possible for each layer to simulate dissolved inorganic nitrogen, phosphate, dissolved and suspended organic matter, dissolved oxygen, and decomposers. The model has been applied to a variety of prototype situations including Lakes George and Sarasota in New York and several lakes in Europe, and it has been found that it gives simulations which are reasonable when compared with available data. Later, the model CLEANER was generalized by Youngberg (1977) so that it could be applied to reservoirs, well-stratified lakes, shallow lakes, and high mountain lakes. Potentially, the model may be applied in studying problems of eutrophication, fisheries management, and thermal pollution. CLEANER was evaluated and adapted in cooperation with scientists in several leading European hydrobiological laboratories, and it has proved to be useful in representing highly diverse limnological conditions. Submodels are included for rotifers, benthic algae, stratification, wave agitation, reservoir interflow, and throughflow. A procedure for sensitivity analysis was developed. The results were helpful in evaluating the model and in setting error ranges on the simulations. They also can be used to indicate the precision required in measuring input data.

On the other hand, some limitations of models must be noted. Some of the problems encountered in the design of a predictive water quality model for a large lake are discussed with reference to Lake Ontario and within the framework of a simple dynamic phosphorus model which incorporates the essential mechanisms of complex plankton models. By fitting the model to a seasonal data base, Simons and Lam (1980) and Babajimopoulos and Bedford (1980) have shown that equally satisfactory simulations are obtained with a different value of parameters and regardless of conditions of annual periodicity imposed on the solution. By comparison of model output with long-term observations, it is demonstrated that seasonal verification studies of dynamic models by themselves are not sufficient to confirm the utility of such models for predicting long-term trends.

The state of the art of water quality-ecological modelling is probably fairly well represented by the phytoplankton productivity models of DiToro et al. (1975) and Thomann et al. (1975), the water quality-ecological models of Chen (1970), Chen and Orlob (1975) and Chen et al. (1975). CLEANER, the BPI-RPI model (Park et al. 1974) adds rigor to biological characterization of impoundments. It appears that model sophistication has gone somewhat beyond understanding of prototype behavior.

Algae-Nutrient Cycle Models It is concluded by Tapp (1978) that eutrophication analysis can be performed with both simple and complex models. The condition of the reservoir was examined for both types and for cases of 90 percent and 99 percent point source removal. Light and temperature are probably the most important parameters determining the process of eutrophication.

Formulations have been developed to model the adaptation of phytoplankton to changing conditions of light and temperature. The photosynthetic response to the interaction between these two parameters is also represented in the work of Groden (1977). Adaptive constructs have been made in the ecosystem model CLEANER, with the result that they increase the long-term applicability of that model under changing conditions, as well as the possibility of applying it to sites with differing conditions. A submodel for internal nutrients, critical to prediction of phytoplankton growth dynamics, has been developed by Desormeau (1978). Constructs were designed to represent storage of nutrients by separate mechanisms for nutrient uptake and assimilation, multiple nutrient limitation by threshold hypothesis, and light dependent photorespiration. The model was calibrated and validated on the basis of data for lakes in Norway and Austria.

The use of the catastrophe theory to obtain a predictive model of eutrophication phenomena is examined and applied to the case of small, highly eutrophic ponds by Duckstein et al. (1979). Phytoplankton dynamics were represented by a nonlinear differential equation derived on the basis of phenomenological considerations. This equation can be reduced analytically to the reasonable form for a cusp catastrophe model. Numerical solution of the dynamic equation is calibrated on the basis of data taken during a phytoplankton bloom in small ponds recorded earlier. Implications for fish ponds are given and possible use of the catastrophe theory approach to model algal succession is discussed.

Some models usable for predicting the effects of changes in environmental factors, such as nutrient inputs in aquatic ecosystems, have also been developed (Slawson and Everett 1976). A methodology for distributing primary productivity measurements among segments of the algal population has been formulated and tested. Such segmentation is necessary to develop a biological population model that allows consideration of biotic diversity, food web stability, and ecosystem complexity. Total algal productivity rates were adequately approximated as the sum of the activity of each algal division.

Some very specific models are also available. A three-compartment mathematical model was developed by Grenney et al. (1974) to represent a phytoplankton population having the capacity for intracellular storage of nitrogen in a nitrate-limited environment. Model coefficients were estimated from data reported in the literature. By incorporating intracellular storage in the model, an organism's ability to compete at any given time was based not only on current environmental conditions, but also on the past history of nutrient uptake. The model was used to demonstrate the effect of environmental fluctuations on the competitive ability of two phytoplankton populations.

A mathematical model was developed to simulate the impact of storm loads on phosphorus, fecal coliform, and dissolved oxygen concentrations in eutrophic lakes and determine the need for storm overflows (Freedman et al. 1980). Results obtained by the model showed that combined sewer and storm loads have a significant impact on lake fecal coliform, but little effect on phosphorus and dissolved oxygen concentrations. Observed variations in dissolved oxygen concentrations of the lake under consideration were caused by changes in chlorophyll-*a*, light, and wind. As a consequence of the modelling analysis, a limited control program for combined sewer overflows was designed, which included only disinfection and removal of objectional solids. Reduction in storm loads of nutrients and BOD would not provide any significant improvements in water quality of analyzed lakes and were not recommended.

A model for evaluating the fate of toxic substances in fish was developed by Leung (1978). Values of important parameters are included and simulated values for methoxychlor and DDT are compared to available data.

Some models have been formulated to simulate fish biomass dynamics, as well as interacting fish species on the basis of food competition (Hackney and Minns 1974; Kitchell et al. 1974: Thornton and Lessem 1978).

Aquaculture Management Models A very limited number of papers are available that deal with modelling for optimum and maximum yield, the implications of these concepts for commercial and recreational fisheries and their national and international applications. Few, if any, appear of much value for pond aquaculture.

Most hatchery management decision problems are multilevel and very complex, involve interactions between different areas of analysis, and are often incorrectly defined (Tomlinson and Brown 1979). Techniques of analysis most commonly applied, such as operations research, cost benefit analysis, econometrics and Bayesian analysis are valuable, but limited. In complex decision problems, decision makers often try to minimize potential regrets and maximize utilities along certain limited and separable measures of benefits at the expense of overall optimality.

Regression analyses have been used for over twenty years to develop useful statistical relationships between estimates of fish catch and various members of a set of abiotic and biotic variables. Empirical predictions of fish yield, based on regression analyses, exist for large North American lakes (Matuszek 1978). Economic models of fisheries are available for a number of different regions of the world: the Gulf of Mexico (Galveston Bay, Texas), Zambia, Canada, and some tropical lakes in Africa and India (Sheehan and Russell 1978; Melack 1976; Toews and Griffith 1977; Grant and Griffin 1979). What all of these models have in common is their empirical structure. Usually they are based on regression analysis with some additional parameters such as primary productivity, standing crop, or gross photosynthesis. Their use in pond aquaculture appears limited.

Special Models A number of useful special models of impoundments were encountered. These included models for routing runoff through the Great Lakes, simulating conservative water quality in reservoir systems, estimating sediment accumulation in reservoirs, evaluating candidate reservoir systems, calculating bank storage adjacent to reservoirs, and predicting the effects of landslides into reservoirs. None of these were considered of much value in the present research.

Comment The ecological models that have been applied to lakes and stratified reservoirs, like LAKECO, CLEANER, MS. CLEANER, etc., have been constructed for the most part following the classical Latka-Volterra equations, with mass and energy conservation principles observed. Adequate hydro-mechanical descriptions are usually assumed rather than simulated. Nevertheless, these models appear to provide a sufficiently solid base for construction of an aquaculture pond model, once the specifics of pond behavior are sufficiently well understood. Field data supporting conceptual notions of pond circulation, nutrient-nekton interactions, the importance of anoxia, toxic effects, etc. are presently scarce.

DISCUSSION OF FINDINGS

The literature review represented by this report resulted in identification of more than 230 references relevant to the principal topic, "Modelling the Hydromechanical and Water Quality Responses of Aquaculture Ponds." However, the great majority of these, as might be expected, focused on details of the behavior of lakes and reservoirs and their hydrochemical and biological properties, and very few articles dealt specifically with mathematical modelling of aquaculture systems, per se.

While the source material will be valuable in the development of pond models, it appears that much is yet to be learned concerning the quantitative responses of small water bodies that are deliberately designed for aquaculture. In particular, there is a dearth of information on the hydraulics of ponds, e.g., effects of natural and man-induced circulation, mixing phenomena, stratification, and detention time. Some guidance is available in the experience of model building for large lakes and reservoirs, but questions of scale are raised, e.g., can this experience be transferred to the level of shallow pond systems? Indications are that it can, with appropriate attention to the fundamentals of hydroscience. The background provided in performance studies of detention basins for water and wastewater treatment should be useful as a starting point.

Water quality studies of small lakes, reservoirs and, in some cases, actual aquaculture ponds, provide a fair basis for qualitative characterization of water quality responses of pond systems. However, in this instance, also, experience is limited insofar as quantitation is concerned. Moreover, most studies deal only with a few constituents, for example, DO, BOD, or toxics, under unique circumstances; few consider the quality picture in a comprehensive way; that is, water quality-ecological interactions are not described quantitatively. More particularly, little attention has been given to temporal characterization of water quality in ponds, during the natural growth cycles of the cultivated species.

An area of obvious import to the efficiency of aquaculture is the interaction between the active aerobic strata in the upper part of the pond and the relatively quiescent benthos. Since this lower zone receives the excess of applied nutrient as particulate organic matter plus excreta, it is sure to be highly active biologically. What role does it play, positively or negatively in the production of the primary species? What of the effects of anoxia? Are these effects mitigated by natural convective cycles or must they be adjusted by deliberate manipulation of the ponds? Such questions, which depend on a better knowledge of pond water quality and its relation to the indigenous ecosystem, are in need of answers not yet found in the literature. They will be necessary in development of a suitable pond model.

Finally, as regards the pond ecosystem itself, most attention has been given to the primary species (for obvious reasons) at the expense of a more comprehensive treatment of the entire ecosystem. Certainly there are other species and trophic levels involved, and while these are usually secondary as far as the goal of aquaculture is concerned, the rest of the ecosystem may impact production of the primary species through control of habitat. An example is the growth of algal species which may restrict light penetration, thereby changing the overall heat energy balance, thus affecting stratification and mixing, enhancing benthic activity, giving rise to anoxia, etc. Under certain circumstances, the role of this lower trophic level may strongly influence productivity. The same argument may be made for the roles of bacteria and zooplankton.

Most aquaculture models tend to deal with the primary species directly and with the rest of the ecosystem indirectly, as a modifying effect, often assessed empirically. There is a need, the authors believe, to structure a pond model to include all these elements of the system, including water quality and the hydrodynamics that may under some conditions actually "drive" the model. At least a capability to consider quantitatively all relevant processes and to determine sensitivity is needed. A mathematical model suited for this purpose is indicated. Its development is the logical next step in the line of research.

LITERATURE CITED

Alabaster, J.S. (ed). 1977. Biological Monitoring of Inland Fisheries. Applied Science Publishers Ltd., Barking, Essex, England.

Aleyev, Y.G. 1977. Necton. Dr. W. Junk, The Hague, The Netherlands.

Asbury, C., and R. Coler. 1980. Toxicity of dissolved ozone to fish eggs and larvae. Jour. of Water Poll. Cont. Fed. 52:1990.

Babajimopoulos, C., and K.W. Bedford. 1980. Formulating lake models which preserve spectral statistics. Jour. of the Hydraulics Div., ASCE 106(HY1):1-19.

Baca, R.G., and R.C. Arnett. 1976. A finite element water quality model for eutrophic lakes. Proc. Intnl. Conf. in Finite Element in Water Res., July, Princeton Univ., Princeton, NJ. 27 pp.

Banks, R.B. 1975. Some features of wind action on shallow lakes. Jour. Environ. Engineer. Div. ASCE 101(EE5):813-827.

__________. 1976. Distribution of BOD and DO in rivers and lakes. Jour. Environ. Engineer. Div, ASCE 102(EE2):265-280.

Bayly, I.A., and E. Williams. 1973. Inland Waters and Their Ecology. Longman, Australia.

Beadle, L.C. 1974. The Inland Waters of Tropical Africa, An Introduction to Tropical Limnology. Longman Group, London.

Bedford, K.W., and C. Babajimopoulos. 1977. Vertical diffusivities. Jour. Environ. Engineer. Div., ASCE 103(EE1):113-125.

Bella, D.A. 1973. Computer simulation of eutrophication. Oregon State University, Corvallis, Dept. of Civil Engineering, Final Report. 14 pp.

Biswas, S. 1977. Thermal stability and phytoplankton in Volta Lake, Ghana. Hydrobiologia 56(3):195-198.

Blake, B.F. 1977. Lake Kainji, Nigeria: A summary of the changes within the fish population since the impoundment of the Niger in 1968. Hydrobiologia 53(2):131-137.

Boyd, C.E. 1976. Water chemistry and plankton in unfertilized ponds in pastures and in woods. Trans. Amer. Fish. Soc. 105(5):634-636.

__________. 1977. Organic matter concentrations and textural properties of muds from different depths in four fish ponds. Hydrobiologia 53(3):-277-279.

__________, J.A. Davis, and E. Johnston. 1978. Die-offs of the blue-green alga, Anabena variablis, in fish ponds. Hydrobiologia 61(2):129-133.

Boyd, C.E., R.P. Romaire, and E. Johnston. 1978. Predicting early morning dissolved oxygen concentrations in channel catfish ponds. Trans. Amer. Fish. Soc. 107(3):484-492.

__________, and C.S. Tucker. 1979. Emergency aeration of fish ponds. Trans. Amer. Fish. Soc. 108:299-306.

Brandes, R. and F.D. Masch. 1971. A Slowly Varying Conservative Transport Model for Shallow Estuaries. Technical Report HYD 12-7103. Hydraulic Engineering Laboratory, University of Texas, Austin.

Brezonik, P.L., and J.L. Fox 1974. The limnology of selected Guatemalan Lakes. Hydrobiologia 45(4):467-487.

Cairns, J. Jr., A.G. Heath, and B.C. Parker. 1975. Temperature influence on chemical toxicity to aquatic organisms. Jour. Water. Poll. Cont. Fed. 47:267.

Canale, R.P. (ed). 1976. Modeling Biochemical Processes in Aquatic Ecosystems. Ann Arbor Science Publ., Inc., Ann Arbor, MI.

Chapra, S.C. 1977. Total phosphorus model for the Great Lakes. Jour. Environ. Engineer. Div., ASCE 103(EE2):147-161.

__________, and S.J. Tarapchak. 1976. A chlorophyll-a model and its relationship to phosphorus loading plots for lakes. Water Resources Res. 12(6):1260-1264.

Chen, C.W. 1970. Concepts and utilities of (an) ecologic model. Jour. San. Div. ASCE 95(SA5):Paper No. 7602, October.

__________, M. Lorenzen, and D.J. Smith. 1975. A Comprehensive Water Quality-Ecologic Model for Lake Ontario. Rep. to Great Lakes Env. Res. Lab., NOAA, by Tetra Tech, Inc., October. 202 pp.

__________, and G.T. Orlob. 1975. Ecologic Simulation of Aquatic Environments, pp. 475-588. In Systems Analysis and Simulation in Ecology, Vol. III, Academic Press, Inc., New York.

Cheng, R.T., R.M. Powell, and T.M. Dillon. 1976. Numerical models of wind driven circulation in lakes. Appl. Math. Modeling 1:141-159. December.

Codell, R.B. 1973. Digital Computer Simulation of Thermal Effluent Dispersion in Rivers, Lakes, and Estuaries. Army Missile Research Development and Engineering Lab., Redstone Arsenal, Alabama. Tech. Rep. RS-74-16, November 5. 227 pp.

Cooper, R.C., D. Jenkins, and L.Y. Young. 1976. Aquatic Microbiology Laboratory Manual. Association of Environmental Engineering Professors, Dept. of Civil Engineering, University of Texas, Austin.

Dasmann, R.F., J.P. Milton, and P.H. Freeman. 1973. Ecological Principles for Economic Development. John Wiley and Sons, New York.

DeGraeve, G.M., and R.W. Ward. 1977. Acclimation of fathead minnows and lake trout to residual chlorine and bromine chloride. Jour. of Water Poll. Cont. Fed. 49:2172-2178.

Desormeau, C.J. 1978. Mathematical Modeling of Phytoplankton Kinetics With Application to Two Alpine Lakes. CEM Rept. 4, Center for Ecological Modeling, Troy, New York.

Dickson, K.L., J. Cairns, Jr., B.C. Greggs, D.I. Messenger, J.L. Plafkin, and W.H. van der Schalie. 1977. Effects of intermittent chlorination on aquatic organisms and communities. Jour. of Water Poll. Cont. Fed. 49:35.

DiToro, D.M., D.J. O'Connor, R.J. Thomann, and J.L. Mancini. 1975. Phytoplankton-Zooplankton Nutrient Interaction Model for Western Lake Erie, pp. 423-474. In B.E. Patten, (ed), Systems Analysis and Simulation in Ecology, Vol. III. Academic Press.

Driver, E.A., and I.G. Peden. 1977. The chemistry of surface water in prairie ponds. Hydrobiologia 53-(1):33-38.

Duckstein, L., J. Casti, and J. Kempf. 1979. Modeling phytoplankton dynamics using catastrophe theory. Water Resources Res. 14(5):1189-1194.

Eccles, D.H. 1974. An outline of the physical limnology of Lake Malawi (Lake Nyasa). Limnol. and Oceanogr. 19(5):730-742.

Edinger, J.E., and E.M. Buchak. 1975. A Hydrodynamic, Two-Dimensional Reservoir Model - The Computational Basis. Rep. to U.S. Army C of E, Ohio River Division, Contr. No. DACW27-74-C-0020 by J.E. Edinger Assoc., Inc., September. 99 pp.

El-Shamy, F.M. 1977. Environmental impacts of hydroelectric power plants. Jour. of Hydraulics Div., ASCE 103(HY9):1007-1020.

Esch, G.W., and R.W. McFarlane. 1976. Thermal Ecology II. Technical Information Center, Energy Research and Development Administration.

Fast, A.W., V.A. Dorr, and R.J. Rosen. 1975a. A submerged hypolimnion aerator. Water Resources Res. 11(2):273-293.

__________, M.W. Lorenzen, and J.H. Glenn. 1976. Comparative study with costs of hypolimnetic aeration. Jour. Environ. Engineer. Div., ASCE 102(EE6):1175-1187.

__________, W.J. Overholtz, and R.A. Tubb. 1975b. Hypolimnetic oxygenation using liquid oxygen. Water Resources Res. 11(2):294-299.

__________, W.J. Overholtz, and R.A. Tubb. 1977. Hyperoxygen concentration in the hypolimnion produced by injection of liquid oxygen. Water Resources Res. 13(2):474-476.

Fillos, J., and H. Biswas. 1976. The release of nutrients from river and lake sediments. Jour. of Water Poll. Cont. Fed. 47:1032.

_______, and W.R. Swanson. 1975. The release of nutrients from river and lake sediments. Jour. of Water Poll. Cont. Fed. 47:1032.

Fischer, H.B., E.J. List, and R.C.Y. Koh, J. Imberger, and N.H. Brooks. 1979. Mixing in Inland and Coastal Waters. Academic Press, NY. 477 pp.

Ford, D.E., and K.W. Thornton. 1979. Time and length scales for the one-dimensional assumption and its relation to ecological models. Water Resources Res. 15(1):113-120.

Forsyth, D.J., and H.S. McCall. 1974. The limnology of a thermal lake: Lake Rotowhero, New Zealand: II. General biology with emphasis on the benthic fauna of chizonomids. Hydrobiologia 44(1):91-104.

Freedman, P.L., R.P. Canale, and J.F. Pendergast. 1980. Modeling storm overflow impacts on eutrophic lakes. Jour. Environ. Engineer. Div. ASCE 100(EE2):335-349.

Fuhs, G.W. 1974. Nutrients and aquatic vegetation effects. Jour. Environ. Engineer Div. ASCE 100(EE2):269-277.

Gakstatter, J.H., A.F. Bartsch, and C.A. Callahan. 1978. The impact of broadly applied effluent phosphorus standards on eutrophication control. Water Resources Res. 14(6):1155-1158.

Gallagher, R.H. 1975. Finite element lake circulation and thermal analysis, pp. 119-131. In R.H. Gallagher, J.T. Oden, C. Taylor, and O.C. Zienkiewicz, (eds), Finite Elements in Fluids, Vol. 1. J. Wiley, New York.

Garrel, M.H., D. Kirschner, and A.W. Fast. 1977. Effects of hypolimnetic aeration of nitrogen and phosphorus in a eutrophic lake. Water Resources Res. 13(2):243-347.

Gebhart, G.E., and R.C. Summerfelt. 1976. Effects of stratification on depth distribution of fish. Jour. Environ. Engineer. Div. ASCE 102(EE6):1215-1228.

Gold, H.J. 1977. Mathematical Modeling of Biological Systems - An Introductory Guidebook. A Wiley Interscience Publication, John Wiley and Sons, Somerset, NJ

Gordon, J.A., and B.A. Skelton. 1977. Reservoir metalimnion oxygen demands. Jour. of the Environ. Engineer. Div. ASCE 103(EE6):1001-1011.

Grant, W.E., and W.L. Griffin. 1979. A bioeconomic model of the Gulf of Mexico shrimp fishery. Trans. Amer. Fish. Soc. 108:1-13.

Grenney, W.J., D.A. Bella, and H.C. Curl, Jr. 1974. Effects of intracellular nutrient pools on growth dynamics. Jour. Water Poll. Cont. Fed. 46:1751.

Groden, T.W. 1977. Modeling Temperature and Light Adaptation of Phytoplankton. CEM Rept. 2, Center for Ecological Modeling, Troy, New York.

Gruendling, G.K., and J.L. Malanchuk. 1974. Seasonal and spatial distribution of phosphates, nitrates, and silicates in Lake Champlain, U.S.A. Hydrobiologia 45(4):405-421.

Hackney, P.A., and C.K. Minns. 1974. A computer model of biomass dynamics and food competition with implications for its use in fishery management. Trans. Amer. Fish. Soc. 103(2):215-255.

Hansen, W. 1962. Hydrodynamical Methods Applied to Oceanographic Problems. Proc. Symp. Math. Hydrodyn. Methods Phys. Oceanogr. Hamburg. 25 pp.

Heath, A.G. 1977. Toxicity of intermittent chlorination to freshwater fish: influence of temperature and chlorine form. Hydrobiologia 56-(1):39-47.

Hickman, M. 1974. Effects of the discharge of thermal effluent from a power station in Lake Wabamun, Alberta, Canada - The eipelic and epipsamic algal communities. Hydrobiologia 45(2-3):199-215.

Hollerman, W.D., and C.E. Boyd. 1980. Nightly aeration to increase production of the channel catfish. Trans. Amer. Fish. Soc. 109:446-452.

Huber, W.C., D.R.F. Harleman, and P.J. Ryan. 1972. temperature prediction in stratified reservoirs. Jour. Hydraul. Div. ASCE, Paper 8839, 98(HY4):645-666.

Hutchinson, G.E. 1975. A Treatise on Limnology. John Wiley and Sons, New York.

Idso, S.B., and J.M. Foster. 1974. Light and temperature relations in a small desert pond as influenced by phytoplankton density variations. Water Resources Res. 10(1):129-132.

Imberger, J., J. Patterson, B. Hebbert, and J.Loh. 1978. Dynamics of a reservoir of medium size. Jour. Hydraul. Div. ASCE, 104(HY5):725-743.

Jana, B.B. 1979. Primary production and bacterioplankton in fish ponds with mono and polyculture. Hydrobiologia 62(1):81-87.

Jassby, A., and T. Powell. 1975. Vertical patterns of eddy diffusion during stratification in Castle Lake, California. Limnol. and Oceanogr. 20(4).

Johnson, N.M., and D.J. Merritt. 1979. Convective and advective circulation of Lake Powell, Utah-Arizona during 1972-1975. Water Resources Res. 15(4):873-884.

Jones, J.R., and R.W. Bachmann. 1976. Prediction of phosphorus and chlorophyll levels in lakes. Jour. of Water Poll. Cont. Fed. 48:2176-2182.

Katz, B.M. 1977. Chlorine dissipation and toxicity in presence of nitrogenous compounds. Jour. Water Poll. Cont. Fed. 49:1627.

Khalaf, A.N., and L.J. MacDonald. 1975. Physico-chemistry conditions in temporary ponds in the new Forest. Hydrobiologia 47(2):301-318.

King, D.L. 1978. Environmental effects of hydraulic structures. Jour. Hydraul. Div. ASCE, 104(HY2):203-221.

King, I.P., W.R. Norton, and K.R. Iceman. 1975. A finite element solution for two-dimension stratified flow problems, pp. 133-156. In R.H. Gallagher, J.T. Oden, C. Taylor, and O.C. Zienkiewicz, (eds), Finite Elements in Fluids, Vol. 1. J. Wiley and Sons, New York.

__________, W.R. Norton, and G.T. Orlob. 1973. A Finite Element Solution for Two-Dimensional Stratified Flow. WRE, Inc., Walnut Creek, Final Rep. March. 80 pp.

Kirchner, W.B., and P.J. Dillon. 1975. An empirical method of estimating the retention of phosphorus in lakes. Water Resources Res. 11-(11):182-183.

Kitchell, J.F., J.F. Koonce, R.V. O'Neill, H.H. Shugart, Jr., J.J. Magnuson, and R.S. Booth. 1974. Model of fish biomass dynamics. Trans. Amer. Fish. Soc. 103(4):786-798.

Koryak, M., L.J. Stafford, and W.H. Montgomery. 1979. The limnological response of a West Virginia multipurpose impoundment to acid inflows. Water Resources Res. 14(4):929-934.

Kothandaraman, V., and R.L. Evans. 1979. Nutrient budget analysis for Rand Lake in Illinois. Jour. of Environ. Engin. Div. ASCE 105(EE3):547-553.

Kranenburg, C. 1979. Destratification of lakes using bubble columns. Jour. Hydraul. Div. ASCE 105-(HY4):333-349.

Kwei, E.A. 1977. Biological, chemical and hydrological characters of coastal lagoons of Ghana, West Africa. Hydrobiologia 56(1).

Lam, D.C.L., and T.J. Simons. 1976. Numerical computations of advective and diffusive transports of chloride in Lake Erie, 1970. J. Fish. Res. Bd. Canada 33(3):537-549.

Larrick, S.R., K.L. Dickson, D.S. Cherry, and J. Cairns, Jr. 1978. Determining fish avoidance of polluted water. Hydrobiologia 61(3):257-265.

Larsen, D.P., and H.T. Mercier. 1975. Shagawa Lake Recovery Characteristics as Depicted by Predictive Modeling. Presented at American Institute of Biological Sciences Mtg., August, Corvallis, OR.

Leendertse, J. 1967. Aspects of a Computational Model for Well-Mixed Estuaries and Coastal Seas. R.M. 5294-PR, The Rand Corp., Santa Monica, CA.

Leung, D.K. 1978. Modeling the bio-accumulation of pesticides in fish. CEM Rept. 5, Center for Ecological Modeling, Troy, New York.

Lewis, D.S.C. 1974. The effects of the formation of Lake Kainji (Nigeria) upon the indigenous fish population. Hydrobiologia 45(2-3):281-301.

Liggett, J.A. 1969. Unsteady circulation in shallow lakes. ASCE Proc. Jour. Hydraul. Div. 95(HY4):Paper 6686: pp. 1273-1288, July. 16 pp.

Lind, D.T., L.L. Smith, and S.J. Broderius. 1977. Chronic effects of hydrogen cyanide on the fathead minnow. Jour. Water Poll. Cont. Fed. 49:262.

Lowe-McConnell, R.H. 1975. Fish Communities in Tropical Freshwaters. Longman, Inc., New York.

MacKenthum, K.M. 1973. Toward a Cleaner Aquatic Environment. EPA, Office of Air and Water Programs, U.S. Government Printing Office, Washington, D.C.

Markofsky, M., and D.R.F. Harleman. 1971. A Predictive Model for Thermal Stratification and Water Quality in Reservoirs. EPA Water Pollution Control, research series, January. 283 pp.

Mattice, J.S., and H.E. Zittel. 1976. Site-specific evaluation of power plant chlorination. Jour. of Water Poll. Cont. Fed. 48:2284.

Matuszek, J.E. 1978. Empirical predictions of fish yields of large American lakes. Trans. Amer. Fish. Soc. 107(3):385-395.

McNeeley, D.L., and W.D. Pearson. 1977. Food habits of channel catfish in a reservoir receiving heated waters. Hydrobiologia 52(2-3):243-249.

Melack, J.M. 1976. Primary productivity and fish yields in tropical lakes. Trans. Amer. Fish. Soc. 105(5):575-580.

Middlebrooks, E.J., D.H. Falkenborg, and T.E. Maloney. 1976. Biosimulation and Nutrient Assessment. Ann Arbor Science Publishers, Inc., Ann Arbor, MI

Mishra, G.P., and A.K. Yadav. 1978. A comparative study of physico-chemical characteristics of river and lake water in Central India. Hydrobiologia 59(3):275-278.

Moretti, P.M., and D.K. McLaughlin. 1977. Hydraulic modeling of mixing in stratified lakes. Jour. Hydraul. Div. ASCE 103(HY4):367-380.

Morgan, W.S.G. 1979. Fish locomotor behavior patterns as a monitoring tool. Jour. Water Poll. Cont. Fed. 51:580.

Mulkey, L.A., and T.N. Sargent. 1974. Waste loads and water management in catfish processing. Jour. Water Poll. Cont. Fed. 46:2193.

Munawar, M. 1974. Limnological studies of freshwater ponds of Hyderabad - India. Hydrobiologia 44(1):13-27.

__________. 1975. Limnological studies on freshwater ponds of Hyderabad, India. IV. The biocenose. Periodicity and species composition of unicellular and colonial phytoplankton in polluted and unpolluted environments. Hydrobiologia 45-(1):1-32.

Niemeyer, G.C. 1978. Solution of coupled nonlinear ecosystem equations. Jour. Environ. Engineer. Div. ASCE 104(EE5):849-861.

Olsen, R.D., and M.R. Sommerfield. 1977. The physical-chemical limnology of a desert reservoir. Hydrobiologia 53(1):117-130.

________, D. Scava, and N.L. Clesceri. 1974. CLEANER - The Lake George Model. IBP, Eastern Deciduous Forest Dome Contribution No. 186. 13 pp.

Orlob, G.T. 1977. Mathematical Modeling of Surface Water Impoundments. Final Report, Project T-0006, Grant No. 6706, Office of Water Research and Technology. 119 pp. June.

Park, R.A., D.Scava, and N.L. Clesceri. 1974. Cleaner - the Lake George model. IBP, Eastern Deciduous Forest Dome Contrib. No. 186, 13 pp.

Patterson, D.J., E. Epstein, and J. McEvoy. 1975. Water Pollution Investigations: Lower Green Bay and Lower Fox River. Rep. to EPA. Contr. No. 68-01-1572, June. 371 pp.

Petr, T. 1974. Distribution, abundance, and food of commercial fish in the Black Volta and the Volta manmade lake in Ghana during the filling period (1964-1968). II. Characidae. Hydrobiologia 45(2-3):303-337.

Pickering, Q.H. 1974. Chronic Toxicity of nickel to the fathead minnow. Jour. Water Poll. Cont. Fed. 46:766.

Platzman, G.W. 1963. The Dynamic Prediction of Wind Tides on Lake Erie. Meteor. Monog. 4, No. 26. 44 pp.

Plumb, R.H., and G.F. Lee. 1975. Response of Lake Superior algae to nutrients and taconite tailings. Jour. Water Poll. Cont. Fed. 47:601.

Powell, T., and A. Jassby. 1974. The estimation of vertical eddy diffusivities below the thermocline in lakes. Water Resources Res. 10(2):191-198.

Rahman, M. 1978. On thermal stratification in reservoirs during the winter season. Water Resources Res. 14(2):377-380.

__________, and N. Marcotte. 1974. On thermal stratification in large bodies of water. Water Resources Res. 10(6):1143-1147.

Rai, H. 1974. Limnological observation on the different rivers and lakes in the Ivory Coast. Hydrobiologia 44(2-3):301-307.

Raney, D.C. 1977. Turbine aspiration for oxygen supplementation. Jour. Environ. Engineer. Div. ASCE 103(EE2):341-352.

Raymond, H.J. 1979. Effects of dams and impoundments on migrations of juvenile chinook salmon and steelhead from the Snake River, 1966 to 1975. Trans. Amer. Fish. Soc. 108:505-529.

Reckhow, K.H. 1979. Quantitative Techniques for the Assessment of Lake Quality. U.S. EPA, Office of Water Planning and Standards, Criteria and Standards Division, Washington, D.C.

Reid, R.O., and B.R. Bodine. 1968. Numerical model for storm surges in Galveston Bay. J. Waterw.:Harbors Div. ASCE 94(WW1):33-57.

_________, and R.D. Wood. 1976. Ecology of Inland Waters and Estuaries. Van Nostrand Co., New York.

Resh, V.H., and J.D. Unzicker. 1975. Water quality monitoring and aquatic organisms: the importance of species identification. Jour. Water Poll. Cont. Fed. 47:9.

Richards, R.T., and M.J. Hroncich. 1976. Perforated-pipe water intake for fish protection. Jour. Hydraul. Div. ASCE 102(HY2):139-149.

Robarts, R.D., and P.R.B. Ward. 1978. Vertical diffusion and nutrient transport in a tropical lake (Lake McIlwaine, Rhodesia). Hydrobiologia 49(1):213-221.

Romaire, R.P., and C.E. Boyd. 1979. Effects of solar radiation on the dynamics of dissolved oxygen in channel catfish ponds. Trans. of Amer. Fish. Soc. 108:473-478.

Ruane, R.J., S. Vigander, and W.R. Nicholas. 1977. Aeration of hydroreleases at Ft. Patrick Henry Dam. Jour. Hydraul. Div. ASCE 103(HY10):1135-1145.

Sager, P.E., and J.H. Wiersma. 1975. Phosphorus sources for lower Green Bay, Lake Michigan. Jour. Water Poll. Cont. Fed. 47:496.

Salanki, J.,and J.E. Ponyi, (eds). 1975. Limnology of Shallow Waters. Symposia Biologica Hungarica Vol. 15, Akademiai Kaido, Budapest.

Schwartzkopf, S.H., and G.L. Hergenzader. 1978. A comparative analysis of the relationship between phytoplankton standing crops and environmental parameters in four entrophic prairie reservoirs. Hydrobiologia 49(3):-261-274.

SCOPE Report 2. 1972. Man-Made Lakes as Modified Ecosystems. International Council of Scientific Unions, Paris.

Sheehan, S.W., and S.O. Russell. 1978. Application of decision theory to salmon management. Water Resources Res. 14(5):976-980.

Shin, E.B., and P.A. Krenke. 1976. Mercury uptake by fish and biomethylation mechanisms. Jour. Water Poll. Cont. Fed. 48:473.

Simons, T.J. 1973. Development of Three Dimensional Numerical Models of the Great Lakes. Canada Centre for Inland Waters, Burlington, Ont. Sci. Series No. 12. 26 pp.

____________, L. Funkquist, and J. Svensson. 1977. Application of a Numerical Model to Lake Vanern. Swedish Meteorological and Oceanographic Inst., NrRH09, 15 pp.

____________, and D.C.L. Lam. 1980. some limitations of water quality models for large lakes: a case study of Lake Ontario. Water Resources Res. 16(1):105-116.

Slawson, G.C., Jr., and L.G. Everett. 1976. Segmented population model of primary productivity. Jour. Environ. Engineer. Div. ASCE 102 (EE1):127-138.

Smedberg, C.T., and R.E. Cannon. 1976. Cyanophage analysis as a biological pollution indicator - bacterial and viral. Jour. Water Poll. Cont. Fed. 48:2416.

Snodgrass, W.J., and C.R. O'Melia. 1975. A Predictive Phosphorus Model for Lakes - Sensitivity Analysis and Application. Environmental Science and Technology. 29 pp.

Sorokin, Yu. I., and E.B. Paveljeva. 1978. On structure and functioning of ecosystems in Salmon Lake. Hydrobiologia 47(1):25-48.

Sreennasan, A. 1974. Limnological studies of an primary production in temple pond ecosystems. Hydrobiologia 48(2):117-124.

Sridharan, N., and G.F. Lee. 1974. Phosphorus studies in Lower Green Bay, Lake Michigan. Jour. Water Poll. Cont. Fed. 46:684.

Stefan, H., and D.E. Ford. 1975. Temperature dynamics in dimictic lakes. Jour. Hydraul. Div. ASCE 101(HY1):-97-114.

__________, and D.E. Ford. 1975. Temperature dynamics in dimictic lakes. Jour. Hydraul. Div. ASCE 102(EE6):-1201-1213.

Stewart, K.M., and S.J. Markello. 1974. Seasonal variations in concentrations of nitrate and total phosphorus, and calculated nutrient loading for six lakes in Western New York. Hydrobiologia 44(1):61-89.

Swensson, U. 1978. A Mathematical Model of the Seasonal Thermocline. Report No. 1002, Dept. of Water Resources Engineering, Lund Institute of Technology, Univ. of Lund, Lund, Sweden.

Taft, E.P., III, and Y.G. Mussalli. 1978. Angled screens and towers for diverting fish at power plants. Jour. Hydraul. Div. ASCE 104(HY5):-623-634.

Tapp, J.S. 1978. Eutrophication analysis with simple and complex models. Jour. Water Poll. Cont. Fed. 50:484.

Tennessee Valley Authority (TVA). 1972. Heat and Mass Transfer Between a Water Surface and Atmosphere. Engineering Lab Report No. 14. 123 pp. Zapp. April.

Thomann, R.V., D.M. DiToro, D.J. O'Connor, and R.P. Winfield. 1973. Mathematical modeling of eutrophication of large lakes. Manhattan College, NY. Environ. Engineer. and Sci. Prog., Ann. Rept. Yr. 1, Apr. 1, 1972-March 31, 1973. 17 pp.

_________, ___________, R.P. Winfield, and D.J. O'Connor. 1974. Mathematical modeling of phytoplankton in Lake Ontario. I. Model Development and Verification. Manhattan College, NY. Final Rept. to Env. Engng. and Sci. Prog. Natl. Env. Res. Center, Grosse Ile, MI. Oct. 1974. 189 pp.

_________, ___________, R.P. Winfield, and D.J. O'Connor. 1975. Mathematical Modeling of Phytoplankton in Lake Ontario 1. Model Development and Verification. EPA Rep. 660/3-75-005, March. 177 pp.

Thomas, W.A., G. Goldstein, and W.H. Wilcox. 1973. Biological Indicators of Environmental Quality: A bibliography of Abstracts. Ann Arbor Science Publishers, Inc., Ann Arbor, MI.

Thornton, K.W., and A.S. Lessem. 1978. A temperature algorithm for modifying biological rates. Trans. Amer. Fish. Soc. 107(2):284-287.

Toews, D.R., and J.S. Griffith. 1977. Empirical estimates of potential fish yield for the Lake Bangweulu system, Zambia, Cent. Africa. Trans. Amer. Fish. Soc. 106(5):417-423.

Tomlinson, J.W.C., and P.S. Brown. 1979. Decision analysis in fish hatchery management. Trans. Amer. Fish. Soc. 108:121-129.

Tucker, C.S., and C.E. Boyd. 1977. Relationships between potassium permanganate treatment and water quality. Trans. Amer. Fish Soc., 106(5):481-488.

________, ________, and E.W. McCoy. 1979. Effects of feeding rate on water quality production of channel catfish, and economic returns. Trans. Amer. Fish. Soc. 108:389-396.

Uzzell, J.C., Jr., and M.N. Ozisik. 1977. Far field circulation velocities in shallow lakes. Jour. Hydraul. Div. ASCE 103(HY4):395-407.

Van Belle, G., and L. Fischer. 1977. Monitoring the environment for ecological change. Jour. Water Poll. Cont. Fed. 49:1671.

Vollenweider, R.A. 1969. Moglichkeiten und Grenzen Elementarer Modelle der Stoffbilanz von Seen. Arch. Hydrobiol. 66:1-36.

Waldrop, W.R., and R.C. Farmer. 1973. Three Dimensional Flow and Sediment Transport at River Mouths. TR 150, Coastal Studies Inst., Louisiana State Univ., Baton Rouge, ONR Proj. NR 388002, Sept. 137 pp.

Walker, W.W., Jr. 1979. Use of hypolimnetic depletion rate as a trophic state index for lakes. Water Resources Res. 15(6):1463-1470.

Water Resources Engineers, Inc. 1967. Prediction of Thermal Energy Distribution in Streams and Reservoirs. Rep. to Cal. Dept. of Fish and Game, June; Rev. Aug. 1968. 90 pp.

Webb, M.S. 1974. Surface temperatures of Lake Erie. Water Resources Res. 10(2):199-210.

Westlake, G.F., H. Kleerekoper, and J. Matis. 1974. The locomotor response of goldfish to a steep gradient of copper ions. Water Resources Res. 10(1):103-105.

Wetzel, R.G. 1975. Limnology. Saunders Co., Philadelphia, PA.

Wiederholm, T. 1980. Use of benthic communities in lake monitoring. Jour. Water Poll. Cont. Fed. 52:537.

Wilhm, J., and N. McLintock. 1978. Dissolved oxygen concentrations and diversity of benthic macro-invertebrates in an artificially destratified lake. Hydrobiologia 7(3-4):513-526.

Young, J.O. 1975. Seasonal and diurnal changes in the water temperature of a temperate pond (England) and a tropical pond (Kenya). Hydrobiologia 7(3-4):513-526.

Youngberg, B.A. 1977. Application of the aquatic model CLEANER to a stratified reservoir system. CEM Rept. 1, Center for Ecological Modeling, Troy, NY.

A SURVEY OF THE MATHEMATICAL MODELS PERTINENT TO FISH PRODUCTION AND TROPICAL POND AQUACULTURE

by

David R. Bernard

INTRODUCTION

Because ponds are small and because biology of tropical aquatic ecosystems is not well understood, the role of models to predict the productivity of tropical ponds has been very limited. Pond ecosystems can easily be altered to create relatively controlled experiments to provide information for management. With larger bodies of water, a resource manager does not have this ability and the information upon which he bases his management must come from models of the resource (Magnuson 1973; Peterman 1975). Because the culturalist need not model what he can describe through experiment on controlled environments, predictive models of pond aquaculture are few relative to the models of aquatic science that describe the productivity of large bodies of water.

The manager of tropical ponds pays a price for his dependence on results from descriptive experiments. These experiments are usually conducted at sites with several ponds as replicates at each site (e.g., Almazan and Boyd 1978; Arce and Boyd 1975; Bishara 1978; Boyd 1976; Haines 1973; Stickney et al. 1979). The efficacy of specific aquaculture practices to boost production is tested by applying these practices to the ponds according to some rational scheme based on aquaculture principles. The experimental results are often reported in tabular form (e.g., Bishara 1978; Boyd 1976; Stickney et al. 1979) or in some kind of statistical format (e.g., Almazan and Boyd 1978; Arce and Boyd 1975; Haines 1973) and then used by someone other than the experimenter and on sites other than those upon which the tests were made. This is one of the ways information is disseminated in the agriculture industry in the United States. By and large, this approach has worked well for U.S. agriculture, but not as well for pond aquaculture. The application of experimental results from descriptive experiments to the other ponds is valid only as long as experimental conditions hold at these new sites. When results of experiments on pond ecosystems are used away from their site of origin, the accuracy of these results often breaks down; for example, the balance of bass-bluegill populations derived through experiments in Alabama does not hold for ponds farther north (Regier 1962). Understanding is missing regarding the mechanisms that drive pond productivity. Without understanding the results of descriptive experiments are too inflexible to conform to new sites and situations without repeating the descriptive experiments at each new site or for each new situation. U.S. Agriculture has overcome this problem through repeating experiments at many different sites under many different sets of environmental conditions. Agriculture is an old science and has had time to develop this approach. Although scientific tropical pond aquaculture does not have this advantage of age, it can use predictive models to speed its development and gain understanding of the mechanisms of pond productivity.

To manage tropical ponds for fish productivity, a dual approach is needed to both easily disseminate experimental results to managers in developing countries and to provide the understanding necessary to measure the precision of these results. In U.S. agriculture, descriptive experiments still provide useful information, because exhaustive laboratory and quantitative analysis help explain the observed precision of experimental results and their applicability to various situations not described by descriptive experiments.

Compared to U.S. agriculture, the science of tropical pond aquaculture is still an infant. In time, tropical pond aquaculture will also mature as a science, and predictive models have an important role to play in that growth.

HISTORY OF PREDICTIVE MODELS IN TROPICAL POND AQUACULTURE

Most of this discussion will concern models that predict the productivity of temperate ponds and lakes. Because little is known about tropical aquatic ecosystems and models specific to tropical ponds are rare, temperate bodies of water must be the focal point of the following discussion. Always keep in mind that lakes are not ponds, temperate is not tropical, and these differences will alter the structure of the predictive models discussed below to some degree.

Predictive Models for the Fish Community

Predictive models of fish production from lacustrine ecosystems can be segregated according to the hierarchy of their focus: ecosystem, stock, or organism (Table 1). Models of whole pond ecosystems (hereafter called empirical models) are directly related to the descriptive experiments discussed in the Introduction where the ecosystem is treated as a "black box" with only inputs and outputs (Figure 1). Given certain inputs and conditions, a certain level of fish production is expected. In empirical models, each pond is assumed to be a replica of the ecosystem. Building these models is akin to statistically building an experimental design with an ANOVA or a regression analysis (Liang et al. 1981). Regression analysis will provide some estimate of the experimental precision which is not directly available from an ANOVA or from a tabular report of experimental results.

Predictive models of stock production often separate stock dynamics into growth, mortality, reproduction, and recruitment of stock members (Figure 2). A stock is the portion of the animals of a single species in a single population that can be harvested (from Ricker 1975). Information on the parts of stock models is derived through experiments that are usually independent of measuring fish production in the ponds (e.g., Bishara 1978; Brown 1970). For instance some experiments may determine the effects of fertilizer on fish growth (Stickney et al. 1979), another experiment may investigate how pH affects mortality rates of fish (Menendez 1976). Although production in stock models is mostly a function of age of the fish (e.g., Deriso 1980; Jensen 1973), environmental influences can be incorporated into these models (e.g., Bakun and Parrish 1981; Gatto and Rinaldi 1976; McKelvey et al. 1980). When information on fish growth, fish survival, and environment are combined, the result can be an estimate of production to compare against *in situ* measurements of production. Stock models that focus on fish tacitly assume that fish and other pond organisms little affect each other's dynamics. When this assumption is false, precision of experimental results declines. Also, with more parts than the empirical models, stock models have more parameters to estimate (Figure 3). But because stock models require a better understanding of the internal mechanisms of pond production, they are more relevant to a wider variety of ponds that are empirical models. For instance, growth by a fish in one pond will be similar to growth of a fish of the same species in another pond, an insight that can be used in pond after pond to estimate fish production without experiments on new ponds.

In stock models, recruitment, growth, survival, and reproduction somewhat explain the variation in production seen in the descriptive experiments from which empirical models are derived.

Table 1

THE HIERARCHY OF EMPIRICAL, STOCK, AND MECHANISTIC MODELS TO PREDICT FISH PRODUCTION FROM TROPICAL PONDS

Empirical Models

$$P = f(\underline{X})$$

Stock Models

$$P = f\{g_1(\underline{X}), g_2(\underline{X}), \ldots g_m(\underline{X})\}$$

Mechanistic Models

$$P = f[g_1\{h_{11}(\underline{X}), h_{12}(\underline{X}) \text{ --- } h_{1n}(\underline{X})\}, g_2\{h_{21}, h_{22}, \text{ --- } h_{2n}\}, \ldots$$

$$g_m\{h_{m1}, h_{m2}, \text{ --- } h_{mn}\}]$$

where

P = production

f, g, h = functions in the models

$\underline{X}$ = a vector of inputs (e.g. x_1 = weight of fish stocked, x_2 = weight of nitrogen fertilizer used, etc.)

Unfortunately, the processes in the stock models are still too simple to incorporate a wide range of aquaculture practices and still gain much understanding of the underlying biological and physical mechanisms of pond productivity.

The predictive models that focus on organisms (hereafter called mechanistic models) describe biological and environmental processes on a finer scale than do the stock models (Figure 4). Again, experiments to model aquaculture principles are independent of the pond ecosystem, and their modelled results are summed to provide estimates of productivity. Growth is modelled as the antagonism between respiration and ingestion (e.g., Caulton 1978; Ursin 1967); mortality is the sum of disease, predation, starvation, and competition (e.g. Smith 1976); and reproduction and recruitment are functions of growth, fecundity, and survival (e.g., Jackson 1979). Although mechanistic models require estimates for many parameters and many independent experiments to get them, these models do link pond practices to pond production by modelling the principles of pond aquaculture. Even more so than do stock models, mechanistic models can explain the fish production seen in the descriptive experiments as functions of aquaculture practices. But mechanistic models are complex syntheses of what is known of the pond ecosystem, and are therefore more difficult to build and more difficult to use in management than are the simpler empirical and stock models. More than the other types, mechanistic models are abstractions of what the modeller feels are the important principles in the ecosystem. Therefore, testing the validity of these models by comparing their predictions of productivity against those from field experiments is imperative (Kerr 1976).

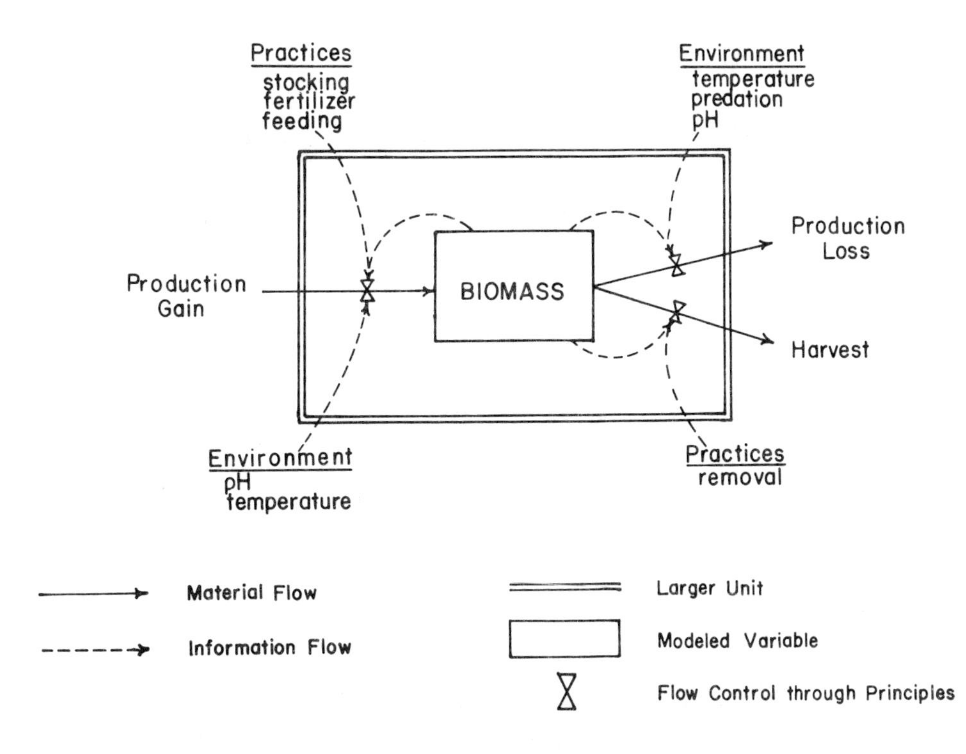

Figure 1

Schematic of a hypothetical empirical model with symbols according to Patten (1975). Biomass flows into the larger unit (the pond ecosystem) through production and flows out through harvest and mortality. Aquaculture practices and environment (physical, chemical, and biological) control inflows and outflows of production through aquaculture principles.

Predictive Models for the Plankton Community

Predictive models of plankton production in ponds, like those for fish production, are classified according to their resolution (Table 1 and Figure 2). Published empirical models predict productivity as a function of morphoedaphic factors (e.g., Ryder et al. 1974), nutrient loading (e.g., Dillon and Rigler 1974a, 1974b), or a "super-organism" whose dynamics depend on light, temperature, and nutrients (e.g., Fee 1973a, 1973b; Jassby and Platt 1976). Stock models of plankton production express phenomena that concern groups of organisms and sum up their dynamics in the model (e.g., Patten and Van Dyne 1968) or in something less than the sum if competition and or predation occur within the plankton community (e.g., Lynch 1977). The tradeoffs between using stock and empirical models are the same as in modelling fish production, only more so because of the greater diversity in the plankton community. The finer the resolution of the model, the more numerous are its parameters, the greater the understanding it provides, and the more difficult it is to use.

Obviously, the fine resolution of the mechanistic model demands that many taxa be modelled. Instead of modelling taxa, modelling functional groups, such as planktivores, benthic predators, etc., can reduce the bookkeeping in these models (e.g., Lehman et al. 1975). However, such shortcuts can go but so far. Because competition, predation, respiration and other processes occur for plankton as for fish (e.g., Dodson et al 1976; Lynch 1977), these process must be included in the mechanistic models.

Most predictive models concern phytoplankton and not zooplankton. Although the same modelling philosophies can be used for both types of plankton, zooplankton are far more difficult to model as a group, because their dynamics are affected by the trophic levels below and above them. The dynamics of primary

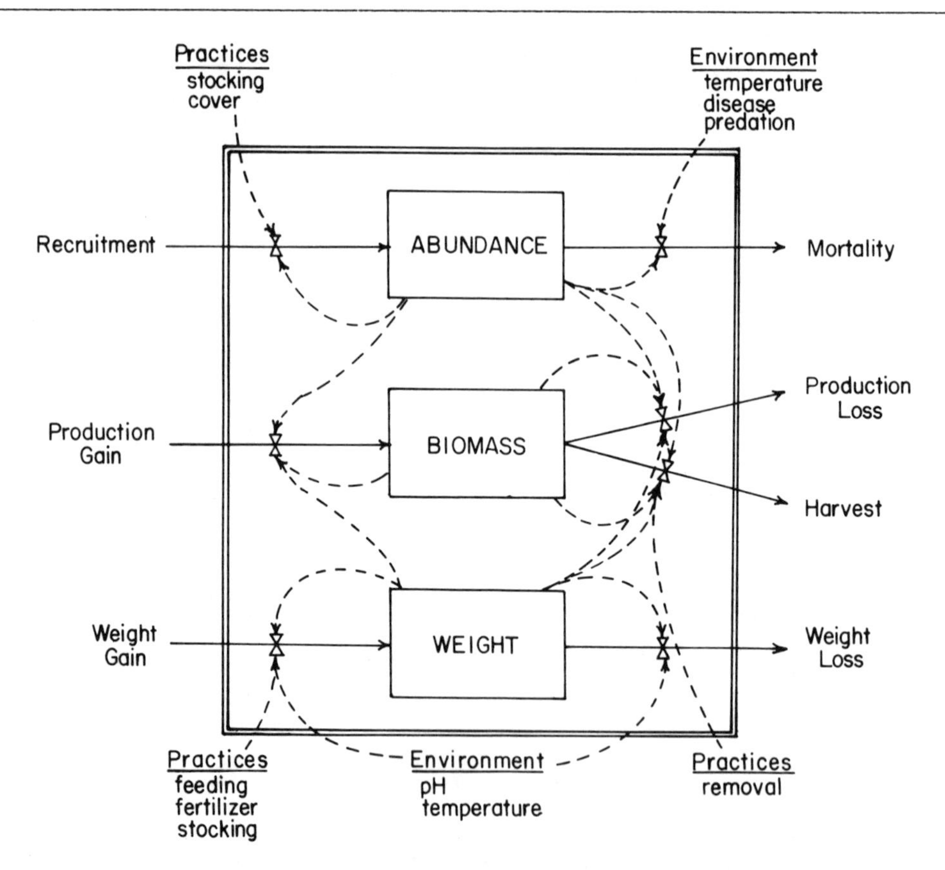

Figure 2.

Schematic of a hypothetical stock model. Symbols are defined in Figure 1. The large, double-lined box represents the conceptual boundaries of the pond ecosystem.

production drive zooplankton production, while vertebrate predators alter zooplankton diversity. Empirical models have generally been poor predictors of zooplankton dynamics (see Buzas 1971). Stock and mechanistic models are extremely rare, although a few principles of zooplankton productivity have been modelled [e.g., stock models: Caswell (1972), Taylor and Slatkin (1981); mechanistic models: Confer and Blades (1975), Peters (1975)].

Because benthos production is difficult to model when it is separated from those factors that affect it, few strictly benthos models exist. Vertebrate predators and omnivores affect benthic production as do allochotonous organic materials (e.g., Hall and Hyatt 1974). However, some principles of pond aquaculture relative to stock and mechanistic models of benthos have been modelled (e.g., Frost 1972; Hall and Hyatt 1974; Ware 1973).

RECOMMENDATIONS

Both empirical and mechanistic models should be used to study tropical pond aquaculture. Empirical models will provide the materials for overseas management of tropical ponds in the developing countries. Also, the mechanistic models can be validated against the descriptive experiments upon which the empirical models are based. Mechanistic models will provide insight to the precision in the empirical models and direct future

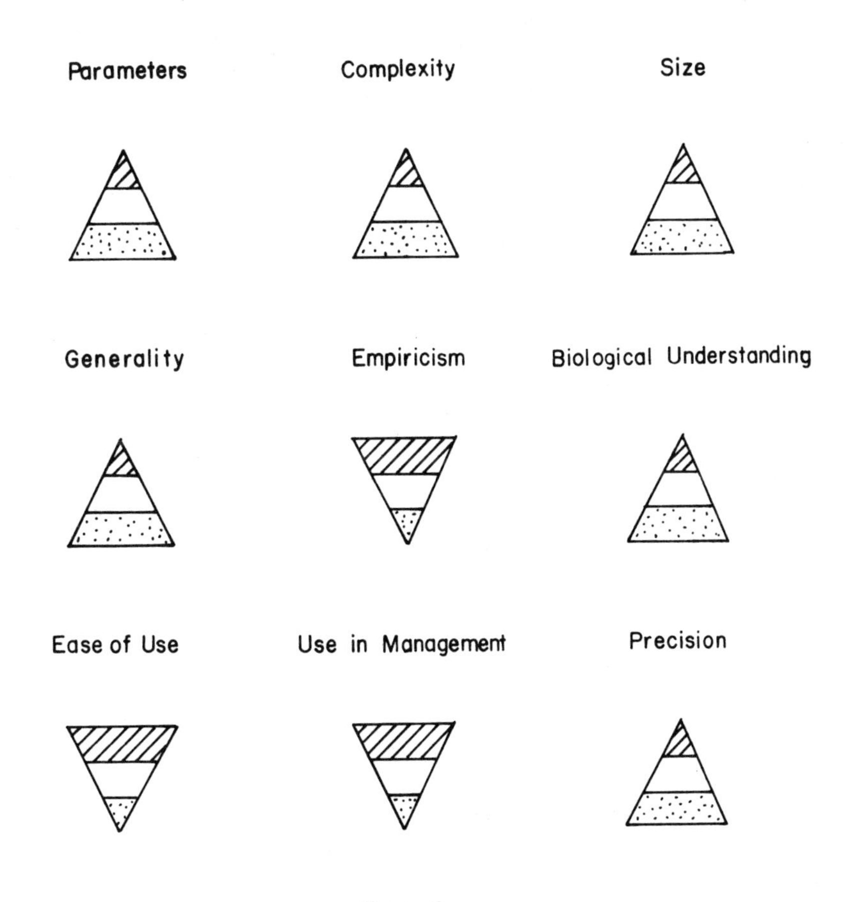

Figure 3.

Relative merits of three approaches to modeling production in tropical ponds: empirical models (the lined area), stock models (the blank area), and mechanistic models (the stippled area). Area within triangles correspond directly to the spectrum of few to many, little to much, small to large, etc.

experiments to improve that precision. Because the advantages and disadvantages of stock models are repeated if both empirical and mechanistic models are used, use of stock models is not necessary.

Because the empirical models are site-specific, they must be constructed for each of the overseas projects on management of tropical ponds. These models should be based on descriptive experiments like those in temperate pond aquaculture (e.g., Haines 1973), but unlike many of these experiments, simply reporting their results is not enough. Results must be rigorously analyzed with ANOVAs and/or regression analysis to provide estimates of experimental precision. The pond manager in the developing country must know the probability of getting undesired results from any particular set of aquaculture practices. Great care should be exercised to develop an experimental design which will provide an analysis of precision through adequate replication at each site.

The manager in the developing country should be presented with this modelling in its simplest possible form (see Peterman 1975), either as tables of inputs (fertilizers, stocking rates, etc.) with the expected levels of production, or as a simple model on a handheld calculator or on a small computer that will provide the same service. If the empirical model is to be used in developing countries for management, it should be stochastic to represent the precision of aquaculture practices to produce desired yields.

The mechanistic models and their results should not be used for management in developing countries but should be used to improve empirical models (see Larkin 1978). The mechanistic models should explain the same data with the same inputs as do the empirical models, but unlike the simpler models, should link production directly to aquaculture practices through modelling aquacultural principles. And because mechanistic models are based on processes, these models can and should be tested with inputs other than those used in the field experiments, then validated against new descriptive experiments with these inputs. Mechanistic models should be deterministic and should be used to explore the imprecision in the empirical models through this procedure.

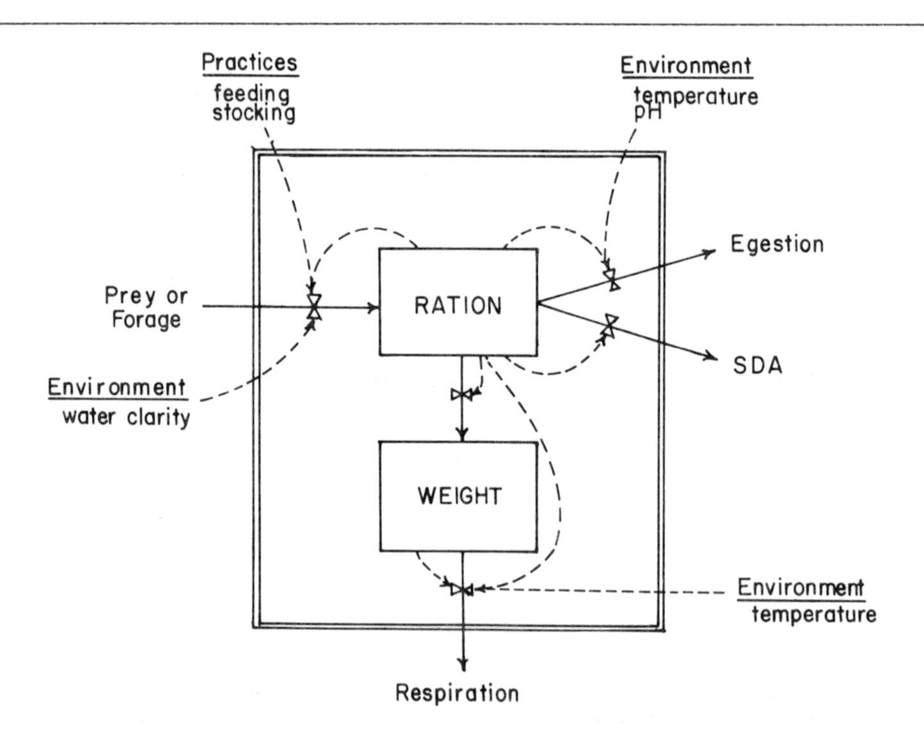

Figure 4.

Schematic of a submodel of growth in a hypothetical mechanistic model. The large, double-lined square corresponds to the variable weight in Figure 2. Symbols are defined in Figure 1.

Both modelling efforts should describe the important parts of the ecosystem and the physical environment in tropical ponds. Some comprehensiveness is implicit in the empirical model; it includes fish, plankton, benthos, and physical environment in a single "black box". In the mechanistic model, these processes and factors must be inserted into the model. Unlike many terrestrial ecosystems, the different trophic levels in aquatic systems can be interdependent and virtually impossible to model separately (Patten 1975).

The physical and chemical environment in ponds drives the ecosystem and is somewhat driven by it. Therefore, the mechanistic models must include models of the physical and chemical environment of ponds.

The Agency for International Development's Aquaculture Program specifies four types of ponds as study sites: coolwater ponds at high elevation, brackish ponds, ponds with their production augmented with energy and/or fertilizers, and ponds with little or no such inputs. Each type of pond requires its own modelling effort. Because the empirical models will be time- and site-specific, they must be built for each site. On the other hand, mechanistic models are not bound to particular sites, but to particular principles. If these principles are universal among the studied ponds, then all mechanistic models will have similar components. As different mechanistic models from different sites are compared, some components of these models common to all four types of ponds can be combined and the modelling effort simplified (see Holling 1966).

LITERATURE CITED

Almazan, G., and C.E. Boyd. 1978. Plankton production and Tilapia yield in ponds. Aquaculture 15:75-77.

Arce, R.G., and C.E. Boyd. 1975. Effects of agricultural limestone on water chemistry, phytoplankton productivity, and fish production in soft water ponds. Trans. Am. Fish. Soc. 104(2):308-312.

Bakun, A.H., and R.H. Parrish. 1981. Environmental inputs to fishery population models for eastern boundary current regions, pp. 67-104. In G.D. Sharp (ed), Workshop on the Effects of Environmental Variation on the Survival of Larval Pelagic Fishes. Workshop Report No. 28. Intergovernmental Oceanographical Commission, UNESCO, Paris.

Bishara, N.F. 1978. Fertilizing fish ponds II - growth of Mugil cephalus in Egypt by pond fertilization and feeding. Aquaculture 13:361-367.

Boyd, C.E. 1976. Nitrogen fertilizer effects on production of Tilapia in ponds fertilized with phosphorus and potassium. Aquaculture 7:385-390.

Brown, B.E. 1970. Exponential decrease in a population of fathead minnows. Trans. Am. Fish. Soc. 99:807-809.

Buzas, M.A. 1971. Analyses of species densities by the multivariate general linear model. Limnol. and Oceanogr. 16(4):667-670.

Caswell, H. 1972. On instantaneous and finite birth rates. Limnol. and Oceanogr. 17(5):787-791.

Caulton, M.S. 1978. The importance of habitat temperatures for growth in the tropical cichlid Tilapia rendalli Boulenger. J. Fish. Biol. 13:99-112.

Confer, J.L., and P.I. Blades. 1975. Omnivorous zooplankton and planktivorous fish. Limnol. and Oceanogr. 20(4):571-579.

Deriso, R.B. 1980. Harvesting strategies and parameter estimation for an age-structure model. Can. J. Fish. Aquat. Sci. 37(2):268-282.

Dillon, P.J., and F.H. Rigler. 1974a. The phosphorus-chlorophyll relationship in lakes. Limnol. and Oceanogr. 19(5):767-773.

_____________, and F.H. Rigler. 1974b. A test of a simple nutrient budget model predicting the phosphorus concentration in lake water. J. Fish. Res. Board Can. 31:1771-1778.

Dodson, S.I., C. Edwards, F. Wiman, and J.C. Normandin. 1976. Zooplankton: specific distribution and food abundance. Limnol. and Oceanogr. 21(2):309-313.

Fee, E.J. 1973a. A numerical model for determining integral primary production and its application to Lake Michigan. J. Fish. Res. Board Can. 30:1447-1468.

__________. 1973b. Modelling primary production in water bodies: a numerical approach that allows vertical inhomogeneities. J. Fish. Res. Board Can. 30:1469-1473.

Frost, B.W. 1972. Effects of size and concentration of food particles on the feeding behavior of the marine planktonic copepod Calanus pacificus. Limnol. and Oceanogr. 17(6):-805-815.

Gatto, M., and S. Rinaldi. 1976. Mean value and variability of fish catches in fluctuating environments. J. Fish. Res. Board Can. 33:189-193.

Haines, T.A. 1973. Effects of nutrient enrichment and a rough fish population (carp) on a game fish population (smallmouth bass). Trans. Am. Fish. Soc. 102(2):346-354.

Hall, K.J., and K.D. Hyatt. 1974. Marion Lake (IBP) - from bacteria to fish. J. Fish. Res. Board Can. 31:893-911.

Holling, C.S. 1966. The strategy of building models of complex ecological systems, pp. 195-214. In K.E.F. Watt (ed), Systems Analysis in Ecology. Academic Press, New York. 276 pp.

Jackson, U.T. 1979. Controlled spawning of largemouth bass. Prog. Fish-Cult. 41(2):90-95.

Jassby, A.D., and T. Platt. 1976. Mathematical formulation of the relationship between photosynthesis and light for phytoplankton. Limnol. and Oceanogr. 21(4):540-547.

Jensen, A.L. 1973. Relation between simple dynamic pool and surplus production models for yield from a fishery. J. Fish. Res. Board Can. 30:998-1002.

Kerr, S.R. 1976. Ecological analysis and the fry paradigm. J. Fish. Res. Board Can. 33(2):329-335.

Larkin, P.A. 1978. Fisheries management - an essay for ecologists. Ann. Rev. Ecol. Syst. 9:57-73.

Lehman, J.T., D.B. Botkin, and G.E. Likens. 1975. The assumptions and rationales of a computer model of phytoplankton population dynamics. Limnol. and Oceanogr. 20(3):343-364.

Liang, Y., J.M. Melack, and J. Wang. 1981. Primary production of fish yields in Chinese ponds and lakes. Trans. Am. Fish. Soc. 110:346-350.

Lynch, M. 1977. Zooplankton competition and plankton community structure. Limnol. and Oceanogr. 22(4):775-777.

Magnuson, J.J. 1973. Application of theory and research in freshwater and marine fisheries programs - a discussion of the papers given by S.H. Smith, R.O. Anderson, K.R. Allen and G.B. Talbot. Trans. Am. Fish. Soc. 102(1):194-195.

McKelvey, R., D. Hankin, K. Yanosko, and C. Snygg. 1980. Stable cycles in multistage recruitment models: an application to the Northern California Dungeness crab (Cancer magister) fishery. Can. J. Fish. Aquat. Sci. 37:2323-2345.

Menendez, R. 1976. Chronic effects of reduced pH on brook trout (Salvelinus fontinalis). J. Fish. Res. Board Can. 33:118-123.

Patten, B.C. 1975. A reservoir cove ecosystem model. Trans. Am. Fish. Soc. 104(3):596-619.

____________, and G.M. Van Dyne. 1968. Factorial productivity experiments in a shallow estuary: energetics of individual plankton species in mixed populations. Limnol. and Oceanogr. 13(2):309-314.

Peterman, R.M. 1975. New techniques for policy evaluation in ecological systems: methodology for a case study of Pacific salmon fisheries. J. Fish. Res. Board Can. 32:2179-2188.

Peters, R.H. 1975. Phosphorus excretion and the measurement of feeding and assimilation by zooplankton. Limnol. and Oceanogr. 20(5):858-859.

Regier, H.A. 1962. Some aspects of the ecological management of warmwater fish in New York farm ponds. Ph.D. Thesis. Cornell Univ. Ithaca, New York. 420 pp.

Ricker, W.E. 1975. Computation and interpretation of biological statistics of fish populations. Fish. Res. Board Can. Bull. No. 191. 382 pp.

Ryder, R.A., S.R. Kerr, K.H. Loftun, and H.A. Regier. 1974. The morphoedaphic index, a fish yield estimator - review and evaluation. J. Fish. Res. Board Can. 31:663-688.

Smith, S.L. 1976. Behavioral suppression of spawning in largemouth bass by interspecific competition for space within spawning areas. Trans. Am. Fish. Soc. 105(6):682-685.

Stickney, R.R., J.H. Hesby, R.B. McGeachin, and W.A. Isbell. 1979. Growth of Tilapia nilotica in ponds with differing histories of organic fertilization. Aquaculture 17:189-194.

Taylor, B.E., and M. Slatkin. 1981. Estimating birth and death rates of zooplankton. Limnol. and Oceanogr. 26(1):143-158.

Ursin, E. 1967. A mathematical model of some aspects of fish growth, respiration, and mortality. J. Fish. Res. Board Can. 24(11):2355-2453.

Ware, D.M. 1973. Risk of epibenthic prey to predation by rainbow trout (Salmo gairdneri). J. Fish. Res. Board Can. 30:787-797.

LITERATURE SUMMARY

The following index and bibliography describe examples of predictive models used to study pond and lake productivity. The citations complement those in the preceding article.

The bibliography does not represent all the literature on predictive models, but does give a comprehensive menu of that literature. Only models of pond or lake ecosystems, models of animals or plants from these ecosystems, and models of biological principles germane to these ecosystems are included in the bibliography. Models of nutrient loading are considered ecosystem models and are included. No models of stream, ocean, or estuarine ecosystems are listed except those based on principles that are germane to pond ecosystems. And although the literature on the stock models of fisheries is extensive, citations on these models are few, because ecosystem and mechanistic models are recommended over stock models for research on the productivity of tropical ponds. However, stock models of plankton are included. Often, the mechanistic models referenced in the index describe one, or a few, principles of pond aquaculture and must be joined with complementary models to create a single mechanistic model of pond production. Some citations are included in the bibliography for their general descriptions of aquatic ecosystems and their predictive models.

The index is an attempt to separate the citations in the bibliography by subject matter. Because differences among stock, ecosystem, and mechanistic models are often a matter of perspective, some models in the index may not fit with readers' judgments. Numbers listed in the index correspond to the numbers of individual references in the bibliography.

INDEX

Subject

BIBLIOGRAPHY

1. Allen, J.C. 1974. Balancing predation and competition in cladocerans. Ecology 55(3):622-629.

2. __________. 1975. Mathematical models of species interactions in time and space. Am. Naturalist 109:319-342.

3. Bailey, V.A., A.J. Nicholson, and E.J. Williams. 1962. Interaction between hosts and parasites when some host individuals are more difficult to find than others. J. Theoretical Biol. 3:1-18.

4. Bannister, T.T. 1974a. Production equations in terms of chlorophyll concentration, quantum yield, and upper limit to production. Limnol. and Oceanogr. 19(1):1-12.

5. __________. 1974b. A general theory of steady state phytoplankton growth in a nutrient saturated mixed layer. Limnol. and Oceanogr. 19(1):13-30.

6. Bartram, W.C. 1980. Experimental development of a model for the feeding of neritic copepods on phytoplankton. J. Plankton Res. 3(1):25-52.

7. Beyer, J.E. 1980. Feeding success of clupeoid fish larvae and stochastic thinking. Dana (1):65-91.

8. Blinov, V.V. 1977. Modelling natural mortality in fish of the younger age groups. J. Ichthyology 17:511-516.

9. Boyd, C.E., R.P. Romaire, and E. Johnston. 1978. Predicting early morning dissolved oxygen concentrations in channel catfish ponds. Trans. Am. Fish. Soc. 107(3):484-492.

10. Buzas, M.A. 1971. Analyses of species densities by the multivariate general linear model. Limnol. and Oceanogr. 16(4):-667-670.

11. Caswell, H. 1972. On instantaneous and finite birth rates. Limnol. and Oceanogr. 17(5):787-791.

12. Caulton, M.S. 1978. The importance of habitat temperatures for growth in the tropical cichlid (Tilapia rendalli Boulenger). J. Fish. Biol. 13:99-112.

13. Chapra, S.C. 1979. Applying phosphorus loading models to embayments. Limnol. and Oceanogr. 24(1):163-168.

14. ____________, and K.H. Reckhow. 1979. Expressing the phosphorus loading concept in probabilistic terms. J. Fish. Res. Board Can. 36:225-229.

15. Chewning, W.C. 1975. Migratory effects in predator-prey models. Mathematical Biosciences 23:-253-262.

16. Comins, H.N., and D.W.E. Blatt. 1974. Prey-predator models in spatially heterogeneous environments. J. Theoretical Biol. 48:75-83.

17. Confer, J. L., and Pamela I. Blades. 1975. Omnivorous zooplankton and planktivorous fish. Limnol. and Oceanogr. 20(4):571-579.

18. Cooley, J.M., and C.K. Minnis. 1978. Prediction of egg development times of freshwater copepods. J. Fish. Res. Board Can. 35:1322-1329.

19. Dillon, P.J. 1975. The phosphorus budget of Cameron Lake, Ontario: the importance of flushing rate to the degree of eutrophy of lakes. Limnol. and Oceanogr. 20(1):28-39.

20. ____________, and F.H. Rigler. 1974. A test of a simple nutrient budget model predicting the phosphorus concentration in lake water. J. Fish. Res. Board Can. 31:1771-1778.

21. Duckstein, L., J. Casti and J. Kempf. 1979. Modeling phytoplankton dynamics using catastrophe theory. Water Resources Res. 14(5):1189-1194.

22. Eberhardt, L.L., and R.E. Nakatani. 1968. A postulated effect of growth on retention time of metabolites. J. Fish. Res. Board Can. 25(3):591-596.

23. Eggers, D.M. 1977a. The nature of prey selection by planktivorous fish. Ecology 58(1):46-59.

24. ___________. 1977b. Factors in interpreting data obtained by diel sampling of fish stomachs. J. Fish. Res. Board Can. 34:290-294.

25. Fee, E.J. 1969. A numerical model for the estimation of photosynthetic production, integrated over time and depth, in natural waters. Limnol. and Oceanogr. 14(6):906-911.

26. ________. 1973a. A numerical model for determining integral primary production and its application to Lake Michigan. J. Fish. Res. Board Can. 30:1447-1468.

27. ________. 1973b. Modelling primary production in water bodies: a numerical approach that allows vertical inhomogeneities. J. Fish. Res. Board Can. 30:1469-1473.

28. ________. 1979. The effect of fertilization with phosphorus and nitrogen versus phosphorus alone on eutrophication of experimental lakes. Limnol. and Oceanogr. 25(6):1149-1152.

29. Frank, P.W. 1960. Prediction of population growth form in Daphnia pulex cultures. Am. Naturalist 94(878):357-377.

30. Freedman, H.I. 1975. A perturbed Kolmogorov-type model for the growth problem. Mathematical Biosciences 23:127-149.

31. Gard, T.C. 1981. Persistence for ecosystem microcosm models. Ecol. Modelling 12:221-229.

32. Gerritsen, J., and J. Rudi Strickler. 1977. Encounter probabilities and community structure in zooplankton: a mathematical model. J. Fish. Res. Board Can. 34:73-82.

33. Glass, N.R. 1969. Discussion of calculation of power function with special reference to respiratory metabolism in fish. J. Fish. Res. Board Can. 26:2643-2650.

34. Goldman, J.C., and E.J. Carpenter. 1974. A kinetic approach to the effect of temperature on algal growth. Limnol. and Oceanogr. 19(5):756-766.

35. Holling, C.S. 1959. Some characteristics of simple types of predation and parasitism. The Canadian Entomologist 91(5):-385-398.

36. Howland, H.C. 1974. Optimal strategies for predator avoidance: the relative importance of speed and maneuverability. J. Theoretical Biol. 47:333-350.

37. Imboden, D.M. 1974. Phosphorus model of lake eutrophication. Limnol. and Oceanogr. 19(2):-297-304.

38. Jassby, A.D., and Trevor Platt. 1976. Mathematical formulation of the relationship between photosynthesis and light for phytoplankton. Limnol. and Oceanogr. 21(4):540-547.

39. Kerr, S.R. 1971a. Analysis of laboratory experiments on growth efficiency of fishes. J. Fish. Res. Board Can. 28:801-808.

40. ________. 1971b. Prediction of fish growth efficiency out in nature. J. Fish. Res. Board Can. 28:809-814.

41. ________. 1971c. A simulation model of lake trout growth. J. Fish. Res. Board Can. 28:815-819.

42. Kitchell, J.F., J.F. Koonce, R.V. O'Neill, H.H. Shugart, Jr., J.J. Magnuson, and R.S. Booth. 1974. Model of fish biomass dynamics. Trans. Am. Fish. Soc. 103(4):786-798.

43. Kitchell, J.F., D.J. Stewart, and D. Weinger. 1977. Applications of a bioenergetics model to yellow perch (Perca flavescens) and walleye (Stizostedion vitreum vitreum). J. Fish. Res. Board Can. 34:1922-1935.

44. Konstantinov, A.S. 1980. Comparative evaluation of the intensity of respiration in fishes. J. Ichthyology 20(1):98-104.

45. Lane, P., and R. Levins. 1977. The dynamics of aquatic systems. 2. The effects of nutrient enrichment on model plankton communities. Limnol. and Oceanogr. 22(3):454-471.

46. Mann, K.H. 1969. The dynamics of aquatic ecosystems. Advances in Ecol. Res. 6:1-73.

47. Matuszek, J.E. 1978. Empirical predictions of fish yields of large American lakes. Trans. Am. Fish. Soc. 107(3):385-394.

48. McAllister, C.D., R.J. Lebrasseur, and T.R. Parsons. 1972. Stability of enriched aquatic ecosystems. Science 175:562-564.

49. McNair, J.N. 1981. A stochastic foraging model with predator training effects. II. Optimal diets. Theoretical Pop. Biol. 19:147-162.

50. Moss, B. 1969. Limitation of algal growth in some central African waters. Limnol. and Oceanogr. 14(4):591-601.

51. Mullin, M.M., E.F. Stewart, and F.J. Fuglister. 1975. Ingestion by planktonic grazers as a function of concentration of food. Limnol. and Oceanogr. 20(2):259-262.

52. Nimura, Y. 1980. A new filter feeding model incorporating the critical food concentration. Japanese Soc. Sci. Fish. Bull. 46(7):787-795.

53. Ostrofsky, M.L. 1978. Modification of phosphorus retention models for use with lakes with low areal loading. J. Fish. Res. Board Can. 35:1532-1536.

54. Parker, R.A. 1974. Empirical functions relating metabolic pro-

cesses in aquatic systems to environmental variables. J. Fish. Res. Board Can. 31:1550-1552.

55. Patalas, K. 1980. Comment on "A relation between lake morphometry and primary productivity and its use in interpreting whole-lake eutrophication experiments". Limnol. and Oceanogr. 25(6):1147-1149.

56. Patten, B.C. 1969. Ecological systems analysis and fisheries science. Trans. Am. Fish. Soc. 98(3):570-581.

57. ___________. 1975. Reservoir cove ecosystem model. Trans. Am. Fish. Soc. 104(3):596-619.

58. Peterman, R.M. and M. Gatto. 1978. Estimation of functional responses of predators on juvenile salmon. J. Fish. Res. Board Can. 35:797-808.

59. Peters, R.H. 1975. Phosphorus excretion and the measurement of feeding and assimilation by zooplankton. Limnol. and Oceanogr. 20(5):858-859.

60. Powers, J.E. 1974. Competition for food: an evaluation of Ivlev's model. Trans. Am. Fish. Soc. 103(4):772-776.

61. ___________., and R.T. Lackey. 1975. Interaction in ecosystems: a queueing approach to modeling. Mathematical Biosciences 25:81-90.

62. Romaire, R.P., and C.E. Boyd. 1979. Effects of solar radiation on dynamics of dissolved oxygen in channel catfish ponds. Trans. Am. Fish. Soc. 108:473-478.

63. Rosenzweig, M.L. 1971. Paradox of enrichment: destabilization of exploitation ecosystems in ecological time. Science 171:385-387.

64. Ross, G.G. 1973. A model for the competitive growth of two diatoms. J. Theoretical Biol. 42:307-331.

65. Scavia, D., and S.C. Chapra. 1977. Comparison of an ecological model of Lake Ontario and phosphorus loading models. J. Fish. Res. Board Can. 34:286-290.

66. Shuter, B.J., J.A. Maclean, F.E. J. Fry, and H.A. Regier. 1980. Stochastic simulation of temperature effects on first-year survival of smallmouth bass. Trans. Am. Fish. Soc. 109(1):1-34.

67. Taghon, G.L., R.F.L. Self, and P.A. Jumars. 1978. Predicting particle selection by deposit feeders: a model and its implications. Limnol. and Oceanogr. 23(4):752-759.

68. Taylor, B.E., and M. Slatkin. 1981. Estimating birth and death rates of zooplankton. Limnol. and Oceanogr. 26(1):143-158.

69. Thomann, R.V. 1977. Comparison of lake phytoplankton models and leading plots. Limnol. and Oceanogr. 22(2):370-373.

70. Thornton, K.W., and A.S. Lessem. 1978. A temperature algorithm for modifying biological rates. Trans. Am. Fish. Soc. 107(2):-284-287.

71. Threlkeld, S.T. 1979. Estimating cladoceran birth rates: the importance of egg mortality and the egg age distribution. Limnol. and Oceanogr. 24(4):601-612.

72. Trump, C.L., and W.C. Leggett. 1980. Optimum swimming speeds in fish: the problem of currents. Can. J. Fish. Aquat. Sci. 37:1086-1092.

73. Ursin, E. 1967. A mathematical model of some aspects of fish growth, respiration, and mortality. J. Fish. Res. Board Can. 24(11):2355-2453.

74. Van Dyne, G.M. 1969. The Ecosystem Concept in Natural Resource Management. Academic Press, New York. 383 pp.

75. Walters, C.J., R. Hilborn, M. Peterman, and M.J. Staley. 1978. Model for examining early ocean limitation of Pacific salmon production. J. Fish. Res. Board Can. 35:1303-1315.

76. Ware, D.M. 1972. Predation by rainbow trout (Salmo gairdneri): the influence of hunger, prey density, and prey size. J. Fish. Res. Board Can. 29:1193-1201.

77. _________. 1973. Risk of epibenthic prey to predation by rainbow trout (Salmo gairdneri). J. Fish. Res. Board Can. 30:787-797.

78. Watt, K.E. F. 1959. Studies on population productivity. II. Factors governing productivity in a population of smallmouth bass. Ecol. Monogr. 29:367-392.

79. Weatherley, A.H. 1976. Factors affecting maximization of fish growth. J. Fish. Res. Board Can. 33:1046-1058.

80. Werner, E.E., and D.J. Hall. 1974. Optimal foraging and the size selection of prey by the bluegill sunfish (Lepomis machrochirus). Ecology 55:1042-1052.

81. Windell, J.T., J.W. Foltz, J.F. Kitchell, D.O. Norris, and J.S. Norris. 1976. Temperature and rate of gastric evacuation by rainbow trout (Salmo gairdneri). Trans. Am. Fish. Soc. 105(6):712-717.

AFTERWORD

POND AQUACULTURE - THE FUTURE

It was noted in the foreword that one of the purposes of this collection of papers was to develop information that would be useful in moving pond aquaculture from a highly developed art form to an agricultural technology. The approach to be taken was to consider what is known and what needs to be known to accomplish this goal. Therefore, the objective of this concluding section is to present a brief critique of the state of the art and a statement about future perspectives.

THE LIMITATIONS OF CURRENT KNOWLEDGE

It is apparent that a great deal of information about pond aquaculture is presently available. Much of this information is descriptive as opposed to quantitative. Thus, while the literature may provide general guidelines for the operation of pond culture systems, it is of limited predictive value. Additionally, because the descriptive data cannot be analyzed statistically, many generalizations found in the contemporary literature must be considered speculative.

In addition, a lack of standardization in experimental design, data collection, and analysis renders the existing data base of limited utility in predicting the performance of pond culture systems. Most of the quantitative data are found in technical reports describing a limited number of observations over a limited time period. Consequently, much of the data lacks statistical precision, is site specific, and the reproducibility of the results is subject to question. However, the existing information base can be utilized as a starting point to formulate testable hypotheses about the performance of pond culture systems.

Among the numerous technical questions about pond culture systems there is a unifying question: how do physical, chemical, and biological processes interact to regulate the productivity of these systems? The description of these processes involves the overlapping disciplines of fish production, water chemistry, and physical and biological limnology. It is a widely held view that the key to maximizing the efficiency of pond aquaculture lies in understanding the principles and then developing management practices that exploit these principles to the greatest advantage. This poses something of a dilemma because to a very large degree the physical, chemical, and biological processes are determined by management practices.

If mathematical models are to be used as management and research tools in improving the efficiency of pond culture systems, additional development will be required. Numerous mathematical models have been developed to describe processes in various aquatic systems such as lakes and waste water treatment systems. Although the mathematical models needed for the description of pond culture systems may be adapted from these models, no existing mathematical models are satisfactory for use in the management of pond aquaculture.

RESEARCH NEEDS

The development of aquaculture technologies will require an understanding of the nature of variation in pond productivity and of site specific variation. It is not clear that a conceptual framework for studying the sources of variation presently exists. It is generally agreed that it will be necessary to describe the dynamic processes occurring in ponds to understand fully the sources of variation. However, it would be naive to think that the dynamics can be described simply in mechanistic terms. Although many of the variables contributing to the variance of pond processes may be highly predictable, others may best be regarded as random variables. This is especially true in cases where the interactions of some variables involve two or more equally probable outcomes, or where the inconsistency of natural phenomena such as climatic factors is involved.

In developing a conceptual framework for studying the sources of variance in pond dynamics it must recognized that the number of correlations increases exponentially with the number of variables observed. It is seldom practical to attempt the direct measurement of a large number of pond variables. A cost effective approach to researching the principles of pond culture systems must initially address the limited subset of correlations that appear to be most important in understanding the underlying principles. This list can become more comprehensive as the processes which constrain the efficiency of the systems are defined.

The quality of the data collected is probably more important than the quantity. Undertaking overly ambitious experiments compromises the quality of results by overtaxing the technical work force. In the design of experiments, the correlations to be tested must be specified and data collection limited to those observations needed for these specific tests.

Wherever possible, standardized experimental designs, methods, and data collection and analysis should be employed in aquaculture research activities. Quantitative standardization is essential if research findings are to be utilized over a wide range of applications and locations.

The most direct approach to maximizing the efficiency of pond culture systems is probably to develop predictive equations that can be used to describe the various dynamic interactions regulating the productivity of the ponds. Models should be developed for both supplemental feeding and fertilizer application (as described in the foreword to this book). Pond management practices can then be optimized by combining the features of these models.

Models describing supplemental feeding should include functional relationships between food consumption and growth rate for a variety of species fed a variety of rations in different pond environments. The gross energy budget serves as a useful model in developing these functional relationships.

The present aquaculture data base is wholly inadequate for modelling nutrient cycling associated with fertilizer addition. To develop predictive models for these systems, it will be necessary to collect physical, chemical, and biological data on a variety of pond environments over a wide geographical range and incorporating a variety of water and soil characteristics. These data must be collected on a standardized protocol so that observations within and between ponds can be compared to provide insights into the nature of site specific variation.

Initially, comprehensive physical, chemical, and biological baseline data should be collected for each pond under carefully controlled conditions of pond preparation and management. Compilation of this baseline should include estimation of the critical standing crop and carrying capacity under the conditions observed. Once this baseline is compiled, the pond environment can be perturbed in various ways and the dynamic responses observed. If the goal of standardizing experimental designs and data collection is realized, a very comprehensive aquaculture data base will be available in a relatively short time. This data base can be utilized to develop predictive models for various pond management practices. These models can be combined to describe combinations of fertilization and supplemental feeding and the resulting functions can be optimized to maximize the efficiency of pond culture operations.

To accomplish such an undertaking will require the coordinated effort of many dedicated researchers. Emphasis should be placed on the word 'coordinated', for without the coordinated efforts of workers throughout the world there is little hope that predictive models can be developed. It is hoped that this collection of papers inspires investigators across the world who share the common dedication to understanding the dynamics of pond culture systems to contribute to and share in the expanding data base. By working together collectively and systematically, understanding of pond dynamics can be accelerated and the improvement of aquaculture technologies hastened.

CONTRIBUTORS

David R. Bernard is a Fishery Biologist with the Alaska Department of Fish and Game. He holds a PhD in Fisheries Science from Virginia Polytechnic Institute. Dr. Bernard served on the faculty of Oregon State University before moving to Alaska. His publications cover a range of topics in fish behavior and population dynamics.

James A. Brock has served as aquaculture disease specialist for the Aquaculture Development Program, Department of Land and Natural Resources, State of Hawaii, since 1978. Dr. Brock, who holds a D.V.M. degree from Washington State University, has been an affiliate faculty member of the Department of Animal Sciences, University of Hawaii. His research interests include the application of the principles of preventive medicine to the control of diseases of aquatic species.

William Chang has conducted extensive studies in limnological research and is an experienced water quality analyst. An assistant research scientist at the Great Lakes Research Division at the University of Michigan, Dr. Chang received his PhD in ecology from Indiana University. He has authored or co-authored a number of publications dealing with marine ecology.

John E. Colt is a senior development engineer in the department of Civil Engineering at the University of California, Davis. Dr. Colt's research interests include the design, construction, and optimization of aquatic culture systems, the effects of the environmental parameters on the growth and mortality of aquatic animals, and the use of aquatic plants and animals for wastewater treatment and resource recovery.

James S. Diana has conducted a number of studies on freshwater fishes - their daily movements, habitats, and food habits. He is an associate professor in the School of Natural Resources and associate research scientist at the Great Lakes Research Division, University of Michigan. Dr. Diana received his master's degree in biology from California State University, Long Beach, and a PhD in zoology from the University of Alberta.

Marlene S. Evans has authored or co-authored a variety of publications dealing with zooplankton populations in Lake Michigan. An associate research scientist at the Great Lakes Research Division of the University of Michigan, Dr. Evans received her PhD in zoology and oceanography from the University of British Columbia.

Arlo W. Fast is an associate researcher at the University of Hawaii's Institute of Marine Biology, where his primary field of interest involves fish culture. Dr. Fast received his PhD in limnology, fisheries biology from Michigan State University.

Donald L. Garling, Jr. is associate professor of fisheries and wildlife at Michigan State University. He has authored or co-authored a number of articles reporting on dietary studies of fisheries. Dr. Garling received his master's degree from Eastern Kentucky University and his doctorate from Mississippi State.

Darrell L. King is a professor in the Department of Fisheries and Wildlife and the Institute of Water Research at Michigan State University. Among his major research interests are the interrelationships between physical, chemical, and biological factors in aquatic ecosystems. Dr. King has received a number of academic honors, including and Outstanding Educators of America award.

James E. Lannan, Jr. is an associate professor of fisheries at Oregon State University and Program Director, Title XII Collaborative Research Support Program, Pond Dynamics/Aquaculture. Dr. Lannan has done extensive research in shellfish and salmon fisheries of the Pacific Northwest. He received his PhD from Oregon State University in biology/fisheries.

C. Kwei Lin has conducted extensive research in ecology and physiology of large lakes, particularly the Great Lakes. Presently an associate research limnologist in the Great

Lakes Research Division, University of Michigan, Dr. Lin received his PhD in limnology/botany from the University of Wisconsin, Milwaukee.

Nikola Marjanovic received his PhD in water resources engineering from the University of California, Davis, where he studied under a Fulbright Scholarship. He is currently a research hydrologist with the Institute of Water Resources, Jaroslav Cerni, in Belgrade, Yugoslavia, specializing in reservoir quality management.

Russell A. Moll is associate research scientist in the Great Lakes Research Division, University of Michigan, and Acting Assistant Director of the Michigan Sea Grant Program. Dr. Moll has conducted extensive investigations of plankton and related limnological conditions and has served as chief scientist during cruises on Lake Huron, Lake Michigan, and Lake Superior. He received his PhD in biology from the State University of New York at Stonybrook.

Gerald T. Orlob is a professor of civil engineering at the University of California, Davis, where he teaches and conducts research programs in water quality management, and mathematical modelling of aquatic systems. Dr. Orlob is a registered engineer and professional hydrologist. He received his PhD in hydraulic engineering from Stanford University.

David Ottey received his B.S. and M.S. at the University of Michigan School of Natural Resources. He is presently a PhD candidate in aquatic ecology at the University of Michigan specializing in benthic invertebrates.

Yung C. Shang is an associate economist in the Department of Agricultural and Resource Economics, University of Hawaii. He has conducted studies on the economic impacts of a variety of Pacific Ocean fisheries. Dr. Shang received his master's degree from Southern Illinois University and his doctorate from the University of Hawaii in agricultural economics.

R. Oneal Smitherman is professor of aquaculture at Auburn University's Fisheries Department and International Center for Aquaculture. He has conducted extensive research in a variety of southern fisheries and has carried out overseas assignments in more than a dozen countries. Dr. Smitherman received a master's degree in animal ecology from North Carolina State and a PhD in fisheries management from Auburn.

George Tchobanoglous, professor at the University of California at Davis, is the author or co-author of more than 70 reports and articles and three textbooks on various subjects in sanitary and environmental engineering. Dr. Tchobanoglous received a master's degree in sanitary engineering from the University of California at Berkeley, and a PhD in civil engineering from Stanford.

Eugene L. Torrans is an assistant professor of fisheries at the University of Arkansas at Pine Bluff. He conducts research in fish culture, toxicology, fish physiology and anatomy, feeding behavior of planktivorous fish, and water quality management of commercial culture ponds. Dr. Torrans received his PhD from the University of Oklahoma in fish culture zoology.

David S. White is an associate research scientist at the Great Lakes Research Division, and assistant professor in the School of Natural Resources at the University of Michigan. The recipient of a master's degree in zoology from DePauw University and his PhD in biology from the University of Louisville, Dr. White is author of a number of publications in the field of aquatic ecology.

Randolph Yamada is a fisheries and livestock specialist for Hawaiian Agronomics (International) Inc., Honolulu, where he is responsible for the technical formulation, management, and evaluation of fisheries and livestock development projects and studies for Hawaiian Agronomics in the Asia/Pacific region. Dr. Yamada holds a master's degree in biology from California State University, Hayward, and a PhD in genetics from the University of Hawaii.

INDEX